GEOGRAPHY 94/95

Ninth Edition

Editor

Gerald R. Pitzl
Macalester College

Gerald R. Pitzl, professor of geography at Macalester
College, received his bachelor's degree in secondary social
science education from the University of Minnesota in 1964
and his M.A. (1971) and Ph.D. (1974) in geography from the
same institution. He teaches a wide array of geography
courses and is the author of a number of articles on
geography, Third World development, and computers in
social science education.

Annual Editions
A Library of Information from the Public Press

The Dushkin Publishing Group, Inc.
Sluice Dock, Guilford, Connecticut 06437

Cover illustration by Mike Eagle

This map has been developed to give you a graphic picture of where the countries of the world are located, the relationship they have with their region and neighbors, and their positions relative to the superpowers and power blocs. We have focused on certain areas to more clearly illustrate these crowded regions.

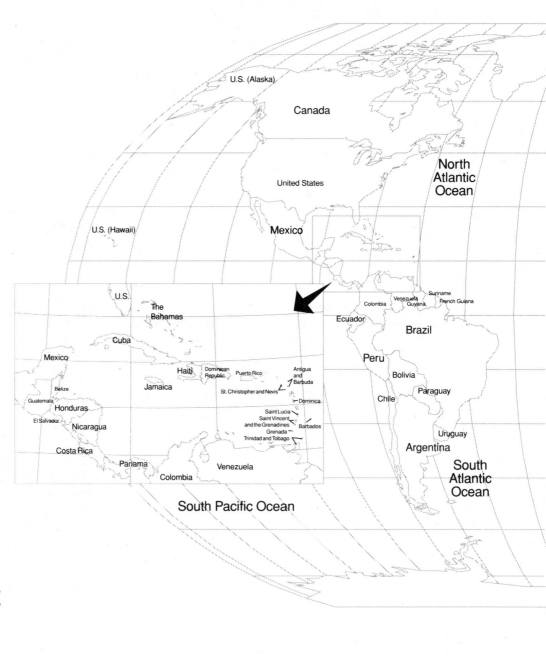

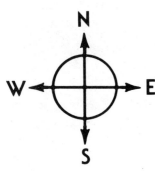

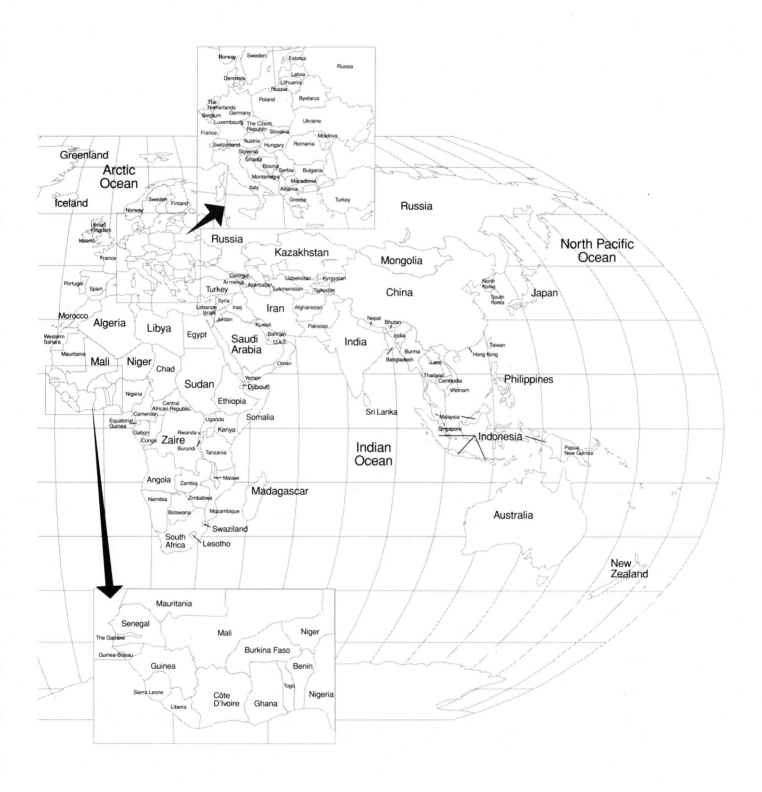

The Annual Editions Series

Annual Editions is a series of over 60 volumes designed to provide the reader with convenient, low-cost access to a wide range of current, carefully selected articles from some of the most important magazines, newspapers, and journals published today. Annual Editions are updated on an annual basis through a continuous monitoring of over 300 periodical sources. All Annual Editions have a number of features designed to make them particularly useful, including topic guides, annotated tables of contents, unit overviews, and indexes. For the teacher using Annual Editions in the classroom, an Instructor's Resource Guide with test questions is available for each volume.

Printed on Recycled Paper

VOLUMES AVAILABLE

Africa
Aging
American Foreign Policy
American Government
American History, Pre-Civil War
American History, Post-Civil War
Anthropology
Biology
Business Ethics
Canadian Politics
Child Growth and Development
China
Comparative Politics
Computers in Education
Computers in Business
Computers in Society
Criminal Justice
Drugs, Society, and Behavior
Dying, Death, and Bereavement
Early Childhood Education
Economics
Educating Exceptional Children
Education
Educational Psychology
Environment
Geography
Global Issues
Health
Human Development
Human Resources
Human Sexuality
India and South Asia
International Business
Japan and the Pacific Rim

Latin America
Life Management
Macroeconomics
Management
Marketing
Marriage and Family
Mass Media
Microeconomics
Middle East and the Islamic World
Money and Banking
Multicultural Education
Nutrition
Personal Growth and Behavior
Physical Anthropology
Psychology
Public Administration
Race and Ethnic Relations
Russia, Eurasia, and Central/Eastern Europe
Social Problems
Sociology
State and Local Government
Third World
Urban Society
Violence and Terrorism
Western Civilization, Pre-Reformation
Western Civilization, Post-Reformation
Western Europe
World History, Pre-Modern
World History, Modern
World Politics

Library of Congress Cataloging in Publication Data
Main entry under title: Annual editions: Geography. 1994/95.
    1. Geography—Periodicals. 2. Anthropo-geography—Periodicals. 3. Natural resources—Periodicals. I. Pitzl, Gerald R., *comp.* II. Title: Geography.
910'.5   ISBN 1–56134–275–0

Ninth Edition

Printed in the United States of America

To the Reader

In publishing ANNUAL EDITIONS we recognize the enormous role played by the magazines, newspapers, and journals of the *public press* in providing current, first-rate educational information in a broad spectrum of interest areas. Within the articles, the best scientists, practitioners, researchers, and commentators draw issues into new perspective as accepted theories and viewpoints are called into account by new events, recent discoveries change old facts, and fresh debate breaks out over important controversies. Many of the articles resulting from this enormous editorial effort are appropriate for students, researchers, and professionals seeking accurate, current material to help bridge the gap between principles and theories and the real world. These articles, however, become more useful for study when those of lasting value are carefully *collected, organized, indexed,* and *reproduced* in a *low-cost format,* which provides easy and permanent access when the material is needed. That is the role played by *Annual Editions.* Under the direction of each volume's *Editor,* who is an expert in the subject area, and with the guidance of an *Advisory Board,* we seek each year to provide in each *ANNUAL EDITION* a current, well-balanced, carefully selected collection of the best of the public press for your study and enjoyment. We think you'll find this volume useful, and we hope you'll take a moment to let us know what you think.

The articles in this ninth edition of *Annual Editions: Geography* represent the wide range of topics associated with the discipline of geography. The major themes of spatial relationships, regional development, the population explosion, and socioeconomic inequalities exemplify the diversity of research areas within geography.

The book is organized into five sections, each of which contains articles relating to themes within the discipline. Articles address the conceptual nature of geography and the global and regional problems in the world today. The latter theme reflects the geographer's concern with finding solutions to these serious global and regional problems. Regional problems, such as food shortages in the Sahel and the greenhouse effect, not only concern geographers, but interest researchers from other disciplines as well.

The association of geography with other disciplines is important because expertise from a number of different fields will be necessary in finding solutions to some difficult problems. Geography should be involved in the search for solutions because the discipline has always been integrative. That is, geography uses evidence from many sources to answer the basic questions, "Where is it?" "Why is it there?" and "What is its relevance?" The articles emphasize the interconnectedness not only of places and regions in the world, but of thrusts to solutions to problems as well. No single discipline will have all of the answers to the problems facing the world today; the complexity of the issues is simply too great.

The articles in the first section of the book discuss particular aspects of geography as a discipline and provide examples of the topics presented in the remaining four sections. The middle three sections represent major themes in geography. The last section addresses important problems faced by geographers and others.

The articles will be useful to both teachers and students in their study of geography. The anthology is designed to provide detail and case study material to supplement the standard textbook treatment of geography. The goals of this anthology are to introduce students to the richness and diversity of topics relating to places and regions on Earth's surface, to pay heed to the serious problems facing humankind, and to stimulate the search for more information on topics of interest.

I would like to express my gratitude to Barbara Wells-Howe for her help in preparing this material for publication. Her typing, organization of materials, and many helpful suggestions are greatly appreciated. Without her diligence and professional efforts, this undertaking could not have been completed. Special thanks are also extended to Ian Nielsen for his continued encouragement during the preparation of this new edition. A word of thanks must go, as well, to all those who recommended articles for inclusion in this volume and who commented on the overall organization of the volume. Peter O. Muller, Sarah W. Bednarz, and Robert S. Bednarz were especially helpful in that regard. Please continue to share your opinions by filling out the article rating form on the last page of this book.

Gerald R. Pitzl

Gerald R. Pitzl
Editor

Contents

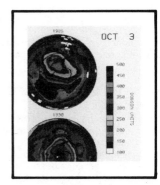

Unit 1

Geography in a Changing World

Nine articles discuss the discipline of geography and the extremely varied and wide-ranging themes that define geography today.

The concepts in bold italics are developed in the article. For further expansion please refer to the Topic Guide and the Index.

Unit 2

Land-Human Relationships

Six articles examine the relationship between humans and the land on which we live. Topics include the devastation of the oceans, the destruction of the rain forests, desertification, pollution, and the effects of human society on the global environment.

The concepts in bold italics are developed in the article. For further expansion please refer to the Topic Guide and the Index.

Unit 3

The Region

Nine selections review the importance of the region as a concept in geography and as an organizing framework for research. A number of world regional trends, as well as the patterns of area relationships, are examined.

The concepts in bold italics are developed in the article. For further expansion please refer to the Topic Guide and the Index.

Unit 4

Spatial Interaction and Mapping

Eight articles discuss the key theme in geographical analysis: place-to-place spatial interaction. Human diffusion, transportation systems, urban growth, and cartography are some of the themes examined.

The concepts in bold italics are developed in the article. For further expansion please refer to the Topic Guide and the Index.

Unit 5

Population, Resources, and Socioeconomic Development

Eight articles examine the effects of population growth on natural resources and the resulting socioeconomic level of development.

The concepts in bold italics are developed in the article. For further expansion please refer to the Topic Guide and the Index.

The concepts in bold italics are developed in the article. For further expansion please refer to the Topic Guide and the Index.

Topic Guide

This topic guide suggests how the selections in this book relate to topics of traditional concern to students and professionals involved with the study of geography. It is useful for locating articles that relate to each other for reading and research. The guide is arranged alphabetically according to topic. Articles may, of course, treat topics that do not appear in the topic guide. In turn, entries in the topic guide do not necessarily constitute a comprehensive listing of all the contents of each selection.

TOPIC AREA	TREATED IN:	TOPIC AREA	TREATED IN:
Accessibility	27. Rhine-Main-Danube Canal 28. Chunnel	Environment	15. Water Tight 27. Rhine-Main-Danube Canal
Agriculture	8. Green Revolution 22. Lash of the Dragon 23. Low Water in the American High Plains 35. Landscape of Hunger	Erosion	6. Beaches on the Brink
		Ethnocentrism	30. Cultural Commitments and World Maps
Alluvium	22. Lash of the Dragon	Forecasters	40. 50 Trends Shaping the World
Apartheid	38. South African Apartheid	Gender	9. Geography and Gender
Balkanization	4. Balkans	Geographic Information Systems (GIS)	see Computer Mapping
Cartography	see Mapmaking		
Climatic Change	11. Current Catastrophe: El Niño 13. Climate Change	Geography	2. American Geographies 38. South African Apartheid
Communication	25. Transportation and Urban Growth	Geopolitical	18. Middle East Geopolitical Transformation
Computer Mapping	31. Sense of Where You Are 32. New Cartographers	Global	16. Rise of the Region State 40. 50 Trends Shaping the World
Conservation	15. Water Tight	Global Positioning System (GPS)	32. New Cartographers
Core-Periphery	17. Africa's Geomosaic Under Stress	Global Warming	6. Beaches on the Brink 7. What's Wrong With the Weather? 14. Exploring the Links Between Desertification and Climate Change
Cultural Geography	24. Key to Understanding the Former Soviet Union		
Deforestation	12. Deforestation Debate	Green Revolution	8. Green Revolution
Demographic Transition	33. Aging of the Human Species	Greenhouse Effect	7. What's Wrong With the Weather? 13. Climate Change
Desertification	14. Exploring the Links Between Desertification and Climate Change 21. Aral Sea Basin	Groundwater	23. Low Water in the American High Plains
Development Gap	37. Why Africa Stays Poor	History of Geography	1. Four Traditions of Geography
Devolution	4. Balkans 5. Will Russia Disintegrate Into Bantustans?	Industrialization	9. Geography and Gender
		Land Degradation	37. Why Africa Stays Poor
Drought	11. Current Catastrophe: El Niño 23. Low Water in the American High Plains	Landscape	2. American Geographies
		Loess	22. Lash of the Dragon
El Niño	7. What's Wrong With the Weather? 11. Current Catastrophe: El Niño		

Geography in a Changing World

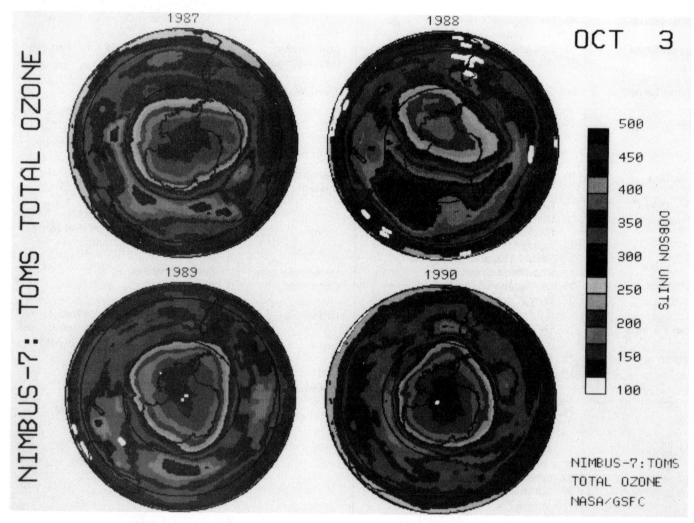

NIMBUS-7: TOMS TOTAL OZONE

1987 1988 OCT 3

1989 1990

500
450
400
350
300
250
200
150
100

DOBSON UNITS

NIMBUS-7:TOMS
TOTAL OZONE
NASA/GSFC

What is geography? This question has been asked innumerable times. It is a question that has not elicited a universally accepted answer, even from those who are considered to be members of the geography profession. The reason lies in the very nature of geography as it has evolved through time.

Geography is an extremely wide-ranging discipline, one that examines appropriate sets of events or circumstances occurring in specific places. Its goal is to answer certain basic questions.

The first question, "Where is it?" establishes the location of the subject under investigation. The concept of location is very important in geography, and its meaning extends beyond the common notion of a specific address or the determination of the latitude and longitude of a place. Geographers are more concerned with the relative location of a place and how that place interacts with other places both far and near. Spatial interaction and the determination of the connections between and among places are important themes in geography.

Once a place is "located," in the geographer's sense of the word, the next question is, "Why is it here?" For example, why are people concentrated in high numbers on the North China Plain, in the Ganges River Valley in India, and along the eastern seaboard in the United States? Conversely, why are there so few people in the Amazon Basin and the Central Siberian Lowlands? Generally, the geographer wants to find out why particular distribution patterns occur and why these patterns change over time.

The element of time is another extremely important ingredient in the geographical mix. The discipline is most concerned with the activities of human beings, and human beings bring about change. As changes occur, new adjustments and modifications are made in the distribution patterns previously established. Patterns change, for instance, as new technology brings about new forms of communication and transportation, and as once-desirable locations decline in favor of new ones. For example, people migrate from once-productive regions such as the Sahel when a disaster such as drought visits the land. Geography, then, is greatly concerned with discovering the underlying processes that can explain the transformation of distribution patterns and interaction forms over time. Geography itself is dynamic as it adjusts as a discipline to handle new situations in a changing world.

Geography is truly an integrating discipline. The geographer will assemble evidence from many sources in order to explain a particular pattern or ongoing process of change. Some of this evidence may even be in the form of concepts or theories borrowed from other disciplines. The first four articles of this unit provide insight into both the conceptual nature of geography and the development of the discipline over time.

Throughout its history, four main themes have been the focus of research work in geography. These themes or traditions, according to William Pattison in "The Four Traditions of Geography," link geography with earth science, establish it as a field that studies land-human relationships, engage it in area studies, and give it a spatial focus. Although Pattison's article first appeared over 20 years ago, it is still referred to and cited frequently today. Much of the geographical research and analysis engaged in today would fall within one or more of Pat-

tison's traditional areas, but new areas are also opening for geographers. In a particularly thought-provoking essay, the eminent author Barry Lopez discusses local geographies and the importance of a sense of place. The theme of place is echoed in a personal narrative recounting experiences on a trip to Japan in an article by Baher Ghosheh. Travel is not recommended, however, to the former Yugoslavia where war and ethnic strife contribute to an unsettling landscape. "The Balkans" details the proposed partition of that region.

Research into apparently abnormal weather shifts in recent years is dealt with in the article, "What's Wrong with the Weather?" An example of erratic weather and a much higher evidence of storminess is found in "Beaches on the Brink."

The next article provides a historical and geographical analysis of the genetic and technological developments associated with the Green Revolution. The Green Revolution refers to attempts to solve the regional problem of food shortage—an effort that involves leaders from scientific, political, and academic backgrounds. Finally, the article "Geography and Gender" looks at the ways that gender has brought about variations in societies.

Looking Ahead: Challenge Questions

Why is geography called an integrating discipline?

How is geography related to earth science? Can you give some examples of land-human relationships? What are area studies? Why is the spatial concept so important in geography? What is your definition of geography?

Why do history and geography rely on one another?

How have humans affected the weather? The environment in general?

How is individual behavior related to the understanding of environment?

Is change a good thing? Why is it important to anticipate change?

Do you understand what interconnectedness means in terms of places? Can you give examples of how you as an individual interact with people in other places? How are you "connected" to the rest of the world?

What do you think the world will be like in the year 2010? Are you pessimistic or optimistic about the future? Is there anything you can do about the future?

The Four Traditions of Geography

William D. Pattison

Late Summer, 1990

To Readers of the *Journal of Geography:*

I am honored to be introducing, for a return to the pages of the *Journal* after more than 25 years, "The Four Traditions of Geography," an article which circulated widely, in this country and others, long after its initial appearance—in reprint, in xerographic copy, and in translation. A second round of life at a level of general interest even approaching that of the first may be too much to expect, but I want you to know in any event that I presented the paper in the beginning as my gift to the geographic community, not as a personal property, and that I re-offer it now in the same spirit.

In my judgment, the article continues to deserve serious attention—perhaps especially so, let me add, among persons aware of the specific problem it was intended to resolve. The background for the paper was my experience as first director of the High School Geography Project (1961–63)—not all of that experience but only the part that found me listening, during numerous conference sessions and associated interviews, to academic geographers as they responded to the project's invitation to locate "basic ideas" representative of them all. I came away with the conclusion that I had been witnessing not a search for consensus but rather a blind struggle for supremacy among honest persons of contrary intellectual commitment. In their dialogue, two or more different terms had been used, often unknowingly, with a single reference, and no less disturbingly, a single term had been used, again often unknowingly, with two or more different references. The article was my attempt to stabilize the discourse. I was proposing a basic nomenclature (with explicitly associated ideas) that would, I trusted, permit the development of mutual comprehension **and** confront all parties concerned with the pluralism inherent in geographic thought.

This intention alone could not have justified my turning to the NCGE as a forum, of course. The fact is that from the onset of my discomfiting realization I had looked forward to larger consequences of a kind consistent with NCGE goals.

As finally formulated, my wish was that the article would serve "to greatly expedite the task of maintaining an alliance between professional geography and pedagogical geography and at the same time to promote communication with laymen" (see my fourth paragraph). I must tell you that I have doubts, in 1990, about the acceptability of my word choice, in saying "professional," "pedagogical," and "layman" in this context, but the message otherwise is as expressive of my hope now as it was then.

I can report to you that twice since its appearance in the *Journal,* my interpretation has received more or less official acceptance—both times, as it happens, at the expense of the earth science tradition. The first occasion was Edward Taaffe's delivery of his presidential address at the 1973 meeting of the Association of American Geographers (see *Annals AAG,* March 1974, pp. 1–16). Taaffe's working-through of aspects of an interrelations among the spatial, area studies, and man-land traditions is by far the most thoughtful and thorough of any of which I am aware. Rather than fault him for omission of the fourth tradition, I compliment him on the grace with which he set it aside in conformity to a meta-epistemology of the American university which decrees the integrity of the social sciences as a consortium in their own right. He was sacrificing such holistic claims as geography might be able to muster for a freedom to argue the case for geography as a social science.

The second occasion was the publication in 1984 of *Guidelines for Geographic Education: Elementary and Secondary Schools,* authored by a committee jointly representing the AAG and the NCGE. Thanks to a recently published letter (see *Journal of Geography,* March-April 1990, pp. 85–86), we know that, of five themes commended to teachers in this source,

The committee lifted the human environmental interaction theme directly from Pattison. The themes of place and location are based on Pattison's spatial or geometric geography, and the theme of region comes from Pattison's area studies or regional geography.

From *Journal of Geography,* Vol. 89, No. 5, September/October 1990, pp. 202-206. Reproduced with permission of the National Council for Geographic Education.

Having thus drawn on my spatial area studies and manland traditions for four of the five themes, the committee could have found the remaining theme, movement, there too—in the spatial tradition (see my sixth paragraph). However that may be, they did not avail themselves of the earth science tradition, their reasons being readily surmised. Peculiar to the elementary and secondary schools is a curriculum category framed as much by theory of citizenship as by theory of knowledge: the social studies. With admiration, I see already in the committee members' adoption of the theme idea a strategy for assimilation of their program to the established repertoire of social studies practice. I see in their exclusion of the earth science tradition an intelligent respect for social studies' purpose.

Here's to the future of education in geography: may it prosper as never before.

W. D. P., 1990

Reprinted from the *Journal of Geography*, 1964, pp. 211-216.

In 1905, one year after professional geography in this country achieved full social identity through the founding of the Association of American Geographers, William Morris Davis responded to a familiar suspicion that geography is simply an undisciplined "omnium-gatherum" by describing an approach that as he saw it imparts a "geographical quality" to some knowledge and accounts for the absence of the quality elsewhere.[1] Davis spoke as president of the AAG. He set an example that was followed by more than one president of that organization. An enduring official concern led the AAG to publish, in 1939 and in 1959, monographs exclusively devoted to a critical review of definitions and their implications.[2]

Every one of the well-known definitions of geography advanced since the founding of the AAG has had its measure of success. Tending to displace one another by turns, each definition has said something true of geography.[3] But from the vantage point of 1964, one can see that each one has also failed. All of them adopted in one way or another a monistic view, a singleness of preference, certain to omit if not to alienate numerous professionals who were in good conscience continuing to participate creatively in the broad geographic enterprise.

The thesis of the present paper is that the work of American geographers, although not conforming to the restrictions implied by any one of these definitions, has exhibited a broad consistency, and that this essential unity has been attributable to a small number of distinct but affiliated traditions, operant as binders in the minds of members of the profession. These traditions are all of great age and have passed into American geography as parts of a general legacy of Western thought. They are shared today by geographers of other nations.

There are four traditions whose identification provides an alternative to the competing monistic definitions that have been the geographer's lot. The resulting pluralistic basis for judgment promises, by full accommodation of what geographers do and by plain-spoken representation thereof, to greatly expedite the task of maintaining an alliance between professional geography and pedagogical geography and at the same time to promote communication with laymen. The following discussion treats the traditions in this order: (1) a spatial tradition, (2) an area studies tradition, (3) a man-land tradition and (4) an earth science tradition.

Spatial Tradition

Entrenched in Western thought is a belief in the importance of spatial analysis, of the act of separating from the happenings of experience such aspects as distance, form, direction and position. It was not until the 17th century that philosophers concentrated attention on these aspects by asking whether or not they were properties of things-in-themselves. Later, when the 18th century writings of Immanuel Kant had become generally circulated, the notion of space as a category including all of these aspects came into widespread use. However, it is evident that particular spatial questions were the subject of highly organized answering attempts long before the time of any of these cogitations. To confirm this point, one need only be reminded of the compilation of elaborate records concerning the location of things in ancient Greece. These were records of sailing distances, of coastlines and of landmarks that grew until they formed the raw material for the great *Geographia* of Claudius Ptolemy in the 2nd century A.D.

A review of American professional geography from the time of its formal organization shows that the spatial tradition of thought had made a deep penetration from the very beginning. For Davis, for Henry Gannett and for most if not all of the 44 other men of the original AAG, the determination and display of spatial aspects of reality through mapping were of undoubted importance, whether contemporary definitions of geography happened to acknowledge this fact or not. One can go further and, by probing beneath the art of mapping, recognize in the behavior of geographers of that time an active interest in the true essentials of the spatial tradition—*geometry* and *movement.* One can trace a basic favoring of movement as a subject of study from the turn-of-the-century work of Emory R. Johnson, writing as professor of transportation at the University of Pennsylvania, through the highly influential theoretical and substantive work of Edward L. Ullman during the past 20 years and thence to an article by a younger geographer on railroad freight traffic in the U.S. and Canada in the *Annals* of the AAG for September 1963.[4]

One can trace a deep attachment to geometry, or positioning-and-layout, from articles on boundaries and population densities in early 20th century volumes of the *Bulletin of the American Geographical Society,* through a controversial pronouncement by Joseph Schaefer in 1953 that granted geographical legitimacy only to studies of spatial patterns[5] and so onward to a recent *Annals* report on electronic scanning of cropland patterns in Pennsylvania.[6]

One might inquire, is discussion of the spatial tradition, after the manner of the remarks just made, likely to bring people within geography closer to an understanding of one another and people outside geography closer to an understanding of geographers? There seem to be at least two reasons for being hopeful. First, an appreciation of this tradition allows one to see a bond of fellowship uniting the elementary school teacher, who attempts the most rudimentary instruction in directions and mapping, with the contemporary research geographer, who dedicates himself to an exploration of central-place theory. One cannot only open the eyes of many teachers to the potentialities of their own instruction, through proper exposition of the spatial tradition, but one can also "hang a bell" on research quantifiers in geography, who are often thought to have wandered so far in their intellectual adventures as to have become lost from the rest. Looking outside geography, one may anticipate benefits from the readiness of countless persons to associate the name "geography" with maps. Latent within this readiness is a willingness to recognize as geography, too, what maps are about—and that is the geometry of and the movement of what is mapped.

Area Studies Tradition

The area studies tradition, like the spatial tradition, is quite strikingly represented in classical antiquity by a practitioner to whose surviving work we can point. He is Strabo, celebrated for his *Geography* which is a massive production addressed to the statesmen of Augustan Rome and intended to sum up and regularize knowledge not of the location of places and associated cartographic facts, as in the somewhat later case of Ptolemy, but of the nature of places, their character and their differentiation. Strabo exhibits interesting attributes of the area-studies tradition that can hardly be overemphasized. They are a pronounced tendency toward subscription primarily to literary standards, an almost omnivorous appetite for information and a self-conscious companionship with history.

It is an extreme good fortune to have in the ranks of modern American geography the scholar Richard Hartshorne, who has pondered the meaning of the area-studies tradition with a legal acuteness that few persons would challenge. In his *Nature of Geography,* his 1939 monograph already cited,[7] he scrutinizes exhaustively the implications of the "interesting attributes" identified in connection with Strabo, even though his concern is with quite other and much later authors, largely German. The major literary problem of unities or wholes he considers from every angle. The Gargantuan appetite for miscellaneous information he accepts and rationalizes. The companionship between area studies and history he clarifies by appraising the so-called idiographic content of both and by affirming the tie of both to what he and Sauer have called "naively given reality."

The area-studies tradition (otherwise known as the chorographic tradition) tended to be excluded from early American professional geography. Today it is beset by certain champions of the spatial tradition who would have one believe that somehow the area-studies way of organizing knowledge is only a subdepartment of spatialism. Still, area-studies as a method of presentation lives and prospers in its own right. One can turn today for reassurance on this score to practically any issue of the *Geographical Review,* just as earlier readers could turn at the opening of the century to that magazine's forerunner.

What is gained by singling out this tradition? It helps toward restoring the faith of many teachers who, being accustomed to administering learning in the area-studies style, have begun to wonder if by doing so they really were keeping in touch with professional geography. (Their doubts are owed all too much to the obscuring effect of technical words attributable to the very professionals who have been intent, ironically, upon protecting that tradition.) Among persons outside the classroom the geographer stands to gain greatly in intelligibility. The title "area-studies" itself carries an understood message in the United States today wherever there is contact with the usages of the academic community. The purpose of characterizing a place, be it neighborhood or nation-state, is readily grasped. Furthermore, recognition of the right of a geographer to be unspecialized may be expected to be forthcoming from people generally, if application for such recognition is made on the merits of this tradition, explicitly.

Man-Land Tradition

That geographers are much given to exploring man-land questions is especially evident to anyone who examines geographic output, not only in this country but also abroad. 0. H. K. Spate, taking an international view, has felt justified by his observations in nominating as the most significant ancient precursor of today's geography neither Ptolemy nor Strabo nor writers typified in their outlook by the geographies of either of these two men, but rather Hippocrates, Greek physician of the 5th century B.C. who left to posterity an extended essay, *On Airs, Waters and Places.*[8] In this work, made up of reflections on human health and conditions of external nature, the questions asked are such as to confine thought almost altogether to presumed influence passing from the latter to the former, questions largely about the effects of winds, drinking water and seasonal changes upon man. Understandable though this uni-directional concern may have been for Hippocrates as medical commentator, and defensible as may be the attraction that this same approach held for students of the condition of man for many, many centuries thereafter, one can only regret that this narrowed version of the man-land tradition, combining all too easily with social Darwinism of the late 19th century, practically overpowered American professional geography in the first generation of its history.[9] The premises of this version governed scores of studies by American geographers in interpreting the rise and fall of nations, the strategy of battles and the construction of public improvements. Eventually this special bias, known as environmentalism, came to be confused with the whole of the man-land tradition in the minds of many people. One can see now, looking back to the years after the ascendancy of environmentalism, that although the spatial tradition was asserting itself with varying degrees of forwardness, and that although the area-studies tradition was also making itself felt, perhaps the most interesting chapters in the story of American professional geography were being written by academicians who were reacting against environmentalism while deliberately remaining within the broad man-land tradition. The rise of culture historians during the last 30 years has meant the dropping of a curtain of culture between land and man, through which it is asserted all influence must pass. Furthermore work of both culture historians and other geographers has exhibited a reversal of the direction of the effects in Hippocrates, man appearing as an independent agent, and the land as a sufferer from action. This trend as presented in published research has reached a high point in the collection of papers titled *Man's Role in Changing the Face of the Earth*. Finally, books and articles can be called to mind that have addressed themselves to the most difficult task of all, a balanced tracing out of interaction between man and environment. Some chapters in the book mentioned above undertake just this. In fact the separateness of this approach is discerned only with difficulty in many places; however, its significance as a general research design that rises above environmentalism, while refusing to abandon the man-land tradition, cannot be mistaken.

The NCGE seems to have associated itself with the man-land tradition, from the time of founding to the present day, more than with any other tradition, although all four of the traditions are amply represented in its official magazine, *The Journal of Geography* and in the proceedings of its annual meetings. This apparent preference on the part of the NCGE members *for defining geography in terms of the man-land tradition* is strong evidence of the appeal that man-land ideas, separately stated, have for persons whose main job is teaching. It should be noted, too, that this inclination reflects a proven acceptance by the general public of learning that centers on resource use and conservation.

Earth Science Tradition

The earth science tradition, embracing study of the earth, the waters of the earth, the atmosphere surrounding the earth and the association between earth and sun, confronts one with a paradox. On the one hand one is assured by professional geographers that their participation in this tradition has declined precipitously in the course of the past few decades, while on the other one knows that college departments of geography across the nation rely substantially, for justification of their role in general education, upon curricular content springing directly from this tradition. From all the reasons that combine to account for this state of affairs, one may, by selecting only two, go far toward achieving an understanding of this tradition. First, there is the fact that American college geography, growing out of departments of geology in many crucial instances, was at one time greatly overweighted in favor of earth science, thus rendering the field unusually liable to a sense of loss as better balance came into being. (This one-time disproportion found reciprocate support for many years in the narrowed, environmentalistic interpretation of the man-land tradition.) Second, here alone in earth science does one encounter subject matter in the normal sense of the term as one reviews geographic traditions. The spatial tradition abstracts

certain aspects of reality; area studies is distinguished by a point of view; the manland tradition dwells upon relationships; but earth science is identifiable through concrete objects. Historians, sociologists and other academicians tend not only to accept but also to ask for help from this part of geography. They readily appreciate earth science as something physically associated with their subjects of study, yet generally beyond their competence to treat. From this appreciation comes strength for geography-as-earth-science in the curriculum.

Only by granting full stature to the earth science tradition can one make sense out of the oft-repeated addage, "Geography is the mother of sciences." This is the tradition that emerged in ancient Greece, most clearly in the work of Aristotle, as a wide-ranging study of natural processes in and near the surface of the earth. This is the tradition that was rejuvenated by Varenius in the 17th century as "Geographia Generalis." This is the tradition that has been subjected to subdivision as the development of science has approached the present day, yielding mineralogy, paleontology, glaciology, meterology and other specialized fields of learning.

Readers who are acquainted with American junior high schools may want to make a challenge at this point, being aware that a current revival of earth sciences is being sponsored in those schools by the field of geology. Belatedly, geography has joined in support of this revival.[10] It may be said that in this connection and in others, American professional geography may have faltered in its adherence to the earth science tradition but not given it up.

In describing geography, there would appear to be some advantages attached to isolating this final tradition. Separation improves the geographer's chances of successfully explaining to educators why geography has extreme difficulty in accommodating itself to social studies programs. Again, separate attention allows one to make understanding contact with members of the American public for whom surrounding nature is known as the geographic environment. And finally, specific reference to the geographer's earth science tradition brings into the open the basis of what is, almost without a doubt, morally the most significant concept in the entire geographic heritage, that of the earth as a unity, the single common habitat of man.

An Overview

The four traditions though distinct in logic are joined in action. One can say of geography that it pursues concurrently all four of them. Taking the traditions in varying combinations, the geographer can explain the conventional divisions of the field. Human or cultural geography turns out to consist of the first three traditions applied to human societies; physical geography, it becomes evident, is the fourth tradition prosecuted under constraints from the first and second traditions. Going further, one can uncover the meanings of "systematic geography," "regional geography," "urban geography," "industrial geography," etc.

It is to be hoped that through a widened willingness to conceive of and discuss the field in terms of these traditions, geography will be better able to secure the inner unity and outer intelligibility to which reference was made at the opening of this paper, and that thereby the effectiveness of geography's contribution to American education and to the general American welfare will be appreciably increased.

1. William Morris Davis, "An Inductive Study of the Content of Geography," *Bulletin of the American Geographical Society,* Vol. 38, No. 1 (1906), 71.

2. Richard Hartshorne, *The Nature of Geography,* Association of American Geographers (1939), and idem., *Perspective on the Nature of Geography,* Association of American Geographers (1959).

3. The essentials of several of these definitions appear in Barry N. Floyd, "Putting Geography in Its Place," *The Journal of Geography,* Vol. 62, No. 3 (March, 1963). 117-120.

4. William H. Wallace, "Freight Traffic Functions of Anglo-American Railroads," *Annals of the Association of American Geographers,* Vol. 53, No. 3 (September, 1963), 312-331.

5. Fred K. Schaefer, "Exceptionalism in Geography: A Methodological Examination," *Annals of the Association of American Geographers,* Vol. 43, No. 3 (September, 1953), 226-249.

6. James P. Latham, "Methodology for an Instrumental Geographic Analysis," *Annals of the Association of American Geographers,* Vol. 53, No. 2 (June, 1963). 194-209.

7. Hartshorne's 1959 monograph, *Perspective on the Nature of Geography,* was also cited earlier. In this later work, he responds to dissents from geographers whose preferred primary commitment lies outside the area studies tradition.

8. O. H. K. Spate, "Quantity and Quality in Geography," *Annals of the Association of American Geographers,* Vol. 50, No. 4 (December, 1960), 379.

9. Evidence of this dominance may be found in Davis's 1905 declaration: "Any statement is of geographical quality if it contains . . . some relation between an element of inorganic control and one of organic response" (Davis, *loc. cit.*).

10. Geography is represented on both the Steering Committee and Advisory Board of the Earth Science Curriculum Project, potentially the most influential organization acting on behalf of earth science in the schools.

The American Geographies: Losing Our Sense of Place

Americans are fast becoming strangers in a strange land, where one roiling river, one scarred patch of desert, is as good as another. America the beautiful exists— a select few still know it intimately—but many of us are settling for a homogenized national geography.

Barry Lopez

Barry Lopez's most recent book is The Rediscovery of North America *now available at Vintage.*

It has become commonplace to observe that Americans know little of the geography of their country, that they are innocent of it as a landscape of rivers, mountains, and towns. They do not know, supposedly, the location of the Delaware Water Gap, the Olympic Mountains, or the Piedmont Plateau; and, the indictment continues, they have little conception of the way the individual components of this landscape are imperiled, from a human perspective, by modern farming practices or industrial pollution.

I do not know how true this is, but it is easy to believe that it is truer than most of us would wish. A recent Gallup Organization and National Geographic Society survey found Americans woefully ignorant of world geography. Three out of four couldn't locate the Persian Gulf. The implication was that we knew no more about our own homeland, and that this ignorance undermined the integrity of our political processes and the efficiency of our business enterprises.

As Americans, we profess a sincere and fierce love for the American landscape, for our rolling prairies, freeflowing rivers, and "purple mountains' majesty"; but it is hard to imagine, actually, where this particular landscape is. It is not just that a nostalgic landscape has passed away—Mark Twain's Mississippi is now dammed from Minnesota to Missouri and the prairies have all been sold and fenced. It is that it has always been a romantic's landscape. In the attenuated form in which it is presented on television today, in magazine articles and in calendar photographs, the essential wildness of the American landscape is reduced to attractive scenery. We look out on a familiar, memorized landscape that portends adventure and promises enrichment. There are no distracting people in it and few artifacts of human life. The animals are all beautiful, diligent, one might even say well-behaved. Nature's unruliness, the power of rivers and skies to intimidate, and any evidence of disastrous human land management practices are all but invisible. It is, in short, a magnificent garden, a colonial vision of paradise imposed on a real place that is, at best, only selectively known.

To truly understand geography requires not only time but a kind of local expertise, an intimacy with place few of us ever develop.

The real American landscape is a face of almost incomprehensible depth and complexity. If one were to sit for a few days, for example, among the ponderosa pine forests and black lava fields of the Cascade Mountains in western Oregon, inhaling the pines' sweet balm on an evening breeze from some point on the barren rock, and then were to step off to the Olympic Peninsula in Washington, to those rain forests with sphagnum moss floors soft as fleece underfoot and Douglas firs too big around for five people to hug, and then head south to walk the ephemeral creeks and sun-blistered playas of the Mojave Desert in southern California, one would be reeling under the sensations. The contrast is not only one of plants and soils, a different array say, of brilliantly colored beetles. The shock to the senses comes from a different shape to the silence, a difference in the very quality of light, in the weight of the air. And this relatively short journey down the West Coast would still leave the traveler with all that lay to the east to explore—the anomalous sand hills of Nebraska, the heat and frog voices of Okefenokee Swamp, the fetch of Chesapeake Bay, the hardwood copses and black bears of the Ozark Mountains.

No one of these places, of course, can be entirely fathomed, biologically or aesthetically. They are mysteries upon which we impose names. Enchantments. We tick the names off glibly but lovingly. We mean no disrespect. Our genuine desire, though we might be skeptical about the time it would take and uncertain of its practical value to us, is to actually know these places. As deeply ingrained in the American psyche as the desire to conquer and control the land is the desire to sojourn in it, to sail up and down Pamlico Sound, to paddle a canoe through Minnesota's boundary waters, to walk on the desert of the Great Salt Lake, to camp in the stony hardwood valleys of Vermont.

To do this well, to really come to an understanding of a specific American geography, requires not only time but a kind of local expertise, an intimacy with place few of us ever develop. There is no way around the former requirement: If you want to know you must take the time. It is not in books. A specific geographical understanding, however, can be sought out and borrowed. It resides with men and women more or less sworn to a place, who abide there, who have a feel for the soil and history, for the turn of leaves and night sounds. Often they are glad to take the outlander in tow.

These local geniuses of American landscape, in my experience, are people in whom geography thrives. They are the antithesis of geographical ignorance. Rarely known outside their own communities, they often seem, at the first encounter, unremarkable and anonymous. They may not be able to recall the name of a particular wildflower—or they may have given it a name known only to them. They might have forgotten the precise circumstances of a local historical event. Or they can't say for certain when the last of the Canada geese passed through in the fall, or can't differentiate between two kinds of trout in the same creek. Like all of us, they have fallen prey to the fallacies of memory and are burdened with ignorance; but they are nearly flawless in the respect they bear these places they love. Their knowledge is intimate rather than encyclopedic, human but not necessarily scholarly. It rings with the concrete details of experience.

America, I believe, teems with such people. The paradox here, between a faulty grasp of geographical knowledge for which Americans are indicted and the intimate, apparently contradictory familiarity of a group of largely anonymous people, is not solely a matter of confused scale. (The local landscape is easier to know than a national geography.) And it is not simply ironic. The paradox is dark. To be succinct: The politics and advertising that seek a national audience must project a national geography; to be broadly useful that geography must, inevitably, be generalized and it is often romantic. It is therefore frequently misleading and imprecise. The same holds true with the entertainment industry, but here the problem might be clearer. The same films, magazines, and television features

that honor an imaginary American landscape also tout the worth of the anonymous men and women who interpret it. Their affinity for the land is lauded, their local allegiance admired. But the rigor of their local geographies, taken as a whole, contradicts a patriotic, national vision of unspoiled, untroubled land. These men and women are ultimately forgotten, along with the details of the landscapes they speak for, in the face of more pressing national matters. It is the chilling nature of modern society to find an ignorance of geography, local or national, as excusable as an ignorance of hand tools; and to find the commitment of people to their home places only momentarily entertaining. And finally naive.

If one were to pass time among Basawara people in the Kalahari Desert, or with Kreen-Akrora in the Amazon Basin, or with Pitjantjatjara Aborigines in Australia, the most salient impression they might leave is of an absolutely stunning knowledge of their local geography—geology, hydrology, biology, and weather. In short, the extensive particulars of their intercourse with it.

In 40,000 years of human history, it has only been in the last few hundred years or so that a people could afford to ignore their local geographies as completely as we do and still survive. Technological innovations from refrigerated trucks to artificial fertilizers, from sophisticated cost accounting to mass air transportation, have utterly changed concepts of season, distance, soil productivity, and the real cost of drawing sustenance from the land. It is now possible for a resident of Boston to bite into a fresh strawberry in the dead of winter; for someone in San Francisco to travel to Atlanta in a few hours with no worry of how formidable might be crossings of the Great Basin Desert or the Mississippi River; for an absentee farmer to gain a tax advantage from a farm that leaches poisons into its water table and on which crops are left to rot. The Pitjantjatjara might shake their heads in bewilderment and bemusement, not because they are primitive or ignorant people, not because they have no sense of irony or are incapable of marveling, but because they have not (many would say not yet) realized a world in which such manipulation of the land—surmounting the imperatives of distance it imposes, for example, or turning the large-scale destruction of forests and arable land in wealth—is desirable or plausible.

In the years I have traveled through America, in cars and on horseback, on foot and by raft, I have repeatedly been brought to a sudden state of awe by some gracile or savage movement of animal, some odd wrapping of tree's foliage by the wind, an unimpeded run of dew-laden prairie stretching to a horizon flat as a coin where a pin-dot sun pales the dawn sky pink. I know these things are beyond intellection, that they are the vivid edges of a world that includes but also transcends the human world. In memory, when I dwell on these things, I know that in a truly national literature there should be odes to the Triassic reds of the Colorado Plateau, to the sharp and ghostly light of the Florida Keys, to the aeolian soils of southern Minnesota, and the Palouse in Washington, though the modern mind abjures the literary potential of such subjects. (If the sand and flood water farmers of Arizona and New Mexico were to take the black loams of Louisiana in their hands they would be flabbergasted, and that is the beginning of literature.) I know there should be eloquent evocations of the cobbled beaches of Maine, the plutonic walls of the Sierra Nevada, the orange canyons of the Kaibab Plateau. I have no doubt, in fact, that there are. They are as numerous and diverse as the eyes and fingers that ponder the country—it is that only a handful of them are known. The great majority are to be found in drawers and boxes, in the letters and private journals of millions of workaday people who have regarded their encounters with the land as an engagement bordering on the spiritual, as being fundamentally linked to their state of health.

One cannot acknowledge the extent and the history of this kind of testimony without being forced to the realization that something strange, if not dangerous, is afoot. Year by year, the number of people with firsthand experience in the land dwindles. Rural populations continue to shift to the cities. The family farm is in a state of demise, and government and industry continue to apply pressure on the native peoples of North America to sever their ties with the land. In the wake of this loss of personal and local knowledge from which a real geography is derived, the knowledge on which a country must ultimately stand, has [be]come something hard to define but I think sinister and unsettling—the packaging and marketing of land as a form of entertainment. An incipient industry, capitalizing on the nostalgia Americans feel for the imagined virgin landscapes of their fathers, and on a desire for adventure, now offers people a convenient though sometimes incomplete or even spurious geography as an inducement to purchase a unique experience. But the line between authentic experience and a superficial exposure to the elements of experience is blurred. And the real landscape, in all its complexity, is distorted even further in the public imagination. No longer innately mysterious and dignified, a ground from which experience grows, it becomes a curiously generic backdrop on which experience is imposed.

In theme parks the profound, subtle, and protracted experience of running a river is reduced to a loud, quick, safe equivalence, a pleasant distraction. People only able to venture into the countryside on annual vacations are, increasingly, schooled in the belief that wild land will, and should, provide thrills and exceptional scenery on a timely basis. If it does not, something is wrong, either with the land itself or possibly with the company outfitting the trip.

People in America, then, face a convoluted situation. The land itself, vast and differentiated, defies the notion of a national geography. If applied at all it must be applied lightly and it must grow out of the concrete detail of local geographies. Yet Americans are daily presented with, and have become accustomed to talking about, a homogenized national geography. one that seems to operate independently of the land, a collection of objects rather than a continuous bolt of fabric. It appears in advertisements, as a background in movies, and in patriotic calendars. The suggestion is that there can be national geography because the constituent parts are interchangeable and can be treated as commodities. In day-to-day affairs, in other words, one place serves as well as another to convey one's point. On reflection, this is an appalling condescension and a terrible imprecision, the very antithesis of knowledge. The idea that either the Green River in Utah or the Salmon River in Idaho will do, or that the valleys of Kentucky and West Virginia are virtually interchangeable, is not just misleading. For people still dependent on the soil for their sustenance, or for people whose memories tie them to those places, it betrays a numbing casualness, a utilitarian, expedient, and commercial

frame of mind. It heralds a society in which it is no longer necessary for human beings to know where they live, except as those places are described and fixed by numbers. The truly difficult and lifelong task of discovering where one lives is finally disdained.

If a society forgets or no longer cares where it lives, then anyone with the political power and the will to do so can manipulate the landscape to conform to certain social ideals or nostalgic visions. People may hardly notice that anything has happened, or assume that whatever happens—a mountain stripped of timber and eroding into its creeks—is for the common good. The more superficial a society's knowledge of the real dimensions of the land it occupies becomes, the more vulnerable the land is to exploitation, to manipulation for short-term gain. The land, virtually powerless before political and commercial entities, finds itself finally with no defenders. It finds itself bereft of intimates with indispensable, concrete knowledge. (Oddly, or perhaps not oddly, while American society continues to value local knowledge as a quaint part of its heritage, it continues to cut such people off from any real political power. This is as true for small farmers and illiterate cowboys as it is for American Indians, native Hawaiians, and Eskimos.)

The intense pressure of imagery in America, and the manipulation of images necessary to a society with specific goals, means the land will inevitably be treated like a commodity; and voices that tend to contradict the proffered image will, one way or another, be silenced or discredited by those in power. This is not new to America; the promulgation in America of a false or imposed geography has been the case from the beginning. All local geographies, as they were defined by hundreds of separate, independent native traditions, were denied in the beginning in favor of an imported and unifying vision of America's natural history. The country, the landscape itself, was eventually defined according to dictates of Progress like Manifest Destiny, and laws like the Homestead Act which reflected a poor understanding of the physical lay of the land.

When I was growing up in southern California, I formed the rudiments of a local geography—eucalyptus trees, February rains, Santa Ana winds. I lost much of it when my family moved to New York City, a move typical of the

modern, peripatetic style of American life, responding to the exigencies of divorce and employment. As a boy I felt a hunger to know the American landscape that was extreme; when I was finally able to travel on my own, I did so. Eventually I visited most of the United States, living for brief periods of time in Arizona, Indiana, Alabama, Georgia, Wyoming, New Jersey, and Montana before settling 20 years ago in western Oregon.

The astonishing level of my ignorance confronted me everywhere I went. I knew early on that the country could not be held together in a few phrases, that its geography was magnificent and incomprehensible, that a man or woman could devote a lifetime to its elucidation and still feel in the end that he had but sailed many thousands of miles over the surface of the ocean. So I came into the habit of traversing landscapes I wanted to know with local tutors and reading what had previously been written about, and in, those places. I came to value exceedingly novels and essays and works of nonfiction that connected human enterprise to real and specific places, and I grew to be mildly distrustful of work that occurred in no particular place, work so cerebral and detached as to be refutable only in an argument of ideas.

These sojourns in various corners of the country infused me, somewhat to my surprise on thinking about it, with a great sense of hope. Whatever despair I had come to feel at a waning sense of the real land and the emergence of false geographies—elements of the land being manipulated, for example, to create erroneous but useful patterns in advertising—was dispelled by the depth of a single person's local knowledge, by the serenity that seemed to come with that intelligence. Any harm that might be done by people who cared nothing for the land, to whom it was not innately worthy but only something ultimately for sale, I thought, would one day have to meet this kind of integrity, people with the same dignity and transcendence as the land they occupied. So when I traveled, when I rolled my sleeping bag out on the shores of the Beaufort Sea, or in the high pastures of the Absaroka Range in Wyoming, or at the bottom of the Grand Canyon, I absorbed those particular testaments to life, the indigenous color and songbird song, the smell of sun-bleached rock, damp earth, and wild honey, with some crude appreciation of the singular magnificence of each of those places. And the reassurance I felt expanded in the knowledge that there were, and would likely always be, people speaking out whenever they felt the dignity of the Earth imperiled in those places.

The promulgation of false geographies, which threaten the fundamental notion of what it means to live somewhere, is a current with a stable and perhaps growing countercurrent. People living in New York City are familiar with the stone basements, the cratonic geology, of that island and have a feeling for birds migrating through in the fall, their sequence and number. They do not find the city alien but human, its attenuated natural history merely different from that of rural Georgia or Kansas. I find the countermeasure, too, among Eskimos who cannot read but who might engage you for days on the subtleties of sea-ice topography. And among men and women who, though they have followed in the footsteps of their parents, have come to the conclusion that they cannot farm or fish or log in the way their ancestors did; the finite boundaries to this sort of wealth have appeared in their lifetime. Or among young men and women who have taken several decades of book-learned agronomy, zoology, silviculture and horticulture, ecology, ethnobotany, and fluvial geomorphology and turned it into a new kind of local knowledge, who have taken up residence in a place and sought, both because of and in spite of their education, to develop a deep intimacy with it. Or they have gone to work, idealistically, for the National Park Service or the fish and wildlife services or for a private institution like the Nature Conservancy. They are people to whom the land is more than politics and economics. These are people for whom the land is alive. It feeds them, directly, and that is how and why they learn its geography.

In the end, then, if one begins among the blue crabs of Chesapeake Bay and wanders for several years, down through the Smoky Mountains and back to the bluegrass hills, along the drainages of the Ohio and into the hill country of Missouri, where in summer a chorus of cicadas might drown out human conversation, then up the Missouri itself, reading on the way the entries of Meriwether Lewis and William Clark and musing on the demise of the plains grizzly and the sturgeon, crosses west into the drainage of the Platte and spends the evenings with Gene Weltfish's *The Lost Universe,*

her book about the Pawnee who once thrived there, then drops south to the Palo Duro Canyon and the irrigated farms of the Llano Estacado in Texas, turns west across the Sangre de Cristo, southernmost of the Rocky Mountain ranges, and moves north and west up onto the slickrock mesas of Utah, those browns and oranges, the ocherous hues reverberating in the deep canyons, then goes north, swinging west to the insular ranges that sit like battleships in the pelagic space of Nevada, camps at the steaming edge of the sulfur springs in the Black Rock desert, where alkaline pans are glazed with a ferocious light, a heat to melt iron, then crosses the northern Sierra Nevada, waist-deep in summer snow in the passes, to descend to the valley of the Sacramento, and rises through groves of the elephantine redwoods in the Coast Range, to arrive at Cape Mendocino, before Balboa's Pacific, cormorants and gulls, gray whales headed north for Unimak Pass in the Aleutians, the winds crashing down on you, facing the ocean over the blue ocean that gives the scene its true vastness, making this crossing, having been so often astonished at the line and the color of the land, the ingenious lives of its plants and animals, the varieties of its darknesses, the intensity of the stars overhead, you would be ashamed to discover, then, in yourself, any capacity to focus on ravages in the land that left you unsettled. You would have seen so much, breathtaking, startling, and outsize, that you might not be able for a long time to break the spell, the sense, especially finishing your journey in the West, that the land had not been as rearranged or quite as compromised as you had first imagined.

After you had slept some nights on the beach, however, with that finite line of the ocean before you and the land stretching out behind you, the wind first battering then cradling you, you would be compelled by memory, obligated by your own involvement, to speak of what left you troubled. To find the rivers dammed and shrunken, the soil washed away, the land fenced, a tracery of pipes and wires and roads laid down everywhere and animals, cutting the eye off repeatedly and confining it—you had expected this. It troubles you no more than your despair over the ruthlessness, the insensitivity, the impetuousness of modern life. What underlies this obvious change, however, is a less noticeable pattern of disruption: acidic lakes, the

skies empty of birds, fouled beaches, the poisonous slags of industry, the sun burning like a molten coin in ruined air.

It is a tenet of certain ideologies that man is responsible for all that is ugly, that everything nature creates is beautiful. Nature's darkness goes partly unreported, of course, and human brilliance is often perversely ignored. What is true is that man has a power, literally beyond his comprehension, to destroy. The lethality of some of what he manufactures, the incompetence with which he stores it or seeks to dispose of it, the cavalier way in which he employs in his daily living substances that threaten his health, the leniency of the courts in these matters (as though products as well as people enjoyed the protection of the Fifth Amendment), and the treatment of open land, rivers, and the atmosphere as if, in some medieval way. they could still be regarded as disposal sinks of infinite capacity, would make you wonder, standing face to in the wind at Cape Mendocino, if we weren't bent on an errant of madness.

The geographies of North America, the myriad small landscapes that make up the national fabric, are threatened—by ignorance of what makes them unique, by utilitarian attitudes, by failure to include them in the moral universe, and by brutal disregard. A testament of minor voices can clear away an ignorance of any place, can inform us of its special qualities; but no voice, by merely telling a story, can cause the poisonous wastes that saturate some parts of the land to decompose, to evaporate. This responsibility falls ultimately to the national community, a vague and fragile entity to be sure, but one that, in America, can be ferocious in exerting its will.

Geography, the formal way in which we grapple with this areal mystery, is finally knowledge that calls up something in the land we recognize and respond to. It gives us a sense of place and a sense of community. Both are indispensable to a state of well-being, an individual's and a country's.

One afternoon on the Siuslaw River in the Coast Range of Oregon, in January, I hooked a steelhead, a sea-run trout, that told me, through the muscles of my hands and arms and shoulders, something of the nature of the thing I was calling "the Siuslaw River." Years ago I had stood under a pecan tree in Upson County, Georgia, idly eating the nuts, when slowly it occurred to me that these nuts would taste different from pecans growing somewhere up in South Carolina. I didn't need a sharp sense of taste to know this, only to pay attention at a level no one had ever told me was necessary. One November dawn, long before the sun rose, I began a vigil at the Dumont Dunes in the Mojave Desert in California, which I kept until a few minutes after the sun broke the horizon. During that time I named to myself the colors by which the sky changed and by which the sand itself flowed like a rising tide through grays and silvers and blues into yellows, pinks, washed duns, and fallow beiges.

It is through the power of observation, the gifts of eye and ear, of tongue and nose and finger, that a place first rises up in our mind; afterward, it is memory that carries the place, that allows it to grow in depth and complexity. For as long as our records go back, we have held these two things dear, landscape and memory. Each infuses us with a different kind of life. The one feeds us, figuratively and literally. The other protects us from lies and tyranny. To keep landscapes intact and the memory of them, our history in them, alive, seems as imperative a task in modern time as finding the extent to which individual expression can be accommodated, before it threatens to destroy the fabric of society.

If I were now to visit another country, I would ask my local companion, before I saw any museum or library, any factory or fabled town, to walk me in the country of his or her youth, to tell me the names of things and how, traditionally, they have been fitted together in a community. I would ask for the stories, the voice of memory over the land. I would ask about the history of storms there, the age of the trees, the winter color of the hills. Only then would I ask to see the museum. I would want first the sense of a real place, to know that I was not inhabiting an idea. I would want to know the lay of the land first, the real geography, and take some measure of the love of it in my companion before stood before the painting or read works of scholarship. I would want to have something real and remembered against which I might hope to measure their truth.

Spatial Environment and Social Adaptation in Japan - A Traveler's Perspective

Baher A. Ghosheh

Humanity's relationship with and adaptation to the environment have fascinated geographers for centuries. Methodologies have undergone significant changes through the years, but traveling remains a superb tool for learning about people and places. "Going native" allows one to be immersed in a culture and thus develop a fuller understanding of the people and their environment.

My personal experience underscores the fact that traveling is an education. In anticipation of a research trip to Japan, I took four Japanese language courses and read numerous books about every aspect of Japanese life. Consequently, I felt well prepared for my trip. After all, I have traveled to more than twenty countries and have managed to feel at home on every trip! My mind visualized every little detail of Japan based on my preparatory research. As the plane approached Narita Airport, I reviewed the basic facts about the country and practiced the essential phrases.

Preparation and experience

Riding the ultramodern train from Narita to Tokyo, I quickly realized that my "thorough" preparation was inadequate! I knew that I was in for Culture Shock and resigned myself to enjoying it. The taxi ride to the hotel confirmed what I had learned about Japan being a crowded country. What fascinated me was the neat organization and cleanliness of Tokyo.

Like most people, I believe myself to be an excellent driver, though many of my friends will dispute that "fact". Taxi drivers in Japan are a rare breed. They are well dressed, polite, efficient, and skillful. The fact that they wear white gloves and uniforms is a sign of the excellent all-round service one can expect in the "Land of the Rising Sun". Taxi cabs are equipped with a small TV, and some provide FAX services for the business traveler.

My friend Akko was waiting for me in the hotel lobby. I was relieved to see her, for my confidence in my language ability eroded as I heard the Japanese speak at incomprehensible speed. My friend had made reservations for me at a modest hotel. A salary-man's hotel, as it is called in Japan, will provide reasonable lodging at moderate prices. One is advised to make reservations well in advance.

I knew that I was in for Culture Shock and resigned myself to enjoying it.

Being a super hyper traveler, I had not slept in more than thirty hours. Akko suggested that we meet for dinner after two hours. The thought of a hot bath and an hour of relaxation suited me fine. As I opened the door to my hotel room, I was to learn first hand what American business needs to learn about Japan. It is a country with very limited space! The room was very small. My search for the closet was in vain . . . there wasn't one! Looking at the bathtub made me smile. It would be a challenge even for a medium size human to fit into the deep but short tub. The room came with a hot water pot along with tea bags and cups. I sipped a few cups of bitter strong green tea as I awaited my friend's arrival.

Dinner introduced me to another aspect of Japanese life. The very people who set the standards in efficiency, politeness, and hard work during the day turn into wild partiers at night. And yes, those who work together, drink and eat together. For an outsider, it seems that the Japanese drink heavily and eat lightly. Drinking seems to help them cope with the pressure of modern Japanese life. My friend was pleased to see me enjoy a variety of sushi and attempt to down some sashimi — raw fish! I had promised myself to live like the Japanese, and do what they do as much as I could, even if that meant having to swallow some expensive but tasteless raw fish!

Spatial conservation in Japanese life

Tokyo is a lively city, with a population larger than that of New York crammed into an area one-third its size. Riding the subway at rush hour allowed me to observe the Japanese version of human adaptation to the city environment. Subway stations bustled with millions of cummuters going in every direction. Whereas New York subways resemble "survival of the fittest", the Japanese believe in "peaceful coexistence"! They do believe in waiting in line. Pushing and shoving are unheard of. Japanese subways are clean and efficient people-movers.

An obsession with reading does not abandon the Japanese even in an overcrowded subway car. While I felt a shortness of breath, everybody around me folded a magazine or a newspaper several times so that it would fit into his/her limited space and then they proceeded to read! I couldn't help but smile in amazement and amusement at all the business people who calmly read cartoons on what seemed to me an intolerable ride.

Spatial adaptation is an ever-present theme in Japanese life. As necessity is the mother of invention, it is only natural that space-scarce Japan should lead the way in spatial conservation schemes. The neo-determinist geographers face great difficulty in explaining Japan's economic prowess in spite of the extreme environmental constraints. The "just-in-time" delivery system (James Rubenstein, FOCUS, Winter 1988) is but another example of ingenious human adaptation to the limits imposed by the environment. Perhaps it is true that abundance allows

From *Focus*, Winter 1991, pp. 19-21, 36. © 1991 by the American Geographical Society. Reprinted by permission.

for wastefulness while scarcity breeds innovation.

Throughout my trip, I noticed many phenomena related to space or, more precisely, the lack of it. Every train station seemed to have one or more bicycle parking lots. Bikes are often used for grocery shopping and travel to and from the subway.

The parking situation is another indication of limited space in Japan. I have spent a long time marvelling at how anybody could park a car in such small spaces. I would probably lose half the car trying to get out of the garage. It is a wonder large American cars do not sell in Japan?!

The absence of closets in Japanese homes and the multiple uses of a room reflect land scarcity. The living/dining room is swiftly transformed into a bedroom in the evening by pushing the table into a corner and bringing out the futons. Sliding doors serve to divide and/or enlarge a room for specific purposes. Initially, Japanese homes may look like toy houses: they are cute, small, and space efficient. An average Japanese home or apartment is about one-half the size of an American home.

Social adaptations to space are amusing and even surprising. I couldn't help but notice that many people wear white masks. Upon asking for the rationale behind the white masks, the answer was simple. People with a cold wear a mask so as not to spread their germs to the thousands of people they pass by! Since then, I feel guilty every time I sneeze in public. Is it really enough to excuse ourselves for spreading cold germs to innocent people we encounter on the streets and other places, or should we follow the Japanese example and wear masks to protect those whom we care about from our germs!

Traveling and wayfinding

A traveler to Japan must take time to ride the Bullet train, *Shinkansen*. Though expensive, it combines efficiency with the luxury of sight-seeing. The *Shinkansen* connects the main cities of Japan — especially those on the main island of Honshu. Tokyo, Nagoya, Osaka, Hiroshima are the main terminals with connections to Yokohama, Kamakura, Kyoto, Nara and other cities. Work has recently been completed on a tunnel connecting the northern island of Hokkaido to the main island. This underwater tunnel is the world's longest. The tunnel, extending 55 kilometers, took twenty-one years to build at a cost of $3750 millions. Riding the *Shinkansen* allows one to enjoy the richly diverse landscape of Japan. The Tokyo-Osaka trip provides a glimpse of Mount Fuji, and illustrates the maximum utilization of space that characterizes the

JAPAN

Hokkaido Island
Sapporo

Sea of Japan

Sendai
Niigata
Honshu Island

Yokohama — Tokyo
Kyoto Mt. Fuji ▲
Osaka Nagoya Kamakura
Hiroshima Nara

Fukuoka

Nagasaki

Kyushu Island

Km
0 100 200

Pacific Ocean

Miami University GIS Lab

country. Rice paddies dot the landscape, and tennis and other recreational fields are abundant alongside the rail tracks — every inch of land has been put to use!

The *Shinkansen* constitutes a viable alternative to crowded highways. As is the case with all trains in Japan, *Shinkansens* are almost always on time! Though the train has a cafeteria, vendors will periodically pass through the cars selling lunch or dinner boxes. These boxed meals are healthy, relatively cheap, and come in a wide variety to satisfy every taste!

Finding one's way in Japan is a challenge to our culturally founded spatial perceptions. The Japanese are center-oriented. The center is emphasized in the house as the hibachi (heater) is placed there — the whole family huddles around the hibachi to eat, drink, or watch T.V. Streets in Japan have no names. Intersections — central points — are named and have signs that provide adequate directions to the surrounding streets. After World War II the Americans named the streets in Japan as they failed to grasp Japan's spatial orientation. Finding a house is another lesson for the *gaigin* (foreigner) in Japan. Houses are numbered according to age — the time they were built — rather than their relative location in the street. Even the Japanese find it necessary to stop by a *Kooban* (police box) for specific directions. Fortunately, *Koobans* are found on almost every corner and the Japanese police are eager to provide assistance. In

Japan, age commands respect even in the numbering of houses!

The abundance of *Koobans* may be one of the reasons behind the country's low crime rates. Cultural values, in my opinion, are a more important factor in making Japan a safe country. Women feel safe enough to walk alone even at late hours of the night.

In Japan, age commands respect even in the numbering of houses!

A *gaigin* is constantly surprised by the many contradictions that characterize the Land of the Rising Sun. The apparent lack of space is puzzling to a visitor as he sees thousands of spacious temples! Even in Tokyo where a square foot of land will command as much as $50,000, temples abound! God would want people to have more of the scarce land to themselves, I thought to myself. Land and real estate prices are astronomical. The area that holds the Imperial Place in Tokyo commands a higher price than the entire state of California! Most Japanese have no illusions about owning a home — it's just beyond dreaming!

Conformity, tradition and obligation are persistent

The Japanese are fashion-conscious and believe in conformity. Students and workers wear uniforms, start the day by singing the school or company song and performing stretching exercises and low-intensity aerobics before work or classes. American students will be shocked to learn that their Japanese counterparts are responsible for cleaning the school. This is done before the beginning of classes! Students and workers alike are careful to use the proper form of speech in their conversations. A higher level of speech — a more polite or humble form — is used when addressing someone of higher status.

Many traditional beliefs exist alongside the practices of ultramodern Japan. The same Japanese who lead the world in technology and innovation still believe in devils and demons. It is not unusual to find a highly educated Japanese seeking advice from a fortune teller or taking a horoscope very seriously. Temples sell letter-style horoscopes and many believe that blood-types and zodiac signs are a basis for matching people. Ancient religious practices are very much alive.

The Japanese perception of civic duty is indeed admirable. People invariably do what they are expected to do. Saving face and maintaining harmony are the main guidelines for social interaction. Individuality is suppressed: "one, by himself, is but a grain of sand on the beach". The group provides its members with identity, support, and a deep sense of belonging. Those who dare to challenge this group mentality are harshly treated: "the nail that sticks out must be hammered down"!

I was very pleased and honored to accept an invitation to a colleague's home. The Japanese generally entertain elsewhere and the home is strictly preserved for the family; I was the first non-family member to enter their home. My friend Hiro, an international business executive who has travelled throughout Asia, Europe, and the Middle East, provides a good example of the many contradictions that exist in Japan. He is a workaholic who works six fourteen-hour days a week. He leaves the house at 7 am to return at 11 pm. Though he has lived in many countries, my friend seemed conservative, even closed-minded and ethnocentric at times. In discussing interracial marriage, my friend refused to even consider the possibility that either of his two eligible daughters wed a non-Japanese. The Japanese are unique and must maintain their purity, he asserted. Both daughters will marry respectable and compatible Japanese men. A husband must meet certain social, educational, financial as well as personality criteria. My friend has already contracted the services of a matchmaker to find suitable partners for his daughters.

I could not help notice that my friend

was feeling the effect of alcohol, as he had already downed six large bottles of Asahi beer. Two more will get him to his daily quota. Alcohol helps him overcome the many pressures he faces daily. Hiro frankly told me that he sees nothing wrong with drinking or even an occasional on-the-side sexual entertainment. As long as it does not affect family life, a man may pursue physical pleasures outside the home. The wife is too busy caring for and educating the children. She cannot be expected to sexually entertain her husband also! Somehow I got the impression that he would rather have somebody else do the entertaining even if the wife was willing and able to provide it! The seemingly conservative Japanese are, at times, permissive and corrupt. I later realized that Japanese TV regularly shows explicit movies at late hours.

What I appreciated most about Japanese TV is the availability of bilingual programming where a viewer can choose to watch the same foreign show in Japanese or in its original foreign language. This is a most valuable service for people who wish to learn foreign languages. This service is also greatly appreciated by foreigners who long for home — especially those who have not yet mastered the Japanese language.

The telephone system is another example of Japan's superb services. One need not fumble with coins; instead, purchase a declining-balance phone card which may be used to make local as well as long-distance calls. These phone cards come in a variety of denominations and have attractive designs. I bought two 500 yen cards rather than one 1000-yen card. They will make good souvenirs and may be used as book markers. I wish AT&T or some other phone company would provide a similar service in the U.S.

It is not easy not to notice the numerous exotic cards that decorate every phone booth in Japan. Ads with photos of attractive girls promising to perform "special" favors for the advertised price are everywhere. So much for conservative Japan! Businessmen see these services as a way to relieve tensions and many women view them as a way to finance their educations and other expenses.

Walking, I believe, allows one to have a closer look at a city. Falling asleep was never so easy for me, as walking for hours every day took its toll on my body. I was fascinated by the many uses vending machines are put to. One can buy milk, rice, cigarettes, soft drinks, juices, and yes even alcohol from the many vending machines found along most streets. No need to worry about checking ID cards, the Japanese sense of responsibility and civic duty will ensure that only those who are eligible to purchase alcohol actually

do so. American teenagers would love to see this kind of service at home!

The Japanese have an extreme sense of responsibility. Fulfillment of obligations, both social and economic, is the norm. Many will pay with their lives for their failure. Students who fail to make the grade on the university entrance or national exams often commit suicide. Japan has the dubious distinction of having one of the highest teenage suicide rates in the world. In carrying on with the Samurai tradition, some managers of failing businesses, police officers who fail to resolve crimes, and people who cannot meet their obligations take their own lives. Social pressure to perform and conform is enormous in Japan.

A visitor to Japan will be reminded of "Good Ole American Service" — rarely found in the U.S. today but almost everywhere in Japan. The moment a customer steps into a store, he is immediately greeted with "Irrasaimase" - welcome! Every employee in sight will politely and seriously utter the phrase. As a customer steps out of the store, he is thanked for shopping or just stopping by: "Domo arrigato goziamasita" — thanks for what you have done! Even the highest-ranking employees actively partake in welcoming, talking to, and thanking the customers. It is customary for the company president to be at the door as the store opens, to spend at least half an hour welcoming the first customers into the store. Courtesy, efficiency, and good service are expected and granted everywhere one goes. Perhaps the fact that the Japanese use the same term, "Kyaku", for customer and guest is indicative of the kind of treatment a customer gets. Oh yes, there is no tipping in Japan. Employees are to perform work as best they can and are not to expect to be tipped for doing their jobs! Restaurants in the U.S. should send their employees for training in Japan! "The Customer is God" say the Japanese.

Ramen and raw fish

Experiencing Japan means acquiring a taste for the food. One may be surprised by what the Japanese eat but if we were to objectively compare theirs to our food, their food — as reflected by international health statistics — is a much healthier diet. Taste is acquired over time and one will, if he eats anything long enough, develop a liking for it!

Whereas the average Japanese is five inches taller than the previous generation, reflecting today's more nutritiously balanced diet, Japanese food remains largely authentic. Not only is eating local food good for a traveler's health and shows a healthy curiosity, it is also good for one's pocket. Food is relatively more expensive in Japan and imported foods are very expensive there. Nevertheless, a traveler can enjoy a delicious and hearty lunch for a few dollars. Noodle stores — *Ramenya* — are a good bet as they provide a wide variety of steaming noodles with vegetables and some kind of meat. To help the frustrated tourist, ramen shops and other restaurants display wax models of the dishes they serve and the price of each dish — what you see is what you get and for the specified price. Since the noodles are extremely hot, the Japanese slurp loudly as they eat them with chop sticks. It may seem impolite to make such loud noises while eating and it is with anything except ramen, where it is just a practical matter and the loud slurps serve as expressions of gratitude for the chef! One would want to eat the dish hot, and yes feel free to sip the broth from the bowl. A clean bowl is an indication of a tasty broth, as one may remember from the movie "Tampopo" in which a woman spends her life perfecting the taste of the Rames's broth!

The Japanese use little oil or fat in their cooking and rice remains the main staple of the Japanese diet. Another way to experience authentic Japanese living and dining is to stay at a *Ryokan* — Japanese inn. These range in price to fit every budget, usually including two meals, breakfast and dinner. As one checks into a *ryokan*, he is invited to take a bath. The water is very hot and one better test the water before it is too late! A bath kimono is provided as well as hot water and tea.

The inn's small rooms are separated by paper walls. In the evening, the maid will bring out and unfold the futons. For people who suffer from backache, the futons have therapeutic value.

Breakfast at the *ryokan* reflects Japan's traditional life. Residents sit on the floor around one large table, indicating Japan's ever-present group mentality. Breakfast consists of a bowl of steamed rice, a raw egg, some fish, Japanese pickles, and tea. The thought of eating fish and raw egg over rice did not appeal to me. However, in keeping up with my promise of "going native", I broke the raw egg into the provided bowl, added some soy sauce to it, beat the mixture well and poured it on top of the rice. Despite my initial reluctance and apprehension, I can honestly say that I enjoyed breakfast.

There are some foods that would require a longer time to appreciate! Even now I cannot understand why one would pay fifty dollars or more to eat a strip of raw fish. Worse yet, why would one risk his life and spend 100-150 dollars to consume the poisonous *Fugo* fish? Perhaps it is the sport of it or the challenge of risking one's life . . . but it certainly can *not* be the taste!

A word of caution for the hungry traveler, Japanese food is digested quickly and one needs to keep some snacks handy. There are many exotic and tasty snacks.

Things are small and cute in Japan: houses, toys, even dishes. Upon ordering a pizza at a European style plaza in Shinjoku, I was dismayed to realize that the pizza pie, supposed to be my dinner, could not be more than an appetizer to me! The Japanese like to socialize in *Kissatens* - tea houses. The *kissaten* offers coffees from all over the world along with scrumptious but not-so-sweet desserts. A cup of coffee will cost up to five dollars but the *kissaten's* atmosphere is well worth it. One may venture and try uniquely Japanese delicacies such as ginger or green tea ice cream and cake.

After spending three weeks there, my fascination with Japan experienced a tremendous boost. The Japanese politeness, helpfulness, industriousness, dedication, hard work, and consideration is indeed admirable. I was so overwhelmed that I would have gladly given everything I owned to the first Japanese to ask for it.

My trip has also taught me a most valuable lesson, "where there is a will, there is a way". The rise of tiny resource-poor, war-devastated Japan to economic prominence in a few decades speaks not only of Japanese determination, it also points to the tremendous human potential world-wide. Other nations can emulate the Japanese and with hard work and determination, control their destiny and improve their living conditions.

THE BALKANS

Ethnic identity versus the modern nation

Koča Jončić

The war in the former Yugoslavia has gone well into a second year, claiming thousands of lives and displacing more than 3 million people. What has made the Serbs, Croats, and Muslims kill one another? What drives a village that months ago coexisted with the neighboring village to destroy it and rape its inhabitants? Everyone knows that the Serbs started the war and that Serbian leader Slobodan Milosevic organized it. But why? With the partition of Yugoslavia, Serbs feared that they would become simply national minorities in the new states of Croatia, Bosnia, and Macedonia, separated from the rest of the population not only by ethnicity, language, customs, and religion but also by traditions.

The reappearance of the old idea that the creation of Greater Serbia and "ethnic cleansing" would save Serbia from destruction became the only ideological axis. That is how the terrorization of the non-Serb population living in areas inhabited by Serbs started, though there have also been cases in which the non-Serb populations have reacted against the Serbs. When people start killing each other because of their different religions and traditions, there are no winners, and war becomes absurd.

The formation of psychological ethnicity based on religious and national ideals clashes with the modern and rational concept of state as defined by the civilized world. Instead of the state as an entity regulated by international law and institutional consensus, the romantic concept of nation-state dating to the end of the medieval period re-emerged in Yugoslavia. The people of the former Yugoslavia started to view themselves not as citizens of a state in a particular territory but rather as members of a psychological community in that territory. And they demanded that this territory belong exclusively to them, leaving no space for others who think, believe, or live differently.

In this way, Serbian populations living in Croatia or Bosnia proclaimed the formation of independent Serbian states within other states, giving rise to the autonomous Serbian republics in Knin and Krajina in Croatia. Politically, the Serbs took treacherous advantage of the mistakes of opponents who faced two alternatives: give autonomy to the Serbs within their own territories or suppress them. When the other parties refused to compromise, the Serbian revolt against the newly created Croatian and Bosnian governments became a massacre for the non-Serbs, organized by Belgrade, realized by the nationalist paramilitary, and backed by the Yugoslav army.

When the Republic of Bosnia-Herzegovina was proclaimed, many Serbs (who represent about 31 percent of Bosnia's population) boycotted the election and forcibly established authority over 70 percent of the Bosnian territory. These acts left no room for political and diplomatic reasoning. The Serbs took over as much territory as they wanted, the Croats took as much as they could, and the Muslims remain practically without a homeland.

Who could have stopped the Serbs? The first hope grew from within: Serbian democratic intellectuals in Belgrade have exerted pressure on the regime, as in the student revolts of March, 1992. The international embargo of Serbia, on the other hand, has merely heightened pressure against those people who want to live in peace. Opposition forces lost perhaps their last chance to put the country back on the path to peace with Milosevic's victory in the December, 1992, election. Serbs voted in favor of domination and war.

The radical Serbs openly proclaim that their game has just started: When Russian President Boris Yeltsin is overthrown and the nationalists take power in Moscow, they say, "the West will not threaten the Serbs." The United Nations and the European Community are left to take actions to help the victims and establish a cease-fire. The Serbs can be considered the victors in Bosnia, although history will condemn them.

Throughout the war in Bosnia, international military intervention has proved highly problematic. The Yugoslav army, ranked fourth in Europe and well equipped, is not like the army of Iraq's Saddam Hussein. Serbian-controlled territories are much more numerous than before, "ethnically cleansed," and inhabited by Serbs; UN troops could not fight the Serbs in a frontal war as they did Saddam's forces.

One solution that might prevent expansion of the war would be sending troops to Macedonia to threaten Milosevic with intervention in case he starts war in this zone. Milosevic may sacrifice the Serbs of Macedonia to gain Greater Serbia.

The Serbs' next target could be [the Albanian enclave of] Kosovo. The Serbian hysteria in Krajina and Bosnia will spread as a sacred war among Orthodox Serbs, who consider the town of Peć their Jerusalem and the Kosovo plain the motherland of the fight against the Muslim Albanians. The Serbs' plan is already known: They want to cleanse the northern region of Kosovo and resettle it with Serbian refugees from Bosnia. Peaceful dispatch of [UN] troops to Kosovo without the approval of Serbia is considered virtually impossible so long as Kosovo is, from the international point of view, considered a domestic problem of Serbia. But if Kosovo gives up its passive resistance, it will lose its international support.

Thus, the war could engulf all the Balkans.
— *Gramoz Pashko, "Koha Jonë" (biweekly),*
Lesh, Albania.

A Serb on the Roots of War

He Blames His Own—And The West

KLASSEKAMPEN

By Grete Gaulin

Ljubisa Rajic is one of the few who struggle from within against the political degeneration and brutalization of Serbian society. In the swamp of Serbian nationalism and chauvinism, he searches for signs of humanism and tolerance. A magazine editor and a professor of Scandinavian languages, his mission is to draw the attention of the rest of Europe to what is happening inside Serbia. He is involved in most of the republic's opposition organizations.

"These signs of political tolerance, of opposition work, will not last long," he says. "The independent press, the high schools and universities—they are all under [Serbian President Slobodan] Milosevic's control. Recently, 1,500 employees of Serbia's state-owned radio and television system were laid off; the police entered the studios, harassed the journalists, and smashed furniture and equipment. At the same time, the universities and schools lost their independence, the independent trade-union movement was banned, and there was a large-scale purge of judges. I see no

A Menacing Air

STATESMAN

The bar in Belgrade's old town was crowded and noisy, until a group of thugs suddenly appeared at the window, leering and shouting. One of the owners quickly shut the entrance, but a couple of the intruders kicked their way in, smashing the door with their boots. One of them hit the owner several times in the face. Then, as suddenly as the violence had erupted, it stopped, and the thugs ran off into the night.

Once one of the safest and most cosmopolitan cities in the Balkans, if not in all of Europe, Belgrade is now a place with an undercurrent of menace, where bar doormen frisk customers for guns. Crime is increasing and often involves weapons, now easily available.

Many people rarely venture out, partly because they cannot afford to and partly because they no longer feel safe. Gangs compete on the streets, demanding protection money from bar and restaurant owners or a share in the business. Non-payment or refusal to pay is a bad idea; a shooting or perhaps a hand grenade soon follows. But not everyone is suffering in the new Chicago of the Balkans. While the majority struggle to survive 20,000-percent annual inflation, for others, war is good business.

Meanwhile, those on fixed incomes are also victims of the war, almost as much as the 500,000-plus refugees (mainly Serbs) sheltering in Serbia and tiny Montenegro.

Sanctions and the costs of the war have helped cause the implosion of the economy, already weakened by the collapse of Eastern-bloc markets for Yugoslav goods. Hospitals are short of drugs, and damaged medical equipment lies unrepaired, say Belgrade doctors.

Serbs believe that the international media has closed its eyes to Serbia's forgotten victims. "The world . . . pretends that this is only a tragedy for Croats and Muslims," says Radmilla, a Serbian refugee from Mostar, a town in western Bosnia. "I did not want to leave my city and my friends." Now Radmilla lives in Belgrade, where she and other Serbian refugees say that they face prejudice from the local Serbs.

As the capital of the former Yugoslavia, Belgrade was the archetypal melting pot. But now, Muslims avoid the mosque, which has been repeatedly attacked, and Croats keep quiet about their ethnic origin. Former Yugoslav President Josip Broz Tito's dream has turned into a nationalist nightmare.
— *Adam LeBor, "New Statesman and Society" (liberal weekly), London.*

future for this country." Instead, Rajic believes that Milosevic will feel the need to demonstrate that he is in control. "And it will come about without any real opposition from the people, without any real struggle against the coming dictatorship."

How was it possible for Milosevic, a man with fascist tendencies, to win a free election? "Eastern Europe, including Russia, has lived under various forms of dictatorship for many hundreds of years," Rajic explains. "Dictatorships were not something new that came with communism's success . . . . The modernization process is now being forced upon the East European countries in the course of just a few years. Such a sudden upheaval creates intense angst in a political culture where, until now, all had their fixed places and privileges."

Rajic views the Balkan conflict from a historical perspective: "Eastern Europe never completed the formation of nation-states. The political map of Eastern Europe is the result of the breakup of the Russian, Hapsburg, and Ottoman empires. We are nations in puberty. The result is several hundred million hysterical people who stand in front of their national mirrors and see reality through it. That is the reason why people support [Russian President Boris] Yeltsin, [Croatian President Franjo] Tudjman, and Milosevic."

Rajic emphasizes that the people in Eastern Europe have not received any of what they were promised. On the contrary, they have lost work, and wages have been drastically reduced. "The West is seen as an enemy," he says, "who takes away from us something we need."

At the end of the 1980s, Rajic notes, a generational change in Yugoslavia ushered in a new elite who sought to redistribute political and economic power. They sought a geographic and national redistribution as well—thus, the war began.

He believes that there is nothing in their culture that causes people in the Balkans to be more willing to come to blows than people elsewhere. But Rajic recognizes motives of vengeance that underlie the present bloody conflicts: "The Serbs in Croatia and Bosnia feared a recurrence of the massacres of the second world war, and the Muslims feared the same. The Croatian nationalists did not hide that they sought to finish what they had begun during the war."

The West must accept its share of responsibility for the war, Rajic says. "One of the major errors by the West was its assumption that Yugoslavia could not exist because there are so many different ethnic groups. That was the starting point for granting recognition to Croatia and Slovenia and, later, Bosnia. But now, the Western powers seek to keep Bosnia together, which proves the inconsistency of their policy. It is typical for the West to see the conflict in black and white: The American characteristic of defining the 'good guys and bad guys' has indeed come through . . . .

"In many ways, the West wanted the war. It served the goal of preventing establishment of a Muslim state in the Balkans and was useful for welding Europe together and creating a West European army. All the demons could be exported to Yugoslavia, and the United States and the European Community could show their muscle."

When the Western countries sought an end to the bloody conflict, he says, it was because some aspects of the war did "not suit their strategy. The refugee problem creates trouble for Europe; [the Europeans] must consider public opinion and the danger that Islamic countries could have become involved. Had it not been for these factors, the West would have just let us kill each other like animals."

—*Klassekampen* (leftist), Oslo.

The Children Of the Rapes

Young Victims of 'Ethnic Cleansing'

stern

By Daniela Horvath

Emina is a strong, lively little girl, four and a half months old, with flame-red hair. She lives in an orphanage for refugee and war-victim children run by the Roman Catholic charity Caritas in Zagreb, for she is an orphan, although both her parents are alive. Emina was born in the Petrova Women's Clinic in Zagreb. Her mother, 17-year-old Mirsada (not her real name), from a small town in Bosnia, did not want to take her baby home or even see her. When Emina's umbilical cord was cut, a hint of a smile crossed Mirsada's face. It was the first time she had smiled for months, remembers Jarmila Škrinjarić, the clinic psychiatrist. Two days later, Mirsada disappeared from the clinic.

Only one thing is known about Emina's father. He is one of the countless Serbian soldiers, or *chetniks*, who raped Mirsada during the four and a half months she was held captive, along with 200 other Bosnian girls and women, in a rape camp near her hometown. When she was liberated during a Bosnian army counterattack, she was already very pregnant.

After weeks on the road, she ended up in the Zagreb clinic, 27 weeks into her pregnancy. The doctors rejected an abortion as too risky by that time. "In the weeks before she gave birth," reports the psychiatrist, "she had a determined but passive-aggressive feeling toward the life growing inside her." Mirsada spoke of the baby in her belly as a foreign object, as "that thing," or "that curse." She wanted desperately to be free of it.

According to estimates by local and international aid groups, at least 30,000 girls and women—mostly Bosnians and Croats—have been raped by Serbian soldiers and partisans. Many have become pregnant. They are the victims of a perfidiously planned strategy that is part of the "ethnic cleansing" of areas claimed by Serbs. This strategy was first apparent during the fighting in Croatian east Slavonia and has reached insane dimensions in Bosnia. The first "*chetnik* babies" are now growing up. "There are certainly hundreds of them," says Zvonimir Šeparović, a former foreign minister of Croatia who now directs the Documentation Center for Genocide and War Crimes in Zagreb.

The rape victims' "psychosexual trauma is one of the worst traumatic experiences that a person can suffer," says Vera Folnegović, a psychiatrist at the Vrapče Clinic in Zagreb. She has been working to help rape victims—chiefly women, but men and children as well. Most of them show symptoms of post-traumatic stress disorder, such as sleep disturbances and terrible nightmares, identity problems, severe depression, and even suicidal tendencies. "Many of them will be overcome by these traumatic experiences even years from now," says Folnegović. But in wartime, neither money nor trained psychiatric personnel are available.

Jelena Brajša, the director of Caritas in Zagreb, has been sending doctors and nurses to hospitals and refugee camps to find rape victims and their children. She has planned two residences, each with 50 apartments, for rape victims and other refugees.

Brajša, a Roman Catholic, repeats the line taken by Pope John Paul II, who recently appealed to the rape victims to reject abortion. The women, he said, should "transform these acts of violence into acts of love and acceptance by accepting the enemy within them." For now, the government of Croatia, which is church-influenced, does not intend to change the three-month limit on legal abortions. A group of Catholic doctors has organized to deny abortions to rape victims. Many women want to abort their children or give them up for adoption, because their families threaten them: "Don't bring home any *chetnik* babies, or there won't be any place for you here, either."

Many women who cannot get abortions try to rid themselves of the babies in other ways. "They simply give the infant away immediately after birth," says Šeparović of the war-crimes center, "or they kill the babies . . . although we have never been able to prove it."

There is also concern about the material survival of refugee women with children, for they have no jobs, no social connections, no future. "If only more help could be provided in these areas, I am sure that many of them would keep their children," says Brajša.

Psychiatrist Škrinjarić and her colleagues have reasons to think otherwise. For example, there is Amra (name changed), 30, from Goražde, who will soon give birth in the Petrova Clinic. In 1992, *chetniks* invaded the part of the eastern Bosnian city where she lived. She hid with 10 women in her house until a group of masked men broke in one night. Amra and another woman were dragged into the bedroom, stripped, and forced to perform oral sex on several men, who also raped them. Among them were two Serbian neighbors. "After this, we are going to take you to a concentration camp, so you can give birth to the babies we made tonight," bragged the Serbs.

Later that night, Amra was able to get away, but, as she made the weeks-long trip to safety in Zagreb, she realized that she was pregnant and tried to rid herself of the child with herbal potions. She is awaiting labor, heavily sedated. She knows that her husband in Goražde, if he did not starve or freeze to death during the winter, would throw her out if she came home with a "*chetnik* bastard." So she intends to give the child away immediately after birth.

The governments in Croatia and Bosnia would rather not deal with the growing number of *chetnik* babies. Psychologists and family-law experts advise giving them up for adoption outside of the region. Croatia is now recommending that Bosnia cede to it the control over these citizens and set up a means for legal adoptions, perhaps under United Nations auspices. At a meeting about the rape victims, former Foreign Minister Šeparović warned about "Mafia-like gangs" that might try to set up a profitable black market in babies, as they have done in Romania. "The first jackals," he said, "have already begun to circle around these poor victims."

— *"Stern" (newsmagazine), Hamburg.*

Seeking a Lasting Role for Russia

A CHAMPION OF MINORITY RIGHTS?

By PAVEL KANDEL

The crisis in Yugoslavia could now be reaching a culmination that could lead either to the long process of a post-war peace settlement or an expansion of the armed conflict beyond the boundaries of Bosnia and Herzegovina, in which case the conflict would probably become internationalized.

The Serbian side in Bosnia has achieved almost all of its military objectives. The 1.4 million Bosnian Serbs, who made up only 31 percent of the republic's population (4 million), captured two thirds of its territory. Similarly, after the Croatian side (population 820,000, or 18 percent of the total) established effective control over most of the areas in Bosnia that are inhabited by Croats and created their own state there, they had reason to be satisfied.

The Muslims (with a population of 1.9 million, or 41 percent of the total) lost control over most of the republics. This explains both their bellicosity and their active search for armed support from outside.

The continuation of sanctions and of pressure from the world community is necessary for the long-term success of the peace process. After all, the only alternative is direct armed intervention in Bosnia by international forces. If intervention does occur, the change in the balance of power in Bosnia will lead to a resumption of fighting in Krajina and an armed uprising by the Albanians in Kosovo, which will spill over into western Macedonia. All that remains is to guess who will find these developments too great a temptation: Albania, Serbia, Bulgaria, Turkey, or Greece.

Therefore, the international influence in Bosnia must also become more flexible. Regardless of sympathies and antipathies, Yugoslavia (Serbia and Montenegro) must remain the chief target of international pressure: It is the strongest of the combatants, and, considering the dimensions of the Serbian-occupied territories in Bosnia, it will have to make the greatest concessions.

It is equally obvious that a massive presence of international peace-keeping forces is the sole means of enforcing agreements. Only this approach will compensate for the inequality in the adversaries' power and prevent Serbia and Croatia from attempting a de facto division of Bosnia. If they annex parts of Bosnia, the basic principle of the postwar European order—the inadmissibility of changing existing borders by force—will have been trampled. An extremely dangerous precedent will have been set, directly undermining stability in the Balkans, in Central and Eastern Europe, and, if one peers a little further, throughout the former Soviet Union as well.

No matter when or how the Yugoslav crisis ends, no matter who is in power in Belgrade and Moscow, Serbia's top priorities will be to break out of its international isolation and secure rehabilitation in the eyes of the world community, revive its devastated economy, restore relations with its Balkan neighbors, and obtain large-scale economic assistance and "return to Europe." Considering the state of Russia's economy, Russia will prove to be a less significant partner for Serbia than will the European Community (EC) and the U.S., no matter how great Russia's efforts to maintain a "historical alliance."

The only way to avoid a general Balkanization of the Balkans, and then of the entire post-totalitarian world, is to preserve the integrity of the former republics of Yugoslavia, above all Bosnia and Herzegovina. To do so, the great powers and the world community must demonstrate their commitment to the principles of European law and order by proving that it is impossible to violate them with impunity.

True, the Western powers have not always been consistent and completely impartial in upholding the general principles of law and order. This was especially apparent in the overly indulgent attitude toward Croatia, which enjoys the special favor of Germany as well as of other states in Central Europe. The Western powers fell short of their mission, too, on the issue of recognizing the independence of Macedonia. By yielding to the insistence of Greece, which opposes the recognition, the EC countries and the U.S. cheapened their own efforts to stabilize the situation in the Balkans and undermined the already shaky foundations of a weak state.

The temptation is great for Russia to use other countries' inconsistencies to justify its own and to balance the anti-Serbian tilt in the Western powers' policies with a pro-Serbian slant in its own. But this would provoke the Serbian side into actions that Russia could not support, not to mention the consequences for itself. Yet filling the vacant role as the most consistent champion of international law and as the defender of ethnic minorities would allow Russia to show independence in a mission that would be irreproachable from the standpoint of the world community. It would also insure Russian influence in the region.

When it insisted on reunifying the regions inhabited by Serbs in Croatia and Bosnia with Yugoslavia, Serbia gave up the right to prohibit the Albanians in Kosovo and the Hungarians in Vojvodina to unite with Albania and Hungary, respectively. The same went for Croatia. While it sought to annex the parts of Bosnia that are inhabited by Croats, Croatia could not deny its own Serbs the right to unite with Yugoslavia. Preventing the federalization of Bosnia is equivalent to protecting Serbia and Croatia from a similar process. Russia's role as a protector of the rights of ethnic minorities, which also meets its own concerns in the former Soviet Union, heightens the stakes of Albania, Hungary, Croatia, and Bosnia in such a role. Yugoslavia will also see its interest in this Russian role once Belgrade renounces its ethnic territorial claims and takes the path of civilized defense of its compatriots.

Playing the role of the most ardent champion of universally recognized principles would make Russia's position invulnerable in terms of international law and more advantageous than the approach of the Western powers, which have displayed political weakness with respect to Macedonia. Stability in Macedonia is even more important for tranquillity in the Balkans than is achieving peace in Bosnia. Russia's role as defender of Macedonian independence makes it an important partner for Bulgaria, Turkey, Serbia, Greece, and Albania, whose interests intersect there.

Considering Russia's skimpy economic capabilities, the potential for its political presence in the Balkans should be realistically assessed. If only for this reason, the role of most consistent champion of legitimacy is not only the preferable role for Russia but the only practical one.

—*"Nezavisimaya Gazeta" (liberal), Moscow.*

Is a Real Peace Possible?

BREAKING HISTORY'S CYCLE

Cambio16

By MILOVAN DJILAS

There are two features that stand out in the conflicts of the former Yugoslavia, whose most ruthless chapter has been the war in Bosnia-Herzegovina. Without identifying those features, no one can make sense of the fighting or envision its outcome.

The first is the tendency of the peoples of Yugoslavia, or rather of their respective nationalisms, to create nation-states. That current has a long history. At least for the Serbs and Croats, it took form as a political strategy during the second half of the last century, and it was fulfilled in 1918 with the formation of Yugoslavia around the Serbian royal house. Croats and Slovenes consented to live in that state, because it protected them from their imperial neighbors. The Serbs accepted it, too, so that all of them could live in a single country. The kingdom of Yugoslavia was built on the belief, propagated by the most prominent intellectuals of the country, especially Serbs and Croats, that its population was a single nationality.

In 1941, with the first blow dealt by its Nazi-fascist conquerors, the single Yugoslav state fell apart. It was revived by the communists and became a federation of national republics. The same idea was applied within Bosnia, where three nationalities—Muslims, Serbs, and Croats—lived side by side. Bosnia became a republic because it was a melting pot of those three nationalities. But the case of Bosnia provides an example of the harshest cruelty: Because its population lives in a melting pot, the ethnic groups were unable to separate from each other without conquering territory and carrying out "ethnic cleansing."

The second feature is that the war in Bosnia has been led by nationalist movements whose internal organization is nondemocratic, movements that are intolerant and chauvinistic toward neighbors of different religions and nationalities. Real ethnic-religious motives are involved, but they are secondary. Religion is used to conceal political and ideological motivations. What has motivated everyone in Bosnia is the deep-rooted idea of creating nation-states—with a single difference: Serbs and Croats wanted to unite with their respective fatherlands, while the Muslims wanted to create a Bosnian state that they would dominate by their numerical strength and by appealing to Islam.

None of the Bosnian movements is Nazi-fascist, and none of them has a solid ideological, social, or political program. But the groups do not differ very much from fascists in their intolerance and fundamentally racist attitude toward the rest of the Bosnian population. Each of the three Bosnian movements is fueled by national myths and an uncritical view of its past. Each tries to arouse the masses by reviving the crimes they committed against each other in the past. Brutality against the innocent is the hallmark of all three gangs. The most efficient, although not necessarily the worst, are the Serbs, given the fact that they have conquered more territory.

The character, methods, and objectives of the three national movements shape their irrational conduct: rejecting negotiation unless forced to it, blindness in the face of the evidence, and mistrust of even the powers that support them or from whom they hope to get backing and aid.

Is peace possible in Bosnia? Real peace will not be possible except through a long process carefully monitored by United Nations forces or NATO.

Unless one understands the nature of this war and of the movements that have led it, one cannot set clear goals for the forces of peace to pursue. Unless a strong peace agreement is endorsed by those with the capacity to enforce it, the war of extermination in Bosnia will last for decades. It will spread, reach into other regions, undermine the edifice of the UN, and dash to the ground the principles of human rights and the international laws of war.

—*"Cambio 16" (newsmagazine), Madrid. Milovan Djilas is a writer and former vice president of Yugoslavia.*

Will Russia disintegrate into Bantustans?

MAP OF THE RUSSIAN FEDERATION

Map supplied by Bogdan Szajkowski.

Bogdan Szajkowski

Bogdan Szajkowski is a Senior Lecturer in Politics at the University of Exeter specialising in social and political conflicts and in ethnic and nationality issues in the former Soviet Union and Central and Eastern Europe. He is the author of numerous books and articles on the former Communist countries. His most recent book, Encyclopedia of Conflicts and Flashpoints in the Former Soviet Union and Central and Eastern Europe, *will be published by Longman.*

Will the Soviet Union Survive until 1984? was the title of a somewhat prophetic book by Andrei Amalrik, published in 1970. Had he lived until 1991, he would have witnessed the final disintegration of the Soviet Union. Today, only two years later, paraphrasing Amalrik we must ask: Will Russia be fragmented into numerous smaller units, the equivalent of South Africa's old Bantustans? The signs are that the processes are set for the 'Bantustanisation' of the once-powerful Federation.

The past four years have seen the emergence of a multitude of conflicts and flashpoints in the former Soviet Union. A map prepared by the Office of the Geographer of the United States at the beginning of 1990 listed some 40 ethno-territorial conflicts in the Soviet Union. By March 1991, some 80 conflicts had been identified by a Russian academic, Vladimir A. Kolossov. By February 1992, Kolossov had listed 164 conflicts affecting 70 per cent of the territory of the former Soviet Union.[1] Today both publications are already substantially out of date. My own research suggests over 204 ethno-territorial conflicts in the former Soviet Union.[2]

The Soviet Union officially ceased to exist on 8 December 1991, when the leaders of Russia, Ukraine and Belarus unilaterally abrogated the Union Treaty of 1922 and created the amorphous Commonwealth of Independent States (CIS). This was the first stage in the disintegration of the Bolshevik empire.

Thereafter trends were set for the second stage – the disintegration of the former constituent republics. It is worth bearing in mind that, while the first stage of disintegration proceeded along the lines of existing borders and the titular majority of a particular republic, the second and subsequent stages are delineated along ethnic lines without clearly defined or indeed previously acknowledged (identifiable) borders.

The demands for independence of the so-called Transdniestr Republic and the Gagauz Republic fractured the territorial and political cohesion of the Moldovan Republic and set a precedent for future divisions accompanied by civil wars and militarisation of a number of areas. The disintegration of Georgia and, more recently, of Tajikistan, followed this route. To this list can also be added Karakalpakia, an autonomous republic on the territory of Uzbekistan, whose Supreme Soviet on 10 April 1993 approved a new Constitution under which the territory will become a sovereign parliamentary republic within Uzbekistan.

The third stage is the disintegration of the Russian Federation. The trend for the forth stage of further disintegration of the Federation's components into even smaller entities is already clearly detectable.

Ethno-territorial conflicts

Although the Soviet Union was a multinational state, only 67 nations out of the 103 recorded in the 1989 census had their own autonomous areas. As early as 1918, Lenin set out the framework for the ethno-territorial division of the Soviet state. According to him, there could be no norm which would ensure the right of all ethnic groups to their own autonomous territories. Rather, autonomous and ordinary districts should be united for economic

From *The World Today,* Vol. 49, Nos. 8-9, August/September 1993, pp. 172-176. © 1993 by the Royal Institute of International Affairs.

purposes in large autonomous regions (*krays*). Consequently, internal divisions of the former Soviet Union were purely administrative; ethnic demarcations seldom corresponded to the ethnic composition of a particular area. Frequent changes in the political-territorial organisation were used mainly for the centralised control and direction of the economy and society. The residues of this Leninist policy are still with us today.[3]

Between 1941 and 1957 repeated changes in the national-territorial organisation of the Soviet Union were made. In 1941-44, seven peoples accused of collaborating with the German occupiers were deprived of their autonomous status and deported to Siberia, Kazakhstan and Kyrgyzstan. The claims of the deported peoples (14 altogether) for the restoration of the boundaries of their states now have a legal basis in addition to their historical and moral foundations. In 1990, the Supreme Soviet of the Russian Federation adopted a special resolution on justice for deported peoples. One of the main points envisages the reconstitution of their national-territorial units with the boundaries which existed on the day of their deportation. But how, in practical terms, is that to be implemented, and what would be the political consequences? What rights do the titular peoples have to their designated territories if their boundaries are legitimised only by Soviet power, which no longer exists?[4]

The catastrophic decline of the Russian economy has had substantial negative consequences for Russia's state sovereignty. The recently published data on the socio-economic situation during the first quarter of 1993 make grim reading indeed.[5] The 19 per cent drop in industrial production during the first quarter, compared with the same period in 1992, has been accompanied by a 193 per cent inflation rate, compared with December last year. The percentage of unprofitable enterprises in all sectors of the national economy rose to 21, compared with 17 per cent last December. The highest proportional share of unprofitable enterprises (between 41-47 per cent) was recorded in the republics of Tuva and Sakha (Yakutia), the Magadan *oblast* and the Chukotka *okrug*. By the end of March 1993, one per cent (1.1m persons) of the total labour force of Russia had been registered as unemployed. Some 38 per cent of the unemployed are young people under 30 years of age. One in every three residents of Russia now has a per capita income below the minimum subsistence level. At the same time there has been a sharp increase in crime – 12 per cent up on the first quarter of 1992 – with only 45 per cent of reported crimes solved.

Communist collectivism and ethnicity

The common denominator for potentially the most explosive conflicts is the intertwining between Communist collectivism and ethnicity. One of the most important aspects of the operation of Communism was the collective nature of the system. Individual rights (including human, civil and property rights) were subjugated to the collective and controlled by the Communist party-state. The system not only negated the individual but, more important, used the oppressive apparatus in order to enforce compliance with collective (party-state) values, structures and procedures. Communist collectivism reinforced group rather than individual identity, but at the same time offered a comfortable net of social and political arrangements. There were few if any choices to be made, the answers were all but supplied, little

if any exercise of individual responsibility was required. The persistence of the political culture of collectivism remains one of the main obstacles to the effective transformation of the former Communist societies. It is also the main factor in the re-emergence of the ethnic conflicts.

There are both objective and subjective elements in the concept of ethnicity.[6] The objective elements cover characteristics which are actually held in common – kinship, physical appearance, culture, language, religion and so on. Some combination of these characteristics, but not necessarily all, would have to be present for a group of people to qualify as an ethnic grouping. The subjective elements rest on the feeling of community. What is important here is the representations which a group has of itself – regardless of whether those representations are actually correct or not. 'The myth can be potent, and it is the group's representations of itself that are important.'[7]

I should like to stress the importance of the subjective elements. Ethnic groups can only be understood in terms of boundary creation and maintenance. In such cases a common culture is not a defining characteristic of an ethnic grouping; it may, in fact, come into existence as a result of a particular grouping asserting its own position. Cultural features are used by ethnic groupings to mark the groupings' boundaries. Similarly, notions of kinship can be projected and/or constructed so as to give greater body to the feelings of commonality within the grouping. The retreat into ethnic socio-political boundaries and values offers safety at turbulent times. In post-Communist Russia, as elsewhere in the former Communist countries, it has become one of the most poignant socio-political forms of organisations and threats.

The decline of presidential authority

The continuous power struggle in the centre and, in particular, the confrontation between the Russian President on the one hand and the Supreme Soviet and the Congress of the People's Deputies on the other, has already had a very adverse effect on the regions.

One of the more recent examples comes from the Rostov *oblast*, where the local soviet abolished on 30 April 1993 the post of the representative of the Russian President. The representative and his staff were told to vacate their offices within a week and stop their activities. A serious conflict between the Supreme Soviet of Mordova and President Yeltsin (and thus the Russian Federation) erupted in April 1993 over the right of the Federation's President to interfere in the republic's power structure. On 2 April the republic's Supreme Soviet voted (by 116 votes to 37) to abolish the position of President of the Mordovan Soviet Socialist Republic. The deputies blamed the incumbent, Vasiliy Guslyannikov, for current economic hardships and accused him of abusing his position and attempting to create one-man rule. In turn, Boris Yeltsin on 8 April issued a decree confirming the powers of Guslyannikov. The decree has been seen in Mordova as a violation of Article 78 of the Constitution of the Russian Federation and Article 3 of the Federation Treaty which state that federal power may not intervene in the organisation of the republics' power structures. On 20 April Mordova's Supreme Soviet, ignoring the presidential decree, dismissed the government and created a new Council of Ministers.[8]

The growing disenchantment of the regions with the Russian

Federation and President Yeltsin's policies were also reflected in the voting figures during the referendum of 25 April 1993.[9] In 10 of the 19 republics – Adygeya, Bashkortostan, Altay, Dagestan, Ingushetia, Kabarda-Balkaria, Karachay-Cherkessia, Mari-El, Mordova and Chuvashia – Yeltsin failed to win a vote of confidence from the majority of voters.[10] It is interesting to note that several major *oblasts* and *okrugs* voted against the President. In the European part of Russia, voters in Belgorod, Bryansk, Kursk, Lipetsk, Orel, Penza, Pskov, Ryazan, Saratov, Smolensk, Tambov and Ulyanov *oblasts* expressed lack of confidence in Yeltsin. Beyond the Urals, voters in Altay *kray*, Admur and Chita *oblasts* and the Aga-Buryat and Ust-Orda Buryat autonomous *okrugs* also failed to deliver a vote of confidence.[11]

Crisis of statehood

The population's confidence in the authority of the state is extremely low. Laws that have been adopted are inoperative. There is increasingly evidence of a crisis of authority and of deepening antagonism between the executive and representative bodies.

As a consequence of the Russian Federation's inability to develop its own concept of state formation and bringing the federal mechanisms into operation, authorities in some of the republics and in *krays* and *oblasts* have been quite successful in building up their power structures based on the efficient interaction of local sources of power. Against the backdrop of the constant weakening of presidential and federal powers and the increasing turmoil in Moscow, local administrations have become guarantors of stability and formed the nuclei of state formation. There has been growing evidence that local soviets are slowly paralysing presidential power and breaking down the unity of executive power.

By now, many of Russia's regions have elected their own heads of administration. Previously, these had been appointed by President Yeltsin. The new heads have become responsible to the local electorate and are primarily influenced by local factors and conditions. Their legitimacy is based mainly on local constituencies rather than on central, federal authorities. If they are to survive in their posts they must above all respond to local demands for greater economic and political autonomy. The resolution of the local agenda – social/ethnical problems, border adjustments and so on – are often at variance with the interests of the Federation and its structures. The elected heads of local administration are unlikely to support the federal authorities (including the President) for long.

In many respects we are seeing the repetition of the 'Gorbachev delusion'. Here was a man confident that he was running *perestroika*, but his *perestroika* operated only in the centre and was executed through presidential decrees. Meanwhile, the peripheries and local party bosses strengthened their own powers and developed and slowly put into operation their own ideas reflecting local needs and aspirations.

The inoperability of the Federation Treaty

The Russian Federation technically consists of 18 union republics and 69 other subjects of the Federation (6 *krays*, 51 *oblasts*, 1 autonomous *oblast* and 11 autonomous *okrugs*). Eighteen of the 20 republics identified in the Federation Treaty[12] and invited to sign the Treaty did, in fact, put their signatures on the document on 31 March 1992.[13] Tatarstan and Checheno-Ingushetia refused to sign. Subsequently, Checheno-Ingushetia split into two separate entities. The Ingush Republic, created by the decision of the Russian Parliament on 4 June 1992, has so far not signed.

The Federation Treaty offered, at least in principle, the opportunity to conclude additional agreements on the re-allocation and mutual delegation of powers. More than a year after its signing, hardly any of the Treaty's provisions have been implemented.[14] The proclamation of norms has not been followed by appropriate additional legal provisions which would allow the exercise of rights granted in the Treaty. According to the Chairman of the Soviet of Nationalities of the Supreme Soviet, Ramazan Abdulatipov, a majority of the subjects of the Federation are dissatisfied with the way the Treaty is being executed.

The central and most contentious issue is that of the status of the components of the Federation and consequently the rights and obligations of the Union republics vis-à-vis the Federation, and similarly the rights of *krays*, *oblasts* and *okrugs* vis-à-vis the republics and the Federation.

The Treaty appears to hold the prospect for all the 87 subjects of the Federation to be given the rights and status of Union republics. Many of the *krays* and *oblasts* and several autonomous *okrugs* have been demanding political and economic rights equal to those of the republics. However, neither the federal nor the republican authorities are willing to accede to these demands, increasingly afraid of the loss of economic and political control and the possibility of demands for a greater degree of political independence.

After a year of confusion over the precise rights and obligations of the subjects of the Federation, President Yeltsin has only recently indicated his opposition to *krays* and *oblasts* acquiring the constitutional right to issue their own laws.[15] He has also spoken against the equality of all the subjects of the Federation as regards political rights. Not only do his pronouncements contradict the spirit of the Federation Treaty, but in many cases they come too late since many of the subjects of the Federation have already adopted a variety of their own legal provisions, which they see as being within their sphere of competence.

In the absence of any effective execution of the Treaty provisions, the republics and regions want to replace the federal authority, a demand which is fiercely opposed by the centre. The absence of a clear demarcation of powers between the centre and the regions is contributing to the weakening of state authority and the integrity of the Federation. For as long as the shape of the new federal structure and the prerogatives of its constituent parts remain unclear, problems of constitutional authority and delineation of prerogatives will, more likely than not, lead to a series of escalating conflicts.

The republics of Tatarstan and Yakutia-Sakha have drafted their own Constitutions. That of Tatarstan ignores the existence of the Russian Federation, while the Yakut version allots only defence and boundary protection to the federal level. There are numerous claims for the partition of 'double republics'. Given the incredibly complex pattern of ethnic distributions, no national and/or linguistic boundary can be wholly satisfactory to all parties. The Yakuts, for example, refer to the boundary of Yakutia as it was in the early nineteenth century, Tatarstan to that before 1552. They also express concern for their 'blood brothers living abroad', claiming the right to annex their settlement areas or at least to establish autonomous territories for them.[16]

On 30 April 1993, **Kalmykia** became a presidential republic within the Russian Federation, when deputies voted by an overwhelming majority to dissolve the Supreme Soviet and replace it with a 25-member 'professional' Parliament. They also abolished the local soviets throughout the country. The decision followed the election, on 11 April, of Kirsan Ilyumzhinov, a 30-year-old multimillionaire, as President of the republic. Subsequently, Ilyumzhinov imposed direct rule through a system of personal representatives in whom he vested special powers. The new President has emphasised the need for economic autonomy from Russia. It is, however, hard to imagine that such an autonomy can be achieved without the loosening and eventual severance of federal links with Russia.

The **Tuva** Supreme Soviet defied the Russian Federation on 11 May 1993 and amended the republican constitution to include the right to self-determination and the right to secede from Russia.[17] It was decided that a new constitution would be debated by the Parliament in June. Nationalists in the republic have long argued that Tuva's incorporation into the Soviet Union was no more legal than that of the Baltic states. Given that two-thirds of the population is Tuvin, secession has become an achievable option.

Bashkortostan has been in serious dispute with the Russian Federation for over 18 months now. In the spring of 1992, the republic's Supreme Soviet demanded of the Russian leadership that 30 per cent of Bashkortostan's industrial output should remain in the republic. The republic signed, albeit with serious reservations, the Federation Treaty establishing the Russian Federation. Bashkortostan insisted that a special appendix should be added to the Treaty. In it, the republic proclaimed that land minerals, natural and other resources (including oil, of which Bashkortostan is a major producer) on its territory are the property of its population and not of the Federation. It declared that issues related to the utilisation of its resources will be regulated by Bashkir law and agreements with the federal government. The republic has also proclaimed itself an 'independent participant in international law and foreign economic relations, except areas it has voluntarily delegated to the Russian Federation'.

In April 1993, Bashkortostan's Parliament approved a question to be put to a republic-wide referendum: 'Do you agree that the Republic of Bashkortostan must have economic independence and treaty-based relations with the Russian Federation and Appendix to it, in the interests of all the peoples of the Republic of Bashkortostan?' The wording of the question predetermines the outcome of the voting – few if any of the voters in the republic are likely to object to greater economic independence. In practice it means the freedom to export its products and maintain its own tax system, whereby Bashkortostan remits fixed payments to the Russian Federation budget, keeping the rest for itself. What is more significant, however, is that the republic's authorities intend to place any agreement with Russia on 'treaty-based relations', i.e., relations between states. By asserting at the referendum the need for treaty-based relations, the Bashkir authorities have put pressure on Moscow to admit that Bashkortostan has a special status within the Federation. That precedent can now be followed by any of the Federation's units.

Tatarstan declared its sovereignty on 30 August 1990. On 21 February 1992 the Parliament of Tatarstan decided to hold a referendum on the status of the republic. Four million voters were asked: 'Do you agree that the Republic of Tatarstan is a sovereign state, a subject of international law, building its relations with the Russian Federation and other republics (states) on the basis of fair treaties?' The referendum took place on 21 March 1992, despite the ruling of Russia's Constitutional Court that it was unlawful. The results confirmed the earlier decision on the declaration of sovereignty of the Tatar state.

In November 1992, the Parliament of Tatarstan adopted a new Constitution which clearly defined the powers, sovereignty and independence of the republic. At the same time the deputies insisted on associated membership for Tatarstan of the Russian Federation – something that is not envisaged by the Federation Treaty. After the adoption of the Constitution, Moscow faced the dilemma of whether to sign a treaty with Tatarstan as an equal partner, thus creating a political precedent, or whether to treat the republic as an integral part of Russia, which Tatarstan refused to acknowledge. The second option could have far-reaching economic, military and political repercussions.

The nationalist and secessionist movement in Tatarstan continues to grow in strength. Eleven organisations and movements in the republic advocate the complete independence of Tatarstan. In an appeal issued on 13 April 1993 they called for a boycott of the all-Russia referendum on 25 April, arguing that Tatarstan had never voluntarily been a part of Russia and that the people of Tatarstan did not need a referendum into which the imperial forces wanted to drag them.[18]

On 11 May 1993, during President Mintimer Shaimiev's visit to Budapest, Tatarstan, in pursuit of its independent foreign and economic policy, signed an economic cooperation agreement with Hungary for 1993-98. Under the agreement Tatarstan will deliver 1.5m tons of crude oil per year to Hungary and in return receive industrial and agricultural products. It was the first such agreement negotiated between Tatarstan and a foreign country. In 1992 trade turnover between the two countries exceeded $235m.

The **Tyumen** region, rich in oil and natural gas, refused to sign the Federation Treaty in March 1992 and is now threatened with the secession of two of its autonomous *okrugs*: Khanty-Mansi and Yamalo-Nenets, both of which want to acquire the status of separate republics. Secession of the two *okrugs* would reduce the area of the Tyumen region from 1.4m sq km to a mere 161.000 sq km and deprive it of much of its resources and industry.

The division of the Magadan *oblast* and the creation of the **Chukot** Republic became a reality when, on 11 May 1993, the Constitutional Court of the Russian Federation decided that the separation of the Chukchee autonomous *okrug* from the Magadan *oblast* was in accordance with the Russian Constitution. In 1989 the Chukchee accounted for only 7.3 per cent of the *okrug*'s population, while Russians and Ukrainians made up 83 per cent. In September 1990, the *okrug*'s soviet proclaimed itself an autonomous republic, and in March 1991 it decided to separate from the *oblast*. Magadan's authorities contended that such a decision could be taken after a referendum had been held. The Court's decision opens the way for the secession of numerous other *okrugs* throughout the Russian Federation.

Conclusions

The argument in this article is based on two broad assumptions. The first is that the residues of Communism will remain for a long time to come. It has proved relatively easy to carry out structural transformation in the former Soviet Union in order to achieve the edifices of liberal democracy. However, their functioning is more often than not at variance with liberal democratic principles and values. These will be able to take root only with generational change. The symbiotic relationship between Communist collectivism and ethnicity will continue to dominate the wider political agenda. It is the most difficult aspect to tackle because it reflects the basic, and in some sense perhaps irrational, feelings of individual and group insecurity.

At the same time, however, it has in political and strategic terms become the avenue for the expression of political, economic and social aspirations which had been denied so far. The substantial credibility gap which exists between the old structural (federal) arrangements and the demands of an essentially new post-Communist situation can only be bridged by drastic action: either by the dismantling of the old structure or through their fundamental modification. So far there has been little, if any, evidence of either.

Russia still wants to remain a federation rather than, for example, a confederation, a commonwealth or a community of nations. The old Tsarist slogan 'Russia is indivisible' is used as a rallying point by new democrats and old Communists alike. In one important respect the new Federation Treaty is even more reactionary than the 1922 Union Treaty, which contained at least a token provision for secession from the Union. The new Treaty does not. According to it, the territory of the Federation is integral and inalienable. The spectre of the disintegration of Russia is indeed threatening, but it is a progressive reality. The way this reality is dealt with in the long term will determine the stability of international relations.

The second assumption is that there are two incompatible processes taking place in Western Europe, on the one hand, and the former Soviet Union and Central and Eastern Europe, on the other. For four decades now the West European agenda has been dominated by integration in political, economic and strategic terms. This has been a long and arduous process based, first, on the clear identification of separate interests and, second, on the development of common strategies and goals. The East is now only at the stage of identifying separate interests. Integration may follow in due course, but if it is forced or artificially accelerated it will inevitably be full of cracks and consequent instabilities.

Perhaps one of the most important lessons to be learned from the historical experience of the former Soviet Union, and from the tragic events in former Yugoslavia, is that the federal organisation of the state and the multinational structure of its population are quite different things. There is an urgent need to re-examine our well-accepted analytical and methodological tools such as the concept of the nation-state, sovereignty, self-determination, nation and borders.

NOTES

1. Vladimir A. Kolossov, *Ethno-Territorial Conflicts and Boundaries in the Former Soviet Union* (University of Durham, International Boundaries Research Unit, 1992), p. 3.
2. Bogdan Szajkowski, *Encyclopedia of Conflicts and Flashpoints in the Former Soviet Union and Central and Eastern Europe* (London: Longman, December 1993).
3. Vladimir A. Kolossov, *op. cit.*, p. 10.
4. *Ibid.*, p. 12.
5. For details, see 'The socio-economic situation and the development of economic reforms in the Russian Federation in the first quarter of 1993', *Ekonomika i Zhizn*, No 17, May 1993.
6. Bogdan Szajkowski and Tim Niblock, 'Islam and Ethnicity in Eastern Europe'. Paper presented to the International Conference on Moslem Minorities/Communities in Post-bipolar Europe' at the Saints Cyril and Metodij University, Skopje, Macedonia. April 1993, pp. 2-3.
7. Eliezer Ben-Rafael and Stephen Sharot, *Ethnicity, Religion and Class in Israeli Society* (Cambridge: Cambridge University Press, 1991), p. 6.
8. BBC SWB, SU/1676 B/5, 30 April 1993.
9. Of the 107.3m eligible voters in the Russian Federation, 69.2m participated in the referendum. Of them, 58.7 per cent had confidence in President Yeltsin; 53 per cent approved of his economic reforms; 31.7 per cent wanted early presidential elections; and 43.1 per cent favoured early parliamentary elections. Under the conditions set out by the Congress of People's Deputies and the Constitutional Court, the last two questions were not passed since they attracted less than half of the potential votes.
10. Only 2.7 per cent of voters in Ingushetia, 14.3 per cent in Dagestan, 25.9 per cent in Karachay-Cherkessia and 35.8 per cent in Kabarda-Balkaria voted 'yes' in answer to the question: 'Do you trust the President?'. BBC SWB, SU/1675 B/2, 29 April 1993; and BBC SWB, SU/1680 B/3, 5 May 1993.
11. BBC SWB, SU/1675 B/2, 29 April 1993. Interestingly, the referendum also showed considerable dissatisfaction with Yeltsin and his policies among the Russians living outside the Russian Federation. For example, of the eligible Russian citizens residing in Estonia, only 27.9 per cent of those voting expressed confidence in the President, with 71.3 per cent voting against. 72.6 per cent rejected the reforms; some 70.3 per cent supported early presidential elections, with 28.3 per cent against; and 50.3 per cent backed early parliamentary elections, with 48.7 per cent against. Of the 4,525 Russian citizens in Latvia who participated in the referendum, 21 per cent voted 'yes' and 78 per cent 'no' on the question of confidence in the President; 19.5 per cent voted 'yes' and 80 per cent 'no' in support of reform policy; 79 per cent voted 'yes' and 19 per cent 'no' on presidential elections; and 40 per cent voted 'yes' and 59 per cent 'no' on the question of fresh elections to the Russian Parliament. BBC SWB, SU/1674 C/5, 28 April 1993.
12. The Federation Treaty replaced the Union Treaty of 29 December 1922, which was abrogated by Russia, Belarus and Ukraine on 8 December 1991 when the three countries created the Commonwealth of Independent States (CIS).
13. The Treaty was signed by the Russian Federation, the Soviet Socialist Republic of Adygeva, the Republic of Bashkortostan, the Buryat Soviet Socialist Republic, the Republic of Gornyy Altay, the Republic of Dagestan, the Kabardin-Balkar Republic, the Republic of Kalmykia-Khalmg Tangch, the Republic of Karachay-Cherkessia, the Republic of Karelia, the Komi Soviet Socialist Republic, the Republic of Mari El, the Mordova Soviet Socialist Republic, the North Ossetian Soviet Socialist Republic, the Republic of Sakha (Yakutia), the Republic of Tuva, the Udmurt Republic, the Republic of Khakassia and the Chuvash Republic.
14. See, for example, an interview with the Chairman of the Soviet of Nationalities of the Supreme Soviet, Ramazan Abdulatipov. BBC SWB, SU/1656 B/5, 6 April 1993.
15. See Yeltsin's address to the Council of Heads of Administration of *Krays, Oblasts* and Autonomous *Okrugs* within Russia on 28 May 1993. BBC SWB, SU/1701 B/1, 29 May 1993.
16. Even in Kazakhstan, with its extremely mixed population and relative tolerance, the legislators wish to extend citizenship to all Kazakhs living 'abroad'.
17. Tuva enjoyed at least nominal independence between 12 August 1921 and 11 October 1944, when it was incorporated into the Soviet Union.
18. *Izvestia*, 13 April 1993.

BEACHES ON THE BRINK

Retreat of the shoreline is nature's response to a rising sea level. Recent storms have renewed the debate over attempts to hold it in place.

Ruth Flanagan

Ruth Flanagan is an associate editor at Earth magazine. While reporting this story, she visited North Carolina and the storm-ravaged coasts of New Jersey and Long Island.

The most striking sight along the island of Westhampton Beach has traditionally been architectural: the contemporary summer homes of well-heeled New Yorkers lined up along the dunes. But after a severe storm blasted the town last December, it was nature, not culture, that commanded attention. The storm sliced through the slender barrier island, carving an inlet hundreds of yards wide that severed the community's only road. The sea swallowed much of the beach, stranding houses in the surf. More than 80 collapsed on their spindly pilings. After each one fell, it took a mere 20 minutes for the waves to chew up the clapboard and spit it all out as flotsam.

The same northeaster claimed millions of cubic yards of sand—and millions of dollars in property—along 200 miles of the coast from the Hamptons in New York to Long Beach Island in southern New Jersey. It also threw into relief fundamental questions about the future of coastal development, particularly on the barrier islands that fringe most of the U.S. Atlantic and Gulf coasts. Since the 1800s, most of these sandy strips have been narrowing. Though gentle swells in spring and summer restore sand, overall the islands have been slimming down year after year.

Many people fear the islands are in deep trouble. The very term "coastal erosion" has become a rallying cry, inspiring herculean attempts to immobilize retreating shorelines. People have tried trapping sand, pumping sand, even building walls that transform chunks of barrier islands into giant sandboxes. Most of these measures do protect property. But in the long run, none stop shoreline erosion. And some do much more harm than good.

"What we're doing as a nation is trying to maintain the status quo and keep everything in place," says Norbert Psuty, associate director of the Institute of Marine and Coastal Sciences at Rutgers University. "It's a losing battle."

It's also a confusing battle. Is the goal to save beaches or the property sidled up to the shore? On some barrier islands, such as Miami Beach and Atlantic City, storm damage and shoreline retreat could threaten billions of dollars' worth of beach-front development. In areas like these, many people feel we have no choice but to hold the ocean at bay—even though stabilizing shores with seawalls and other structures destroys natural beach systems. In many regions, however, people are moving toward "softer" solutions, such as replenishing beaches with sand from offshore. A few planners and residents are even starting to discuss what may be politically painful but ultimately necessary option: strategic retreat from the sea.

Human beings have undoubtedly accelerated erosion on many barrier islands. But according to Orrin Pilkey, a coastal geologist at Duke University, the main reason most islands are narrowing worldwide appears to be nature itself: rising sea level. About 20,000 years ago, during the Ice Age, much of the Atlantic Ocean's water was locked up in massive continental ice sheets. As a result, sea level along the U.S. East Coast was about 450 feet lower than it is now. As the climate warmed, the ice sheets began to melt and sea level rose rapidly. About 4,000 years ago, the rise slowed, but in the 1800s it picked up again. Since then the sea has crept up at an average rate of one foot per century. The amount of land being swallowed up differs from place to place, depending on other conditions such as the slope of the land. Even a small sea level rise generally has a marked effect on barrier islands because their slope is gentle.

Yet the sea's advance doesn't threaten the integrity of island systems, Pilkey says. In fact, he believes that by slimming down, the islands are actually securing their long-term survival.

Geologic evidence shows that barrier islands naturally tighten their belts when the sea rises. Then, after tens, hundreds or thousands of years, they become slender

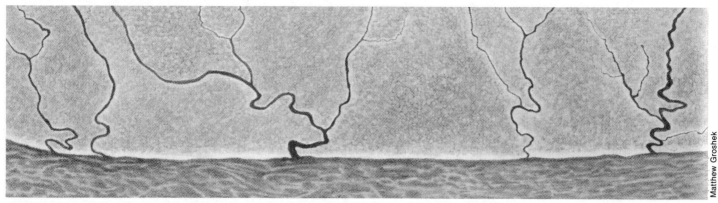

EVOLUTION OF A SHORELINE: Phase 1, 20,000 years ago. Ice Age glaciers are advancing, causing sea level to drop. No barrier islands line the shore, which lies far seaward of today's coast. Rivers flowing through the low-lying coastal plain empty directly into the ocean; there are no estuaries.

enough for even trifling storms to transfer sand clear across their widths. At that critical point the narrowing stops. Storms then remove sand from the front side of the islands and add it to the back. By this process, the islands migrate inland in pace with the rising sea. If the sea rises slowly enough, moving islands don't usually smash into mainland coasts. Instead, the shorelines behind the islands retreat at the same time. The two landmasses move back in step, like partners in a tango.

The encroaching ocean doesn't really damage the recreational beach either, Pilkey says. The swath of sand we plunk our towels on doesn't actually erode in response to sea level rise. It simply moves back, remaining the same width while the island as a whole narrows down. Erosion becomes a problem only when we put up immobile structures. As Pilkey puts it, "Coastal erosion is not a problem for beaches. It's only a problem for buildings."

It's not surprising that barrier island management is a major concern in the United States. Although these strips of sand fringe nearly every flat-lying coastal plain in the world, North America is blessed with the longest barrier island chain of all. These islands extend in a perforated line running parallel to the mainland some 2,700 miles from Long Island all the way to the Yucatan Peninsula. Some are highly developed and bear little resemblance to their natural state. Others remain wild. A few, like Cumberland Island in Georgia, are plump. A few other, like Assateague off the Maryland coast, are wisp-thin and low-lying and are migrating swiftly toward the mainland.

But varied as they are, all barrier islands share a few common features. They form only on gently sloping coastal plains. They require both sufficient sand and strong enough waves to move the sand about. And though sea level rise is now eroding many of them, they all required rising seas to form in the first place.

To demonstrate how barrier islands evolve, Orrin Pilkey takes his students to Shackleford Banks on North Carolina, an uninhabited island a half-mile wide and 7.5 miles long in Cape Lookout National Seashore. Today,

much of the island is covered with a roller coaster of high dunes. The interior is forested with cedars and gnarled live oaks. Salt marsh thrives behind the island in a sheltered lagoon. Shackleford's front side is a long stretch of tawny beach behind which clumps of sea oats cling to hummocks of sand. The island looks sturdy and permanent.

But 20,000 years ago, when ice sheets to the north were still growing, there was no barrier island there or anywhere along the Atlantic Coast. Shackleford was part of a vast, forested plain, including what is now the submerged continental shelf. The plain was flatter than Kansas and extended seaward 40 miles.

Several thousand years later, the ice sheets began melting and the sea began to encroach. River valleys on the plain drowned and widened into estuaries. Only the high areas between the valleys were left jutting above water into the sea, forming protruding headlands. Then, as soon as the headlands emerged, waves began eroding them away. The waves generally approached the shore at an angle, forming longshore currents that carried huge loads of suspended sand picked up from the eroding headlands. When the waves lost energy, they deposited their sand in linear spits across the mouths of the estuaries.

At the same time, dunes were forming on the spits. Fair weather waves pushed sand onto the beach from the inner continental shelf. Winds carried some of this sand to the interior of the island, and plants like Shackleford's sea oats held much of it in place. Storms and hurricanes transported huge volumes of sand to the island's interior through gaps between the dunes. This supply allowed secondary ranks of dunes to form.

Finally, the spits broke free from the mainland and became true islands. Sometimes storms broke the spits into separate islands by carving inlets through them. In other cases, the spits gained their independence by more gradual means: Rising sea level flooded the lowlands behind the dunes, transforming the ridges of sand into

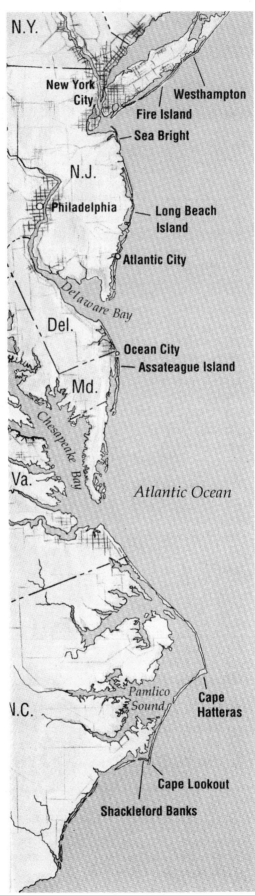

THE NATURAL BEACH

Vegetation helps stabilize the talcum-fine sand just inland of the beach on Cumberland Island in Georgia (above). Farther inland on this wide barrier island near the Florida border are thick forests of moss-draped live oak.

North Carolina's Cape Lookout lighthouse (above) stands on a classic barrier island. The ocean beach is to the right. Sheltered behind dunes are grasses, shrubs and trees. The lagoon lies to the left. Salt marshes are common in calm lagoon waters, such as those behind Bear Island, N.C. (below).

islands and the former lowlands into quiet salt marsh lagoons.

Once they were cast out on their own, the barrier islands could take care of themselves. They no longer needed the eroding headlands to provide them with sand. Instead, waves brought in a steady supply from the shoreface, the submerged part of the beach. The islands sustained themselves for thousands of years in this state of dynamic balance, constantly changing in response to the action of waves and wind.

This doesn't mean that the going has always been easy for the barrier islands. For some periods between 18,000 and 6,000 years ago, sea level rose at the rate of five feet per century—about five times faster than it's rising today, Pilkey says. Some of these rapid rises must have caused many islands to disappear completely. During other periods, barrier islands migrated quickly enough to keep up with the rise. Many of these migrating islands have even preserved records of their travels. On Shackleford, for instance, bits of pearlescent oyster shell rest in the sand on the ocean side of the island. The oysters didn't live along the unprotected oceanfront. They grew on the other side of the island, in the shelter of a quite lagoon. Since then, the island has migrated back, plowing over its former lagoon.

People have been building permanent structures on barrier islands for only a tiny sliver of time, starting with a few scattered settlements in the 1700s. But by a quirk of fate, the development of the islands overlaps with a turning point in their history. After remaining stable for thousands of years, sea level began to rise during the 1800s, according to Pilkey. Thus, permanent settlement came at a time when most islands had already begun slimming and preparing to move.

Intensive development on barrier islands is an even more recent phenomenon. Construction boomed after World War II, spurred on by post-war prosperity, and has continued at a dizzying rate ever since. The most recent surge of development occurred in the past 25 years—a period in which thousands of hotels, condominiums and beach homes have sprung up. Significantly, this boom also coincides with a stint of unprecedented calm weather along the Atlantic coast. Many scientists consider this an unlucky break: The relative rarity of severe storms and hurricanes may have led people to overestimate the safety of barrier islands.

"In the last three decades or so, we've had very very few storms on the East Coast, and it's encouraged people to develop without fear," Rutgers' Norbert Psuty says. Perhaps then it's no coincidence that the number of homes located in coastal and riverine "high hazard areas" rose 40 percent from 1966 to 1989.

Pilkey fears that a general lack of experience with storms and hurricanes has also encouraged risky construction practices that transform relatively safe islands into dangerous ones. He is seriously concerned, for example, about the removal and "notching" of dunes, the

Michael Baytoff

A boardwalk and high-rise gambling halls line the shore in Atlantic city, N.J. (above). The beach here depends on periodic replenishment efforts, in which massive amounts of sand are added to compensate for erosion. In Seaside Park, N.J. (below), houses are densely packed nearly up to the surf line.

Robert Perron

THE DEVELOPED BEACH

On Long Beach Island in New Jersey, summer homes lie behind a barrier of dunes. These homes are less protected from the sea than they may seem. As sea level rises, Long Beach Island may eventually migrate inland right out from under them.

Michael Baytoff

EVOLUTION OF A SHORELINE: Phase 2, 17,000 years ago. Glaciers are melting and sea level is rising. The mouths of rivers drown and widen into estuaries. High areas between the estuaries are left as headlands. Sand eroding from the headlands is deposited in spits that grow steadily longer across the mouths of the estuaries—barrier islands in the making.

building of houses on areas of beach that once were inlets, and the construction of roads that cut direct, perpendicular paths to the beach—standard practice on many barrier islands.

"In a big storm, a surge can cut through those roads like a knife through hot butter," Pilkey says. The roads not only usher in stormwaters, but often cut off residents' only access to escape routes.

Intensive shoreline development has already put many towns on a collision course with the everchanging islands. But when clashes occur, the islands generally suffer as much damage as their human inhabitants. Once communities place immobile structures near the shore, coastal erosion inevitably becomes a problem, and people often feel compelled to stabilize the beaches to protect themselves and their property. Sadly, rigid stabilization generally degrades and sometimes destroys the beaches that attracted development in the first place.

No stabilizing structures illustrate this process quite as dramatically as seawalls. Massive and unyielding, they run along beaches as a barricade against the waves. (Smaller parallel-lying obstructions, such as revetments and bulkheads, are also part of the seawall family.) People have often erected seawalls as a defensive measure following devastating storms or hurricanes. In 1900, for instance, the worst hurricane disaster in U.S. history struck Galveston, Texas, killing 6,000 people. In response, the survivors built the largest protective seawall ever placed on a barrier island.

Today, seawalls armor much of the developed shoreline. Many have done their job admirably, protecting homes and other buildings that otherwise would probably have tumbled into the sea. But seawalls extract a high toll for their services. Obviously, they block easy access to the beach and blot out the view. More importantly, though, they narrow and, according to Pilkey, even eliminate beaches.

Many geologists and engineers consider the seawall at the historic town of Sea Bright, N.J., a kind of object lesson: an example of the final, sorry consequences of hemming in a moving shore. In the late 1800s, Sea Bright was one of several island resort communities along a barrier spit in northern New Jersey that boasted a broad, sandy beach. A railroad running through the town to nearby Sandy Hook helped transport thousands of people each year to the prized bathing grounds in the area. In the early 1900s, a seawall was constructed to protect the railroad. Today, the gray, fortresslike wall in Sea Bright dominates the shoreline, looming 15 to 18 feet high and stretching more than six miles.

Unfortunately, the wall is protecting the town at the beach's expense. As the spit has slimmed down, the beach has narrowed dramatically in front of the wall. This process, called "passive erosion," occurs whenever a static structure is placed on a retreating shore. Psuty and Pilkey believe the wall has also caused "active erosion." Its presence caused the sea floor in front of it to steepen. The steep grade increased the storm-wave energy striking the shoreline, accelerating erosion and making it impossible for incoming waves to deposit sand from the continental shelf. Today, northern Sea Bright has no beach. Just big waves and a very big wall.

In places like Sea Bright where erosion is severe, communities often erect other structures specifically designed to build up the beaches. Groins and jetties are wall-like sand traps that run perpendicular to the shore, interrupting the transport of sand across the coast by longshore currents. Groins are used to widen a particular beach. Jetties (basically long groins) protect harbors and inlets. These devices line thousands of developed barrier shores.

Both groins and jetties can be very effective at building up beaches. Yet their drawbacks are obvious. Wherever they plump up one stretch of beach, they starve others down the coast. They work, essentially, by "robbing Peter to pay Paul." This often leads to a domino effect: Once one groin is in place, another is needed at the beach next door.

Matthew Groshek

Matthew Groshek

EVOLUTION OF A SHORELINE: Phase 3, 14,000 years ago. High water from storms occasionally washes over the spits, breaking their connections with the mainland and carving inlets across the strands. The result: a line of barrier islands guarding the mainland, with quiet salt-marsh lagoons in between. Currents deliver replenishing sand to the forested, dune-lined islands, which are now in dynamic balance.

Jack Dermid

Sand bags (above) offer meager protection from the waves, which have taken a bite out of the beach at South Nags Head, N.C.

Courtesy: Moss Archives Sea Bright, N.J.

THE DAMAGED BEACH

The ocean crashes against a seawall in northern Sea Bright, N.J. (above). In this very spot in 1931 (below), bathers set up their umbrellas on a sandy strip called Highland Beach. The seawall, built in the early 1900s to protect the railroad, hastened the beach's demise by cutting off its supply of replenishing sand.

Matthew Groshek

EVOLUTION OF A SHORELINE: Phase 4, today . . . and the future. With few exceptions, barrier islands are slimming down as sea level rise resumes after a hiatus of several thousand years. Waves and wind push sand from the ocean side of the narrow islands to the lagoon side, taking from the front and adding to the back. In this way, some islands are migrating shoreward. They could eventually crash into the mainland—especially where seawalls and other protective constructions prevent the mainland coast from retreating.

The shifting sands of Assateague Island in Maryland and Virginia demonstrate the tremendous impact jetties can have. In 1933, the Army Corps of Engineers built jetties on either side of an inlet separating the Maryland beach resort of Ocean City from Assateague. The corps' purpose was to keep the inlet open for navigation, but the jetties had an unintended impact. They trapped sand to the north, providing Ocean City with the bonus of a wider recreational beach. But they prevented sand from reaching the northern part of Assateague on the other side of the inlet. As a result, this part of the island has slimmed considerably and moved back toward the mainland. Today, the northern part of the island is migrating rapidly toward a boat channel. This situation is sure to create an interesting management dilemma in the near future.

In the last few decades, however, barrier island communities have been building fewer and fewer structures to stabilize shores. Instead, most have embraced "beach replenishment" or "nourishment" to repair coastal erosion. Replenishment involves adding sand to beaches, both to provide recreational opportunities and to buffer property from storms. Sometimes the sand is dredged from silted harbors; in other cases, it's pumped from offshore bars or other parts of the coastal system. According to a report in the *Journal of Coastal Management,* more than 400 miles of U.S. shoreline were replenished as of 1987.

But even replenishment has its down side. Sand of the right size and texture may be hard to obtain. Dredging sand may also disrupt marine organisms and other parts of the coastal system, Pilkey says. And replenished beaches are often expensive and frustratingly short-lived. "It's all sacrificial sand," explains Diane Abell, a landscape architect at Fire Island National Seashore in New York. "We know a storm will come eventually and wash it all away."

Past replenishment projects have shown widely varying results. According to a report by Pilkey and his colleague Tonya Clayton, Ocean City, N.J., replenished a stretch of beach at a cost of $5.2 million in 1982. Two and a half months and 18 northeasters later, the beach had disappeared. On the other hand, the sand pumped onto 10.5 miles of Miami Beach between 1976 and 1981 (at a cost of $54 million) is still in place. The project even won a preservation award from the American Shore and Beach Preservation Association, Pilkey and Clayton report. It also received the Golden Fleece Award from Senator William Proxmire for wasting the taxpayers' money.

There is one other solution to coastal erosion. We could stop intervening and let the ocean advance. We could stop rebuilding storm-damaged buildings, for instance, and instead move them back as needed in step with the shore. We could let sand go where waves and wind transport it.

Though it's very controversial, the idea of retreat is creeping into the corners of coastal policy. A number of states, such as South Carolina, North Carolina and Maine, have adopted strong regulations prohibiting hard stabilization. At the federal level, the Army Corps of Engineers is now required to weigh the costs and benefits of retreat along with other solutions. Years ago, Pilkey says, many people would have scoffed at such an approach.

In the mid-1980s, Nags Head, N.C., adopted strict standards requiring that new beach homes be constructed on deep lots so that the houses have room to move back. "We consciously adopted the principal of retreat as part of our land-use planning," says Donald Bryan, the town's former mayor.

Of course, retreat is not practical on densely populated islands like Miami Beach or Atlantic City, where vast sums of money are tied up in shoreline development and where there is little, if any, room for buildings and utilities to move back. But even in smaller communities, retreat is sure to be painful. It could shrink the tax base, creating economic and political fallout. Requiring people to move from their homes would be emotionally wrenching as well.

LESSONS FOR THE FUTURE

A ruined home lies in the surf off Fire Island, N.Y., after last December's storm (right). Efforts to control the sea often have unanticipated results. A good example is a jetty protecting an inlet bordering Ocean City, Md., (below). It blocked the flow of sand to Assateague Island (foreground), which has thinned and migrated toward the mainland.

Diane Abell

Judy Johnson

But the rising costs of fighting the sea will undoubtedly force more communities to consider retreat in the near future. For instance, Psuty predicts that the exorbitant cost of replenishing the beach and maintaining the seawall at Sea Bright will ultimately compel people to abandon the area.

Nationally, reimbursing coastal dwellers for damage they incur from storms and hurricanes has helped put pressure on the National Flood Insurance Program. But obviously it is not only coastal disasters that soak up these funds. In July, the severe flooding in the Midwest completely drained the program's coffers. Additional money could come from federal taxpayers in a coastal emergency. This is hardly comforting though, since a single year of bad storms can cost up to $4 billion.

The expense of maintaining barrier islands for development thus raises issues of fairness, such as who pays and who benefits. After bad storms strike, taxpayers may be asked to foot the bill not only for insurance payments but also for beach repairs. Following last December's northeaster, city officials and property owners in Westhampton and nearby Fire Island clamored for aid in closing newly opened inlets, arguing that the channels posed a risk to the islands' ecosystems. The property owners claim that ocean water moving through the inlets threatens shellfish in the bays. But the inlets don't damage the system in general, or the islands in particular—only the people who depend on them to stay the same year after year.

A few things, at least, are certain: Our problems on barrier islands are not going to go away, especially if global warming accelerates the sea's rise. Some scientists believe the warming may also increase the frequency of coastal storms, making barrier islands and spits more vulnerable in the future.

But even without global warming, the sea will continue to encroach. The islands will continue to move. And, in the long run, we may have no choice but to get out of the way.

What's Wrong with the

PHILIP ELMER-DEWITT

JUST WHEN SUMMER SHOULD HAVE been coming in, it snowed last week in Colorado, punctuating several days of unseasonable 32°C (90°F) weather with enough snowfall to close three mountain highways. Paris was hit with a torrential rainstorm—the worst in a decade—that crippled the city, poisoned the Seine with sewer effluent, and clogged the river with 300 tons of dead fish. In one hour in early May, a squall dumped a record 110 mm (4⅓ in.) of rain on Hong Kong, turning steep city streets into rushing rivers and killing five. In the Middle East this January, the wettest, coldest winter in recent memory was capped by a storm that blanketed Amman, Damascus and Jerusalem with much more snow than anyone there had seen for 40 years.

If it continues as it has begun, 1992 could turn out to be almost as bizarre as 1991, a year in which North America's spring arrived in winter, its summer in spring and its winter in autumn. The period from December 1991 to March 1992 has already gone into the National Weather Service's record books as the warmest winter in at least 97 years. It hardly rained at all in rainy Seattle in May. Texas in January was swamped with twice as much precipitation as normal, and Southern California, where it never rains, was socked with floodwaters so powerful they carried cars out to sea. Africa is having its worst drought in 50 years, and eastern Australia, which is supposed to have summer when the northern hemisphere has winter, had to do without this year. Instead of balmy days and bright sunshine, Melbourne racked up a record 12 consecutive days of rain and the coldest January in 137 years, which is as far back as anyone Down Under kept track.

What is going on? Experts say fluctuations from normal readings are, well, normal and that weird weather is the rule, not the exception. But the highs and lows and wets and drys over the past two years have

been so extreme that anxious questions are arising. Could these outbursts of wacky weather be related to those fires from the gulf war? That hole in the ozone layer? The global warming trend that environmentalists have been predicting for so many years?

The questions are more than idle speculation. This week at the Earth Summit in Rio de Janeiro, world leaders will be adding their signatures to a treaty to prevent climate change, a document that was significantly weakened during presummit negotiations, in part because of U.S. contentions that the threat of global warming has been overblown. But the Bush Administration's skepticism must contend with the direct experience of millions of citizens who are worried that when the weather gets as odd as it has been of late, something must be wrong.

Scientists, however, are more cautious than the umbrella-carrying public. Even climatologists who believe that global warming may eventually trigger extreme weather variations like the ones we are experiencing say it is too early to prove a direct connection. The outbreak of freakish weather could also have been partly caused by one or more of several large-scale atmospheric events now under way. The main suspects, in descending order of likelihood:

EL NIÑO. To meteorologists, the weather phenomenon named after the Christ child is not a theory but a recognizable and recurrent climatological event. Every few years around Christmastime, a huge pool of warm seawater in the western Pacific begins to expand eastward toward Ecuador, nudging the jet streams off course and dis-

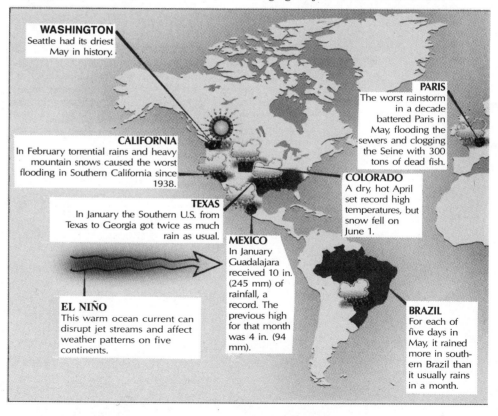

WASHINGTON
Seattle had its driest May in history.

PARIS
The worst rainstorm in a decade battered Paris in May, flooding the sewers and clogging the Seine with 300 tons of dead fish.

CALIFORNIA
In February torrential rains and heavy mountain snows caused the worst flooding in Southern California since 1938.

COLORADO
A dry, hot April set record high temperatures, but snow fell on June 1.

TEXAS
In January the Southern U.S. from Texas to Georgia got twice as much rain as usual.

MEXICO
In January Guadalajara received 10 in. (245 mm) of rainfall, a record. The previous high for that month was 4 in. (94 mm).

EL NIÑO
This warm ocean current can disrupt jet streams and affect weather patterns on five continents.

BRAZIL
For each of five days in May, it rained more in southern Brazil than it usually rains in a month.

Weather?

There's nothing unusual about unusual weather. But global warming, a volcano and a stray ocean current may be making things even freakier.

rupting weather patterns across half the earth's surface. The El Niño that began last year and is now breaking up has been linked to record flooding in Latin America, the unseasonably warm winter in North America and the droughts in Africa.

PINATUBO. The full effects of the eruption of Mount Pinatubo in the Philippines last June—probably the largest volcanic explosion of the 20th century—are starting to be felt this year. The volcano heaved 20 million tons of gas and ash into the stratosphere, where they formed a global haze that will scatter sunlight and could lower temperatures—by half a degree Fahrenheit—for the next three or four years. Smoke from the gulf-war fires, by contrast, never reached the stratosphere and had no measurable effect on the world's weather.

GREENHOUSE GASES. It is known that the level of CO_2, methane and other heat-trapping gases in the atmosphere has increased 50% since the start of the Industrial Revolution. Measurements also indicate that the world's average temperature has increased 1°F over the past 100 years. The rest is conjecture. Computer models suggest that as the buildup of greenhouse gases continues, average temperatures could jump 3°F to 9°F over the next 60 years. Some scientists speculate that even a small rise in average temperatures could lead to greater extremes in weather patterns from time to time and place to place.

The problem with sorting out these influences is that they interact in complex ways and may, to some extent, cancel each other out. Pinatubo's cooling effects could counteract the warming caused by greenhouse gases, at least over the short term. At the same time, El Niño's warming influence seems to have suppressed the early cooling effects of Pinatubo's global haze.

Predicting the weather is, in the best of circumstances, a game of chance. Even with the most powerful supercomputers, forecasters will never be able to see ahead more than a couple of weeks with any accuracy. Climatologist Stephen Schneider of the National Center for Atmospheric Research compares the typical weather forecast to guessing what bumpers a pinball will hit after it has left the flipper. "What's happening now," he says, "is we're tilting the machine in several directions at once."

Of course, there have always been volcanic eruptions, and the tales of El Niño date back at least to the Spanish conquistadors. Old-timers can point to freak weather occurrences that put the Los Angeles floods to shame, like the 1928 storm that bombarded southwestern Nebraska with hailstones the size of grapefruit. Or the blizzard of 1888 that buried the Eastern Seaboard in snowdrifts the size of four-story buildings. "There is a record set somewhere every day," says Steve Zebiak, an atmospheric scientist at Columbia University's Lamont-Doherty Geological Observatory.

What is new is that for the first time some of the influences that shape our weather are man-made. Experts say it could be 20 or 30 years before they know for certain what effect the buildup of greenhouse gases, the destruction of ancient forests or the depletion of the ozone layer have had. Policymakers looking for excuses not to halt those trends will always be able to point to scientific uncertainty. As Schneider puts it, "We're insulting the system at a faster rate than we can understand." The risk is that by the time we understand what is happening to the weather, it may be too late to do anything about it. —*Reported by David Bjerklie/ New York, with other bureaus*

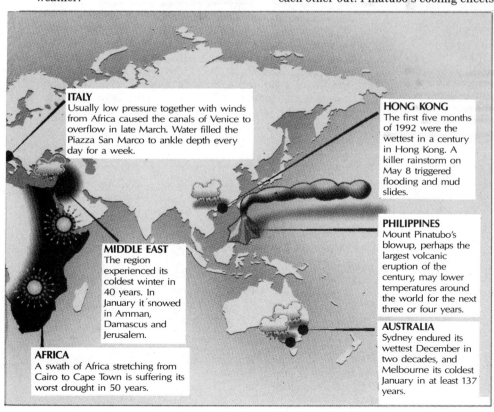

ITALY
Usually low pressure together with winds from Africa caused the canals of Venice to overflow in late March. Water filled the Piazza San Marco to ankle depth every day for a week.

HONG KONG
The first five months of 1992 were the wettest in a century in Hong Kong. A killer rainstorm on May 8 triggered flooding and mud slides.

MIDDLE EAST
The region experienced its coldest winter in 40 years. In January it snowed in Amman, Damascus and Jerusalem.

PHILIPPINES
Mount Pinatubo's blowup, perhaps the largest volcanic eruption of the century, may lower temperatures around the world for the next three or four years.

AUSTRALIA
Sydney endured its wettest December in two decades, and Melbourne its coldest January in at least 137 years.

AFRICA
A swath of Africa stretching from Cairo to Cape Town is suffering its worst drought in 50 years.

TIME Graphic by Joe Lortola

The Green Revolution

Robert E. Huke

Robert E. Huke *was educated at Dartmouth College and Syracuse University, and he is currently Professor of Geography, Dartmouth College, Hanover, NH 03755. His teaching and research interests focus on agriculture, especially in South and Southeast Asia where he has completed a number of village and regional level research projects. A second major interest lies with the development of computer-assisted instruction modules for use in teaching and in training agricultural extension workers.*

The Green Revolution refers to a complex package that includes improved seeds and a wide range of management practices. The new plant types show a strong positive response to fertilizer because of a high leaf area index, short stature, and stiff straw that resists lodging. These plant varieties have resistance to many insect pests and plant diseases. The management package is concerned with timing; rate and method of application of various inputs; appropriate spacing of plants; thorough weeding; careful monitoring and control of pests; and improved harvesting, drying, and threshing methods. Of all the management practices, the control of water is perhaps the most important, because its timely application is essential to efficient utilization of fertilizer and the attainment of high yields (Anderson et al. 1982).

The Green Revolution is not a miracle of modern agricultural technology, but rather an evolution that differs only in time and place from earlier developments in industrialized countries. In both cases, the route to significant yield improvements involved plant materials having high genetic potential farmed with improved management practices that included considerably increased levels of input.

This paper provides a brief historic and geographic

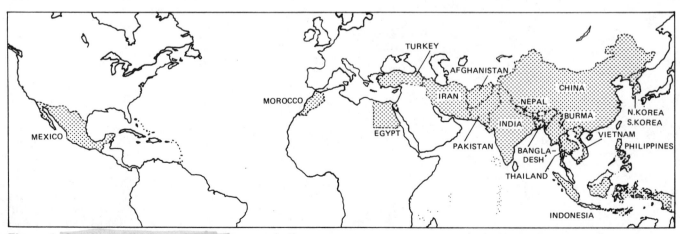

Figure 1. Chief benefiting countries. Source: Author.

From *Journal of Geography,* Vol. 84, No. 6, November/December 1985, pp. 248-254. Reproduced with permission of the National Council for Geographic Education.

view of the genetic and technological developments that led to the Green Revolution. It describes some of the characteristics of the area where these developments were most readily applied and raises questions concerning the unequal impact of the revolution.

Targeted Areas

Yield increases and improved food availability have been particularly important in a group of eighteen heavily populated countries where wheat and/or rice are major food crops. These countries, which are the chief beneficiaries of the Green Revolution, extend across the subtropical part of the world from Korea in the east to Mexico in the west (Figure 1). Fifteen of these countries are contiguous, describing an arc in southern Asia, and are characterized by high rural and farming populations. The benefiting countries include eighteen percent of the earth's land surface, thirty-two percent of its cultivated area, and are home to fifty-six percent of the world's population. Except for China, their population numbers are increasing at a markedly higher rate than that for the world as a whole, and population density is already at extraordinary levels. In mid-1985, there were 290 persons per square mile (750 per square kilometer) compared to forty-nine (127) for the remainder of the world. Population density is almost six times higher than that of the rest of the world; and the nutritional density, or population per square mile of cultivated land, is over 1,500.

The high population-to-land ratio helps to explain other characteristics of the area as well. Grain is raised chiefly on small farms without mechanization and often in monocultivation. The annual crop provides a major portion of the total farm income, and frequently a significant portion of the crop stays on the farm to provide food for the coming year. The farm population dependent on either wheat or rice as a major crop totals at least 1.7 billion people, or 320 million families. Many of these families have no land at all and depend on farm employment for their income. Even those who own or rent land have an average of less than 2.5 acres (about one hectare) per family. The vast majority of these Third World farmers live close to the margin of existence. Surpluses of food and money are minimal.

Unfortunately for these and other underdeveloped countries, the Green Revolution promises more than it can deliver. The term "Green Revolution" implies that a solution has been found for the problem of providing adequate food for a population that grows exponentially. A solution has not been found. The Green Revolution has provided for a quantum leap in food production per unit area, but further increases will be slower, smaller, and probably more expensive. What the Green Revolution has done is to provide time—a breathing space—for people to find a more permanent solution to the population-food problem.

Genetic History: The Green Evolution

The term, "Green Revolution," was originally inspired by work with wheat and rice during the late 1960s, but today is applied to developments in a broad range of food crops throughout the tropics and subtropics. Much of the fundamental research is done at a series of international centers that conduct basic research and train scientists from and in underdeveloped countries. National research centers adapt this research to local needs.

The genes that sparked the Green Revolution were brought to the attention of the post-World War II scientific community by S.C. Salmon who carried seeds of an obscure, short, stiff-strawed, heavy-seeded Japanese wheat (Norin No. 10) to the US in 1946. In 1953 a small packet of second-generation seeds (Norin crossed with a tall North American wheat) was received in Mexico and became the foundation for a breeding program that eventually resulted in a Nobel Peace Prize for Norman Borlaug. In 1962 the first few of Borlaug's Mexican semi-dwarf, rust-resistant seeds were matured in New Dehli. M.S. Swaminathan, then head of the Indian Agricultural Research Institute's Division of Genetics, was so impressed that Borlaug was invited to India where the two scientists launched a program of agricultural innovation in the subcontinent (CIMMYT Economics Program 1983).

Also in 1962, Peter Jennings at the International Rice Research Institute in the Philippines crossed Peta (a tall Indonesian rice variety) with Dee-geo-woo-gen (a short statured variety from Taiwan) (Chandler 1982), and 130 seeds were formed. From these seeds, IR8 (the first of the IRRI modern high-yielding varieties) was identified and named. IR8 was short and sturdy, tillered well, had great seedling vigor, responded well to fertilizer, had moderate seed dormancy, was reasonably resistant to tungro virus, and was essentially insensitive to photoperiod. (Varieties that are insensitive to photoperiod can be planted and harvested in any season without regard to the number of hours of daylight.) IR8 produced record yields almost everywhere it was tested. Unfortunately, it also had several disadvantages. The grain was bold and chalky in appearance, which detracted from its value; it was subject to considerable breakage in the milling process; and the amalose content of its starch was so high as to cause a hardening after cooking and cooling (a distinct handicap for sales to Asian consumers). It was also susceptible to bacterial blight and several races of rice blast. IR8 was soon replaced by a sequence of newer varieties that overcame IR8's disadvantages but maintained most of its yield advantage.

Contact with rice scientists in China was difficult and limited until the 1970s. Nonetheless, a small but steady exchange of genetic material and published research was maintained through Hong Kong and East Pakistan (now called Bangladesh). In China the breeding efforts in rice were divided between the development of improved varieties and hybrids, for which new seeds had to be developed for each planting. The emphasis on hybrids was so strong that by the early 1980s almost twenty percent of China's total rice area was planted to hybrids (Huke 1982). In other rice growing countries, the cost and effort of raising hybrid seed was not considered worth the gains to be realized from hybrid vigor until recently.

Innovations Through Breeding

The breeding of rice and wheat in the mid-twentieth century developed along three distinctive and equally important lines. The first innovation was the introduction of the dwarfing gene that allowed plants to use high levels of fertilizer and achieve yield levels far above those previously possible. Along with this change came the elimination of photoperiod sensitivity. This innovation was of greatest benefit to farmers on the highest quality land that could be irrigated year round. But even for farmers on rainfed land, it allowed increased flexibility and the opportunity to adjust planting dates to the soil moisture levels of individual seasons. It also allowed some spreading out of labor demands in areas where monocultivation was the normal mode of operation and, therefore, moderated the "boom" and "bust" character of the labor market.

1. GEOGRAPHY IN A CHANGING WORLD

The second innovation was the genetic capability to resist attacks by various insect pests and diseases. One reason for the rapid and widespread acceptance of wheat varieties from Mexico in the early 1960s was a strong resistance to various rusts. With rice, early releases were not as resistant to attack. In many parts of Asia, losses to insects and diseases were often severe, especially during the monsoon season. Several early varieties produced yields up to their genetic potential only with heavy and frequent application of chemical insect controls. Farmer resistance to the use of expensive systemics, inappropriate use of sprays, increasing environmental concerns, and the development of resistant genotypes among insect populations hastened the development of grain types with genetic resistance. Newer varieties of both rice and wheat require smaller amounts of pesticides than varieties used in the 1970s.

The third innovation was the development of plants with much shorter growing seasons. Before this innovation, the normal practice in Asia was for farmers to plant rice with the first monsoon rains in June and to harvest in mid-December. A 180-day growing season was common. The growing season, however, was reduced to 150 days with IR8 and 110 days with IR36. The adoption of IR36 allows farmers in areas subject to erratic, early rains to evade moisture problems by planting late, and it allows farmers more favorable conditions to supplement the rice crop with an early planting or post-rice planting of a vegetable or a pulse.

The importance of these innovations may be appreciated with reference to wheat in Bangladesh, where output increased from 100,000 metric tons to over 1,000,000 metric tons in twelve years. When the country was established as an independent nation in 1971, many writers believed the problem of producing sufficient food for over 90,000,000 people on a land area the size of New York State, and whose farm base is subject to annual deep flooding as well as to periodic devastation by typhoons, was close to hopeless. Rice is the major food crop, and wheat a secondary crop used chiefly for making chapatis. The environment for rice production is probably the most difficult of any rice-growing country. Of 24.7 million acres (ten million hectares) planted annually, only ten percent is irrigated; twelve percent is flooded to a depth of greater than three feet (one meter); almost ten percent is dryland, lacking even bunds to pond rainfall; and the remainder is rainfed, frequently subject to massive but temporary flooding. Despite such handicaps, Bangladesh has converted about fifteen percent of its rice area to modern varieties and has improved its national rice yield by almost thirty percent. At the same time, wheat acreages, yields, and production increased sharply (Figure 2). This shift came about only after changes had been made in rice and wheat seeds. The shortened rice growing season provided a window of opportunity for Bangladeshi farmers; and the rust-resistant, short growing season wheat allowed farmers to take advantage of the opportunity. Wheat is now grown as a winter season crop on large areas that previously produced only a rainy season, long-duration rice crop.

It is improbable, however, that wheat expansion would have taken place rapidly were it not for the fact that farmers had prior experience with modern rice varieties and had received help from extension agents. In less than fifteen years, wheat yields more than doubled. The area planted to wheat expanded five times, and wheat production increased ten times (CIMMYT 1982). In the world's most crowded country where food shortages and natural disasters occur frequently, the food situation by mid-1985 in Bangladesh was marginally better than it was a decade earlier. In Bangladesh, modern agricultural technology has already proven that it can provide the time for the country to tackle the population-food problem.

Inputs and Yield

Modern plant varieties give higher yields than do traditional ones. They are planted on the best land, tend to occupy irrigated or at least reliable water control areas, receive

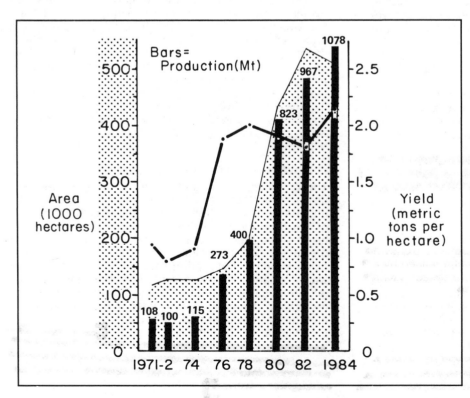

Figure 2. Bangladesh wheat: area, yield, production. Source: US Department of Agriculture (1984).

Table 1.

Percentage Change in Area, Production, and Yield of Rice: 1965-67 to 1982-84

Country	Change in rice area		Change in rice production		Change in rice yield	
	million hectares	percentage	million metric tons	percentage	metric tons per hectare	percentage
China	3.1	10	67.9	73	1.76	57
India	4.0	11	31.5	69	0.66	51
Indonesia	1.6	21	18.8	122	1.72	83
Bangladesh	1.4	15	7.0	46	0.44	27
Thailand	2.8	43	5.2	42	− 0.02	− 1
Burma	0.1	2	7.0	97	1.45	94
Vietnam	0.9	19	5.3	60	0.66	75
Philippines	0.3	10	3.8	93	0.99	75
S. Korea	<.1	0	2.3	45	1.96	47
Pakistan	0.6	43	3.2	160	1.20	83
N. Korea	0.2	33	2.7	117	2.47	66
Total	+ 15.0	+ 14	+ 154.7	+ 74	+ 1.05	+ 52
World less 11 above	+ 3	+ 15	+ 5.3	+ 11	− 0.10	− 4

Source: Palacpac (1982) and US Department of Agriculture (1984).

Table 2.

Percentage Change in Area, Population, and Yield of Wheat: 1965-1984

Country	Change in wheat area		Change in wheat production		Change in wheat yield	
	thousands of hectares	percentage	thousands of metric tons	percentage	metric tons per hectare	percentage
Egypt	47	+ 9	728	+ 57	1.08	+ 45
Morocco	318	+ 19	874	+ 80	.34	+ 52
Afghanistan	294	+ 13	800	+ 36	.20	+ 21
Bangladesh	405	+ 405	960	+ 813	.96	+ 81
China	4106	+ 17	51671	+ 173	1.63	+ 135
India	9685	+ 72	25900	+ 156	.61	+ 50
Iran	2000	+ 50	2582	+ 66	.10	+ 10
Pakistan	2052	+ 39	6284	+ 103	.54	+ 47
Turkey	1574	+ 22	3264	+ 33	.12	+ 9
Total	20481	+ 39	93063	+ 131	.82	+ 66
World less 9 above		− 4		+ 24		+ 29

Source: US Department of Agriculture (1984) and FAO (1967).

higher levels of fertilizer, and are often more carefully weeded than traditional varieties. Plant protection is more commonly extended to modern varieties; the net labor input is higher, and the modern varieties are first adopted by the most innovative farmers. For almost every input controlled by farmers, modern varieties have an advantage over traditional ones.

Changes in area planted, production, and yield for rice and wheat for Green Revolution countries are shown in Tables 1 & 2. In the case of rice, area increased by a modest fifteen percent over eighteen years, reflecting the fact that land was already at a premium and that the best land had already been developed. The new area was land previously avoided for rice because of poor soils, excessive flooding, excessive drainage, remoteness, or some other physical or locational handicap. Despite this modest increase in area, the production of rice rose by seventy-four percent over the same period because of a remarkable increase in yield per unit area.

Output of modern grain varieties increases sharply with the application of the first unit of a new input, but an additional unit of that same input results in a lower additional increase. For example, a first weeding may result in a yield increase of fifteen percent over that of no weeding, but a second weeding may result in only a ten percent additional increase. Eventually, one further weeding will cost more than the value of the resulting yield increase. With chemical fertilizers, the response will be more dramatic. At some level, a point will be reached where one additional unit of input will result in a decrease in yield. Nonetheless, at all levels of fertilizer input, the expected yield of modern varieties exceeds that of traditional varieties. The general formula describing this relationship is:

Final Yield = Base Yield + A(Fertilizer) − B(Fertilizer)^2.

The case of Indonesia provides an example of the differential impact of high nitrogen fertilizer on irrigated rice (Figure 3). The coefficients for (Fertilizer) and

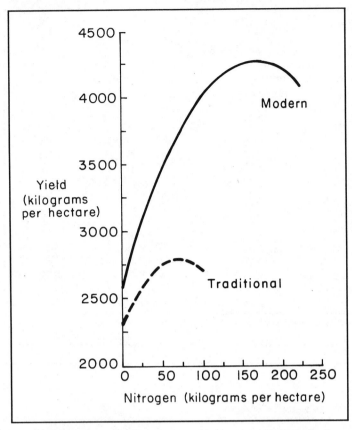

Figure 3. Yield response of nitrogen on irrigated rice in Indonesia. Source: Herdt and Capule (1983).

(Fertilizer)^2 vary with rice variety, water control, soil character, and season but always show a more rapid response on the part of improved grain types. The observation that modern varieties require more fertilizer is only partly correct. Although they do require more fertilizer to reach their maximum potential, modern varieties outyield traditional types even without the application of fertilizer (Barker and Herdt 1985).

Impact of the Green Revolution

From 1965 through 1983, underdeveloped countries out-performed developed countries in terms of the rate of increase in food production (Figure 4). Unfortunately, population growth was so high in underdeveloped countries that their per capita gain was no more than in developed countries, about one percent per year.

On a regional scale, the impact of the improved technology is illustrated in Tables 1 & 2. In the case of rice, the benefiting countries increased their average yield by an impressive fifty-two percent between 1965 and 1983, representing almost 2.5 percent per year. In the same period, their total production went up by seventy-

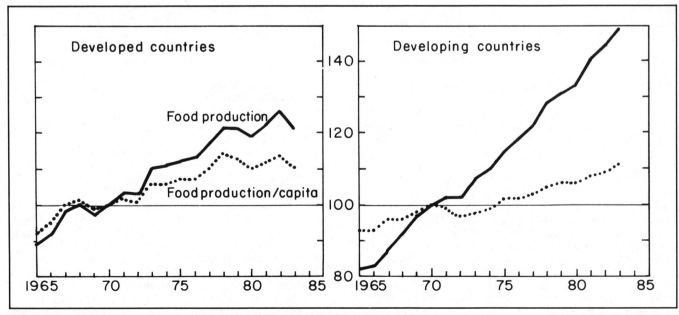

Figure 4. Food indexes (1970 = 100 percent). Source: After FAO (1984).

Table 3.

Rice Area and Nitrogen Used 1979-1980

Country	Rice area planted thousand hectares	Nitrogen kilograms per hectare	Yield rough rice metric tons per hectare
Burma	5,013	11	2.19
Bangladesh	10,308	21	2.02
India	40,200	19	2.01
Pakistan	2,026	52	2.37
Indonesia	8,495	73	3.29
Philippines	3,543	33	2.15
Thailand	8,288	10	1.82
S. Korea	1,314	131	5.90

Source: Martinez and Diamond (1982) and FAS (1983).

four percent. For the rest of the world, yield decreased by four percent, and production advanced by only eleven percent. Higher yields among the beneficiaries were related to improved irrigation, increased use of fertilizers, better weeding, and plant protection, all applied to a stronger genetic stock.

High yielding modern varieties of wheat and rice have not been universally adopted, nor have their full potentials been realized by the farmers using them. On experimental farms, the new seeds commonly produce four times the yield of traditional varieties, but on farmers' fields the advantage is far smaller. The difference is often called the "yield gap" and probably occurs because the environment on the farmers' fields is less ideal than on experimental farms where neither flood nor drought is a problem, where soil conditions are most favorable, where each field is carefully monitored by specialists, and where the level of inputs, especially fertilizer, is well above that used by farmers. On experimental farms, the new varieties maximize their yield on most soil types at levels of nitrogen ranging from 130 to 200 pounds per acre (150 to 225 kilograms per hectare). Depending on the relative price of fertilizer and the market value of rough rice, the optimum level of application to maximize returns without measurable damage to the ecosystem is between ninety and 130 pounds per acre (100 to 150 kilograms per hectare). These levels are seldom achieved in Third World countries.

In South and Southeast Asia, modern wheat seeds have been planted to roughly eighty percent of the wheat area compared to under fifty percent for rice. The lower figure for rice is related to the character of the new seed and the physical geography of Asia's rice lands. The seed produces a plant about three feet (about one meter) tall that matures in roughly 110 days. The plant does poorly when subject to submergence, and it does not have time to recover from drought stress that may occur early in the growing season. Modern rice cannot be grown at all on areas normally subject to more than three feet (about one meter) of standing water during the wet season. Rice varieties also provide only modest yield advantage under saline conditions or in high sulfate soils (IRRI 1985). Large areas of the great delta regions of South and Southeast Asia, therefore, are unsuitable for today's modern rice varieties. With this in mind, the fifty-two percent mean yield increase already achieved is even more remarkable.

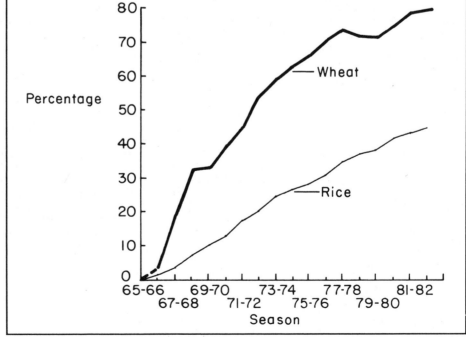

Figure 5. Percentage adoption of modern wheat and rice in Asia. Source: Dalrymple (1978, 1985a, 1985b).

1. GEOGRAPHY IN A CHANGING WORLD

It is difficult to identify and document those factors beyond the physical environment that have contributed directly to unequal benefits, and many questions have been raised. Has the Green Revolution resulted in unfair advantages accruing to the already well-to-do, or have the poor benefited as well? Do farmers with large holdings reap disproportionately more profit than those with small holdings? Do owners of land prosper while tenants and the landless become poorer?

In 1982, The American Association for the Advancement of Science published the results of a symposium devoted to discussing these questions (Anderson et al. 1982). The publication consists of eighteen well-documented papers that provide different answers to these fundamental questions. In a much briefer report, more than 100 sources reporting on modern rice varieties in Asia are analyzed for answers to the same questions (Herdt and Capule 1983). The evidence is strongly conflicting across time, place, and cultural setting and, therefore, the authors conclude that there are no consistent findings relating successful adoption of the Green Revolution package to either farm size or tenure.

In a report on two Indonesian and two Philippine villages, strong evidence was found supporting the environment as the principal determinant of success in adoption.

The four village cases invariably show that farm size was not a factor in the adoption of MV [modern varieties]. In the three villages with environmental conditions suitable for MV, both large and small farmers planted MV in nearly 100% of their paddy areas. On the other hand, both large and small farmers rejected MV when they found the MV to be unsuited to their environment. In neither case was a significant difference observed in the levels of either yield or application of modern inputs such as fertilizers (Hayami and Kikuchi 1982, pp. 212-13).

Conclusion

In the years ahead, there are two groups of challenges facing agricultural scientists and social scientists in regard to the Green Revolution. The first is to continue research on breeding and farm systems. The objective of breeding should be to develop plants that will adapt more readily to the less than ideal environments where most of the poorest farmers of the world live. The objective of farming systems research should be to maximize year-round food production on a given plot of land and to substitute home-grown or home-produced inputs for purchased industrial inputs where possible.

The second objective is to continue to extend the improved technology to those farmers who have not yet participated and to encourage those who have adopted one-half the package to intensify their operations. For example, a majority of farmers in Asia apply very low levels of nitrogen to their crop. In environments as different as Indonesia and South Korea where nitrogen levels are high, yields are significantly above those of neighboring countries. For the other countries, much unexploited potential remains to be developed through fertilizers and other management inputs.

Success in these broad research objectives holds promise for possibly a fifty percent increase in yields.

Such a result would provide the world with two or three decades to find a more permanent solution to the population-food question.

References

Anderson, R.S., ed. et al. 1982. *Science, Politics and the Agricultural Revolution in Asia.* Boulder, CO: Westview Press.

Barker, R., and Herdt, R.V. 1985. *The Rice Economy of Asia.* Resources for the Future Series. Baltimore: Johns Hopkins University Press, forthcoming.

Chandler, R.F., Jr. 1982. *An Adventure in Applied Science: A History of the International Rice Research Institute.* Manila, Philippines: International Rice Research Institute.

CIMMYT. 1982. "Wheat in Bangladesh." *CIMMYT Today No. 15.* Mexico, D.F. Mexico: Centro Internacional de Mejoramiento de Maiz y Trigo.

CIMMYT Economics Program. 1983. *World Wheat Facts and Trends.* Report 2: An Analysis of Rapidly Rising Third World Consumption and Imports of Wheat. Mexico, D.F. Mexico: Centro Internacional de Mejoramiento de Maiz y Trigo.

Dalrymple, D.G. 1978. *Development and Spread of High-Yielding Varieties of Wheat and Rice in the Less Developed Nations.* Report No. 95. Washington, DC: US Department of Agriculture.

———. 1985a. *Development and Spread of High Yielding Rice Varieties in the Developing Countries.* AID Technical Bulletin Series. Washington, DC: United States Agency for International Development.

———. 1985b. *Development and Spread of High Yielding Wheat Varieties in the Developing Countries.* AID Technical Bulletin Series. Washington, DC: United States Agency for International Development.

FAO. 1967. *FAO Production Yearbook 1967.* Rome.

———. 1984. *FAO Production Yearbook 1984.* Rome.

FAS. 1983. *Foreign Agriculture Circular FG-26-83 World Rice Reference Tables.* Washington, DC: US Department of Agriculture.

Hayami, Y., and Kikuchi, M. 1982. *Asian Village Economy at the Crossroads: An Economic Approach to Institutional Change.* Baltimore: Johns Hopkins Press.

Herdt, R.V., and Capule, C. 1983. *Adoption, Spread, and Production Impact of Modern Rice Varieties in Asia.* Manila, Philippines: International Rice Research Institute.

Huke, R.E. 1982. *Rice Area by Type of Culture: South, Southeast and East Asia.* Manila, Philippines: International Rice Research Institute.

IRRI. 1985. *International Rice Research: 25 Years of Partnership.* Manila, Philippines: International Rice Research Institute.

Martinez, A., and Diamond, R.B. 1982. *Fertilizer Use Statistics in Crop Production.* Muscle Shoals, AL: International Fertilizer Development Center.

Palacpac, A.C. 1982. *World Rice Statistics.* Los Banos, Philippines: IRRI, Department of Agricultural Economics.

US Department of Agriculture. 1984. *Agriculture Statistics 1984.* Washington, DC: US Government Printing Office.

GEOGRAPHY AND GENDER

Urban geographers have long been concerned with the way social classes and ethnic divisions show up in the landscape. This article considers the geography of another form of social differentiation that has only recently attracted the attention of geographers – that between women and men. Examining this form of differentiation can help us to understand a wide range of phenomena from the persistence of poverty in the Third World to the processes of large scale industrialisation, suburban growth and gentrification.

Liz Bondi

Dr Liz Bondi lectures in Social Geography at the University of Edinburgh. Her research and teaching interests are gender issues and educational provision. She is a member of the Women and Geography Study Group of the Institute of British Geographers.

INTRODUCTION

In Britain, we accept without question that most secretaries and typists are women, and that most people working on building sites are men. However, in India, we would find that most secretaries and typists are men, and that some construction sites are places where women work. As geographers, we are often interested in the ways in which one place differs from another and the reasons for such differences. Yet, so far, little attention has been given to geographical differences in the work of women and men. This article aims to highlight the importance of such differences to the study of geography by using the concept of *gender roles* (defined below). The first section considers some definitions. The second section examines variations in gender roles between countries and looks, briefly, at the relationship between economic development and gender inequalities. The third section assesses the relationship between gender roles and spatial patterns in urban areas; it focusses on the processes of large-scale industrialisation, suburban growth and gentrification.

DEFINITIONS

Sex and Gender

Sociologists have provided a valuable distinction between the terms 'sex' and 'gender'. The former refers to biological differences between the male and female of a species. It is used of humans insofar as we are merely another biological animal. 'Gender', on the other hand, refers to *social* differences between men and women. The term was originally used to describe the grammatical distinction between feminine and masculine words. This is most familiar in a language like French, which defines all nouns as masculine or feminine. The term has since been broadened to include all socially created distinctions between women and men.

Gender Roles and Sex-typing

Because gender differences are created socially rather than biologically, they vary across space and time. This can be illustrated most clearly by means of the concept of 'gender roles'. *Gender roles* arise from the allocation of tasks to different people on the basis of their sex. Although the content of the roles may vary (as illustrated in the next section) in many societies women are regarded as carers and nurturers whereas men are regarded as religious, political or economic leaders.

The development of gender roles results in *tasks* becoming *'sex-typed'* in that they acquire a strong association with femininity or masculinity. This is true of many familiar social activities. Take paid employment in this country as an example. Most people in Britain today work in jobs dominated very markedly by either women or men (see Figure 1).

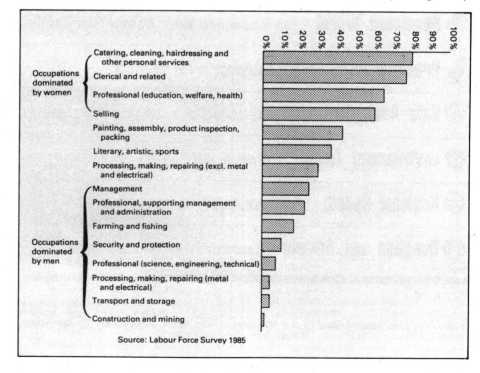

Figure 1   Women as a Percentage of all Employees by Occupational Group

From *Geography Review*, May 1989, pp. 2-6. Reprinted by permission of Philip Allan Publishers, Ltd., England.

1. GEOGRAPHY IN A CHANGING WORLD

Thus, cleaning, nursing and typing are widely recognised as 'women's work', while mining, managing and building are equally widely seen as 'men's work'. The divisions are not watertight, but we make clear the unusual status of exceptions by referring to them as the 'male nurse' (since 'nurse' is assumed to be female) or 'female engineer' (since 'engineer' is assumed to be male). Sex-typing occurs in many other activities. For example, childcare and domestic work, whether paid or unpaid, are generally considered to be tasks for women. And, while both women and men enjoy a range of leisure activities, they often do different things: contrast the participants in a football match and an aerobics class.

VARIATIONS BETWEEN SOCIETIES

This kind of sex-typing occurs in all known societies. However, with the exception of domestic work and childcare, the allocation of activities to women or men varies between societies. One survey of a range of productive tasks in over 400 societies found that every task that was carried out by men in one society was

carried out by women in another society, with the single exception of metal working. For example, in much of sub-Saharan Africa, women work in the fields, growing basic subsistence crops for their families, whereas in much of Latin America, women's agricultural work is confined to tending animals and food processing. Variations are also evident within societies over time. For example, farmwork in this country is almost exclusively performed by men, especially where heavy machinery is involved. Yet, during wartime the same work was undertaken by women. Thus, there is both a history and a geography of gender roles.

Inequality

Interest in gender differences is often prompted by a particular concern with the position of women. This arises because the role of women is generally associated with inferior status, socially, politically and/or economically. For example, in Britain, hourly wage rates for women average less than 75% of those of men. In many societies women are very poorly represented in government: only in a handful of countries (mainly

Scandinavian) do women make up more than 20% of national legislatures. In some countries, notably India and China, the strong desire for male rather than female children is symptomatic of the lowly social status accorded to women. There is, therefore, a geography of gender inequality as well as a geography of gender roles.

Understanding Spatial and Temporal Variations

Geographical and historical variations in gender roles and gender inequalities underpin the argument that many apparently 'natural' differences between women and men are created socially, and are not biologically given. From a geographical perspective, it is important to observe that gender divisions are specific to specific societies. This presents geographers with an important task, namely to understand geographical variations in gender roles and gender inequalities.

The first task here is to map the geography of gender. It is useful at this point to distinguish between the geography of women and the geography

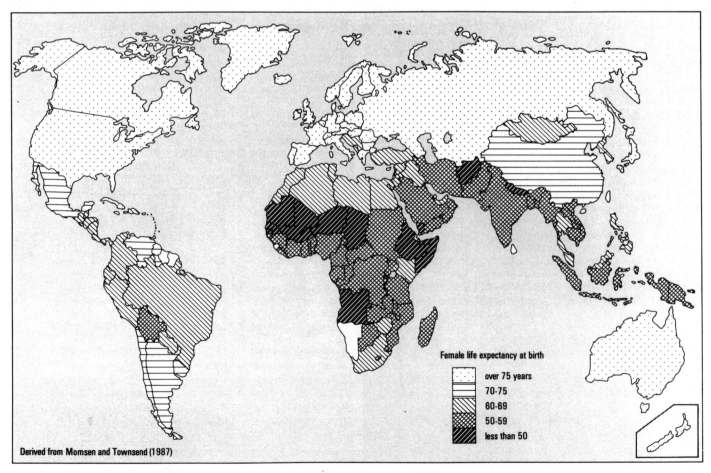

Derived from Momsen and Townsend (1987)

Female life expectancy at birth

- over 75 years
- 70-75
- 60-69
- 50-59
- less than 50

Figure 2 Female Life Expectancy at Birth in the Early 1980s

of gender. The former focuses exclusively on women and could be mapped by comparing their position in different countries. An example is given in Figure 2, which shows that female life expectancy at birth is highest in the developed countries (in North America, Europe, the Soviet Union and Australasia) and lowest in the poorest countries of the Third World (in sub-Saharan Africa and South Asia). This pattern suggests that the geography of women is one aspect of the geography of poverty and development at a world scale: the position of women by and large reflects the position of their countries within the global economic order.

Turning to the geography of gender, this focuses on differences between women and men *within* societies. For example, Figure 3 measures the gap between the literacy rates of women and of men. Contrasts between rich and poor countries are again clear. Thus, women in poor countries tend to suffer low status relative to men in their own countries as well as in comparison with women in other countries. This pattern could be interpreted in two ways: it could be argued to indicate that as development proceeds gender inequalities diminish, or

it could be argued to indicate that wide gender inequalities impede the process of development. There is probably some truth in both arguments (see, for example, **article by Townsend and Townsend in *Geography,* Vol. 73**).

Detailed examination of other gender maps reveals many interesting discrepancies, which are not always easy to explain. Why, for example should patterns of industrial employment vary widely between culturally and economically similar countries: women make up 49% of the workforce in manufacturing industry in Tunisia but only 6% in Egypt and Pakistan. Much work remains to be done in order to answer such questions. (For a fuller discussion, see *Geography of Gender in the Third World*, edited by J.H. Momsen and J. Townsend.)

SPATIAL PATTERNS WITHIN SOCIETIES

A second issue for geographers concerns the relationship between gender divisions and various aspects of spatial organisation within societies. In this connection it is useful to consider how urban geographers

have used models of city structure to describe the geographical separation between different social groups. By mapping social areas in cities we are showing how *social* differences between households come to be expressed in *spatial* patterns of residential segregation. Most attention has focused on differences in ethnic group, social class and stage in the life cycle. As the previous sections have shown, distinctions between women and men provide another kind of social difference in society. The question to be considered here is whether these gender differences lead to patterns of spatial segregation. At first sight, this does not appear to offer a promising line of enquiry: men and women belong to the same households and live under the same roofs so that a spatial separation corresponding to that observed in connection with social class, ethnic group and stage in the life cycle cannot be expected. However, there are other ways in which gender differences might be influential.

It has been argued above that gender roles vary between societies. We might then ask whether different kinds of gender role are associated with different urban forms. This question could be tackled by

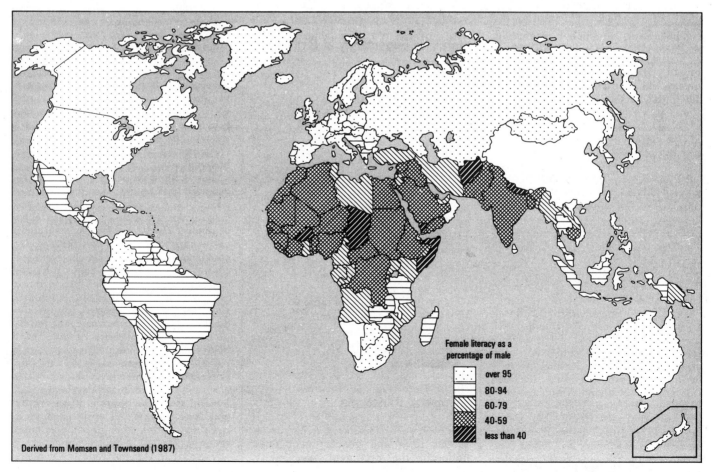

Derived from Momsen and Townsend (1987)

Female literacy as a percentage of male

over 95
80-94
60-79
40-59
less than 40

Figure 3   Female Literacy Rates as a Percentage of Male in the Early 1980s

comparing urban spatial structure in societies with different gender roles. An alternative approach is to examine change within one society. For example, in Britain, social history provides us with a great deal of evidence about changes in gender roles, while urban history provides evidence of changing urban structure. Drawing these two bodies of work together, we can begin to assess the interconnections between gender and urban organisation. A few examples will serve to illustrate the point.

Large-scale Industrialisation

In mid-nineteenth century Britain, the scale of industrial production began to increase markedly. Large factories were established in industrial towns, providing places of work for an increasing proportion of the population. Previously, for the bulk of the population, production had been organised on the basis of individual households, exemplified by the farm in rural areas, and the artisan workshop in urban areas. Consequently, place of residence and place of work were, for most people, identical or adjacent. Further, although most tasks were sex-typed, women and men worked alongside one another, contributing in various ways to the survival of the household. In particular, there was no rigid demarcation between domestic and other kinds of work since most activities contributed more or less directly to subsistence.

These patterns began to change with the emergence of large-scale factory production, which precipitated a spatial separation between home and work on a scale previously unknown. This separation created the possibility of separate spheres of life for women and men. Such a pattern developed most clearly among the expanding middle classes. In these groups, women became increasingly identified with the home and increasingly separated from productive activities. Thus, their roles were redefined to exclude work outside the home. Within the home a new notion of domesticity emerged, strongly associated with the supposedly natural capabilities of women as nurturers and homemakers. By contrast men became the sole breadwinners for their families and found in their homes a refuge from the increasingly stressful world of work.

These developments illustrate that the spatial separation between home and work may have served as a precondition for the emergence of new ideas about the roles appropriate for women and men. In the nineteenth century these new ideas developed among a particular sector of society, namely the middle classes. Within this group, the exclusion of productive work from the home, and the confinement of women to the home, permitted the emergence of a distinctively female domestic realm. *This is important for geographers because it illustrates how changes in spatial organisation can influence and contribute to aspects of social life.*

Suburban Growth

The second example concerns the subsequent influence of this model of female domesticity on urban residential development. Although initially restricted to the middle classes, during the twentieth century more and more working class families aspired to a similar kind of family form, in which the wife was able to devote herself full-time to motherhood and homemaking. Another widespread aspiration was for improvements in housing and particularly for the advantages offered by suburban living. There are connections between the two trends. Suburban living served to accentuate the separation between the realms of work and home both in terms of distance and through the creation of large areas under single land-uses, whether industrial or residential. This, in turn, accentuated the separation between the daily lives of women as homemakers and men as breadwinners. The trend reached a peak in the 1950s, especially in the USA. At this time, many advertisements used images of women looking after children in suburban homes to promote the sale of consumer goods. This was effective because the images of both female domesticity and suburban living appealed to the majority of the population.

This example illustrates how a particular social form, namely a nuclear family with a dependent wife, can operate as a factor contributing to changes in the spatial organisation of urban areas in the form of suburban growth. *Gender roles, therefore, may form part of the explanation of geographical phenomena such as urban structure.*

Gentrification

The third example concerns contemporary changes in urban residential patterns. After decades of outward movement by affluent social groups, a return to small pockets within inner-urban areas is now evident. This process is variously known as 'gentrification', 'urban renaissance', 'inner-city revitalisation' and so on. A variety of factors have been cited in connection with this trend, including changes in the position of women both in the family and in the labour force. With regard to the former, women are tending to marry and have children later, and to

Date of woman's birth	Average number of children per woman
1920	2.00
1925	2.12
1930	2.35
1935	2.42
1940	2.36
1945	2.20
1950	2.06
1960	1.98
1965	1.95
1970	1.97
1975	2.00

Source: Social Trends 18, p. 48 (1988) HMSO.
Note: Data are projected for women born since 1950.

Table 1 Changes in Completed Family Size

have smaller families (see Table 1), with births in more rapid succession. At the same time, women are returning to paid employment more quickly after starting their families (see Table 2). Overall therefore, women today tend to spend a much greater proportion of their lives in paid employment than women of previous generations. An additional factor of significance in connection with gentrification is the increasing success of middle-class women in obtaining well-paid, career jobs. These factors combine to create more relatively affluent, childless households, capable of paying for increasingly expensive private housing. Moreover, many of these households are much more interested in the cultural benefits and convenience of inner urban locations than the imagery of family life associated with the suburbs. Thus, just as suburbia represented the expression of 'ideal' gender divisions for earlier generations, so gentrification may represent an ideal for at least some sectors of contemporary society. *In this way, changes in the role of women may contribute to our understanding of the evolving geography of cities.*

CONCLUSION

The position of women in society is an issue recognised to be important in a growing number of disciplines. This reflects a broader social awareness of, and questioning of, traditional assumptions about gender roles in both developed and developing countries. Geographers have an important part to play in this movement. As this article has illustrated, there is a geography of gender in more than one sense. First, there are geographical variations in gender divisions that need to be mapped and explained. Such studies can contribute to our understanding of both the causes and consequences of Third World development. Secondly, gender divisions may be influenced by or may themselves influence the spatial organisation of society. In this way, a focus on gender has much to offer urban and social geography. The examples provided here are by no means exhaustive but are intended to indicate the importance of a gender

Period in which first child was born	Median no. of years before mother first returns to work (years)
1950—54	9.7
1955—59	8.7
1960—64	7.0
1965—69	5.5
1970—74	4.8
1975—79	3.7

Source: J. Martin and C. Roberts (1984) *Women and Employment. A Lifetime Perspective*, HMSO.

Table 2 Changes in Rate at which Women Return to Work After Starting Their Families

perspective within different geographic specialisms. There is much more to discover and understand.

FURTHER READING

Momsen, J.H. and Townsend, J. (Eds) (1987) *Geography of Gender in the Third World*, Hutchinson, London.

Townsend, J.G. and Townsend, A.R. (1988) 'Teaching gender North—South', *Geography*, Vol. 73, pp. 193—207.

Women and Geography Study Group (1984) *Geography and Gender*, Hutchinson, London.

Land-Human Relationships

The home of humankind is the surface of Earth and the thin layer of atmosphere enveloping it. Here the human populace has struggled over time to change the physical setting and to create the telltale signs of occupation. Humankind has greatly modified Earth's surface to suit its purposes. At the same time, humankind has been influenced greatly by the very environment that it has worked to change.

This basic relationship of humans and land is important in geography. As you will recall, in unit 1 William Pattison identifies this relationship as one of the four traditions of geography. Geographers observe, study, and analyze the ways in which human occupants of Earth have interacted with the physical environment.

This section presents a number of articles that illustrate the theme of land-human relationships. In some cases, the association of humans and the physical world has been mutually beneficial; in others, environmental degradation has been the result.

At the present time, the potential for major modifications of Earth's surface and the atmosphere is greater than at any other time in history. It is crucially important that the consequences of these modifications for the environment be clearly understood before such efforts are undertaken.

The first article in this unit delves deeply into the continued decline in fresh water quality in a number of major rivers in the United States. The article by Pamela Naber Knox details our ability to predict the El Niño phenomenon and its effects on global climate patterns.

Can the rain forests of Earth be saved? What is the climatic impact of their wholesale destruction? The next article discusses a new NASA program to improve the assessments of global deforestation.

Human and environmental problems associated with the greenhouse effect are reviewed in the article by George Sanderson. How closely related are desertification and climate change? This important question is posed in "Exploring the Links Between Desertification and Climate Change." Sandra Postel's article, "Water Tight," discusses water conservation practices in four North American cities.

The articles presented in this section provide a small sample of the many ways in which humans interact with the environment. The outcomes of these interactions may be positive or negative; they may enhance the position of humankind and protect the environment, or they may do just the opposite. Human beings are the guardians of the physical world; they have it in their power to protect, to neglect, or to destroy.

Looking Ahead: Challenge Questions

What are the long-range implications of atmospheric pollution? Can you explain the greenhouse effect?

How can the problem of regional transfer of pollutants be solved?

The manufacture of goods needed by humans produces pollutants that degrade the environment. How can this dilemma be solved?

What would the climatic effect of loss of the tropical rain forests be?

Where in the world are there serious problems of desertification and drought? Why are these areas increasing in size?

What will be the major forms of energy in the next century?

How are you as an individual related to the land?

Can humankind do anything to ensure the protection of the environment?

Unit 2

TROUBLED
WATERS

Two decades after passage of the Clean Water Act, American waterways no longer catch fire with chemical pollution, but they do carry invisible toxic waste that threatens fish and people. Congress tackles the issue this fall, but the answers may be as intricate as rivers themselves.

"Eventually all things merge into one, and a river runs through it."

—*Norman Maclean*

ANNETTE MCGIVNEY

ANNETTE MCGIVNEY *is associate editor of* Backpacker Magazine. *She lives on the Delaware River and sees pollution first-hand.*

For Norman Maclean, Montana's Big Blackfoot River was a pristine and spiritual place where any faithful fly fisherman could enjoy a near-religious experience, partaking in the best that nature had to offer. The trout-filled waters of the Blackfoot shaped Maclean's life and inspired him to write a book, *A River Runs Through It*, filled with romantic descriptions of the Blackfoot, which inspired Robert Redford to buy the screenplay rights and produce a movie about the beautiful river of Maclean's youth. Flyfishing, brotherhood, growing up in Montana— the Hollywood production had all the wholesome goodness of homemade bread, except for one thing. By the time Redford was ready to start filming two years ago, some 16 years after the book was published, the Big Blackfoot lacked the aesthetics necessary to serve as the setting for the movie.

Like many rivers in Montana, and in the rest of the U.S., the mighty Blackfoot has been broken by the resource-guzzling activities of humans and is no longer the proud water that Maclean described. The Montana locals weren't surprised that Redford and his crew had to simulate the Blackfoot, which hasn't supported a healthy fish population for at least 10 years. Its banks are scarred by clearcuts, and foul-looking mine waste slurps against road crossings. "The Blackfoot has been over-mined, over-cut, over-fished, over-recreated, and overlooked," says Becky Garland, president of the Big Blackfoot Chapter of Trout Unlimited, which is helping to revive the river.

Taken out of context, Maclean's words are prophetic — all of our land use practices and wastes have indeed merged into one, and our rivers run through it. Since Cleveland's Cuyahoga River caught fire in 1969, federal and state governments have worked hard to get the junk out of polluted waters. It's rare now to see a river topped with chemical foam, flowing bright red from industrial discharges, or reeking from raw sewage. But in many instances, these waters are not much healthier than they were two decades ago.

The Blackfoot is part of 2,279 miles of rivers and streams in Montana listed in the Environmental Protection Agency's (EPA) 1990 National Water Quality Inven-

> "The public thinks that the rivers are cleaned up because the goo is gone. If water quality is improving, why are all the fish dying?"

tory as having "elevated toxins." Nationwide, more than 28,000 river miles contained a hazardous level of toxins in 1990, and some 26 million fish were killed by pollution. "The public perception is that the rivers are cleaned up, that they're better than ever, because the goo is gone," says Kevin Coyle, president of American Rivers. "But if water quality is improving, then why are all the fish dying? Eventually, our rivers will look incredibly clean and they will be deadly." What worries him most is the invisible pollution that spills undetected from industry, and flows with rain water from every paved street and farmed field into rivers and streams. "Back in the 70s, people thought of water quality

From *E Magazine*, September/October 1993, pp. 30-37. Reprinted with permission from *E*, the Environmental Magazine, subscriptions $20/yr; P.O. Box 6667, Syracuse, NY 13217. (800) 825-0061.

in terms of gooey stuff washing up on the docks," he says. "Today, non-point source pollution is the major problem for rivers, and it's not usually something people can see. Poison runoff can even be aesthetic because the water looks so clear."

When the Clean Water Act (officially called the Federal Water Pollution Control Act) was passed in 1972, the mandate was broad: "to restore and maintain the chemical, physical and biological integrity of the nation's waters." But the immediate task was to stop municipalities and industries from dumping untreated or poorly treated wastes into public waters. And urban rivers like the Delaware, Potomac and Hudson, which once stunk from blocks away, are now popular spots for fishing, canoeing and picnicking. Much of the progress has come from improved wastewater treatment. The federal government has spent $56 billion on new treatment plants since the early 70s.

But controlling point-source discharges is about as far as the nation has gotten in meeting the goals of the Clean Water Act. Under the deadlines set out in 1972, by 1983 all of the nation's waters were to be fishable and swimmable. And by 1985, the discharge of any pollutant into U.S. waters was to have been eliminated. According to EPA's 1990 Water Quality Inventory, 37 percent of rivers and streams assessed did not meet fishing and swimming standards. The figure could be higher, since only 36 percent of the nation's 1.8 million river miles were monitored. And industrial releases of toxic substances into American waters still average 360 million pounds a year.

"Rivers are the catch basins for all our land use practices—farmers spraying their fields, homeowners spraying their lawns, logging, grazing, everything," Coyle says. "And our land use practices are causing aquatic species to go extinct at a rapid rate. The invertebrates are shouting out most loudly, telling us some rivers are dying."

A recent study by The Nature Conservancy found that 65 to 70 percent of bottom-of-the-food-chain aquatic species like crayfish and freshwater mussels are classified as rare to extinct, while only 11 to 14 percent of North American terrestrial species (birds, mammals and reptiles) are considered rare to extinct. Since freshwater mussels are filter-feeders and have little tolerance for water-borne pollutants, they have been seen as the proverbial canaries in the coal mine. "The unprecedented decline of mussels is nothing less

than a red flag, a distress signal from our rivers, streams and creeks," says Nature Conservancy president John Sawhill. "We cannot ignore this warning and the degradation of water quality that it implies."

The Clean Water Act is up for reauthorization this year, and the need for the federal government to come to the rescue of rivers is just as urgent as it was in 1969. A preliminary bill was introduced in the Senate last June by Max Baucus (D-MT), head of the Committee on Environment and Public Works, and John Chafee (R-RI), the committee's ranking minority leader. Following summer hearings, the final bill is expected to be submitted to the Senate for approval early this fall.

"The House isn't as far along as the Senate with a reauthorization bill, but getting Clean Water Act legislation through this session is a top priority for

---

"The industrialization of agriculture over the last two decades has had a tremendous effect on rivers. Farmers spray crops with all kinds of chemicals, and weeds with herbicides instead of tilling the land. And people wonder why the cancer rate is going up in Iowa."

Congress," says Robyn Roberts of the Clean Water Network, a coalition of 450 environmental and community organizations that want a "strengthened" Clean Water Act. Supported by national players like the Sierra Club, American Rivers and the National Wildlife Federation, as well as local grassroots groups, the Network wants a reauthorized Clean Water Act that will "eliminate the use of the nation's waters—and other parts of the environment—as dumping grounds for society's wastes." The Network also wants Congress to federalize water quality programs and make them less permissive, to focus on non-point source pollution, preserve wetlands, expand regulations of toxic discharges, and prevent clean waters from getting polluted. Many of these components already appear in the broad language and numerous amendments to the Clean Water Act, but federal and state agencies are failing to carry them out. "The law seems fine to me as it is written," says Nina Bell, executive

director of Northwest Environmental Advocates (NEA) in Portland, Oregon. "It's just the implementation that stinks."

Keeping Tabs

Over the past decade, local groups like NEA have led the fight against river degradation by spending a great deal of time policing the polluters to stay in compliance with water quality standards. The Willamette River no longer has rafts of sewage floating atop its waters or slime lining its banks. Municipal wastewater treatment plants have been built, industries must obtain discharge permits, and some 200 miles of greenway have been acquired for a large riverfront park system. In 1989, the U.S. Army Corps of Engineers described the Willamette as "one of the cleanest streams of comparable size in the nation," and the EPA often refers to it as an example of a river restoration success story.

The Willamette, however, is far from restored and requires constant management. Water levels are regulated by a series of 13 dams, most of the fish are hatchery stock, and waste discharges are diluted with impounded water. A 148-mile strip of the Upper Willamette has unsafe levels of dioxin, a carcinogenic byproduct of pulp mills and sewage treatment plants that use chlorine. Other toxins violate water quality standards in scattered sections of the river, including DDT, PCBs, chlordane, arsenic and heavy metals. On the Lower Willamette, Portland uses 43 antiquated storm drains called combined sewer outfalls (CSOs) which dump raw sewage into the river when it rains. "I wouldn't take my child swimming in the Willamette," Bell says. "There's too much sewage. You can see toilet paper floating by after a rain."

Even more frustrating than the pollution is the lack of information about it. "The fundamental problem is we don't have enough monitoring. The states aren't required by EPA to collect certain water quality data, so, like with the Willamette, information is very sketchy," Bell says. The Oregon Department of Environmental Quality has admitted as much: "The river's current water quality is not completely known. Much of the existing information on the river describes conventional water quality measurements, such as levels of dissolved oxygen. Insufficient information exists on other topics, such as the impact of toxins on the overall health of the river."

The Clean Water Act requires states to designate uses for rivers—usually fishing and/or swimming—and then report to the

EPA on whether or not those uses are met. The EPA offers guidelines, but the decision on which rivers to monitor and what testing methods to use is up to the states. Their information gets compiled by the EPA every two years for the National Water Quality Inventory, which typically only reports on about one third of the nation's river miles. And since every state has its own testing methods, the EPA can't combine all of the information into a national overview. "In times of limited resources, you have to monitor where there are suspected problems," says Elizabeth Jester Fellows, branch chief for monitoring in EPA's Office of Wetlands, Oceans and Watersheds. "Since only about one third of rivers are monitored in a two-year period, we are recommending that the states rotate their testing each period so more rivers are covered."

The EPA also encourages states to shift their focus from testing for the obvious—specific chemicals from point-source discharges—toward monitoring for non-point source pollution. But in 1991, the federal government gave less than $70 million to states for ambient water monitoring — assessing the quality of the aquatic environment rather than studying the effect of specific discharges. Federal monies for new wastewater treatment plants in 1991 totaled $6 billion. "Increased funding for monitoring is often viewed by Congress as a resource sink," says Fellows, "because it's hard to prove the money has been used effectively."

Richard Sparks, an aquatic ecologist with the State of Illinois, says, however, that river water quality won't improve unless the feds step up to the plate. "There needs to be more leadership from the federal government. States set up programs in response to federal requirements and grant opportunities. Most states are too strapped financially just to start up new water quality programs on their own," he says. "Our monitoring is supposed to provide a kind of 'state of the aquatic environment' report for the EPA, but I consider many of the measures the states use to gauge this as irrelevant." The EPA does recommend that states incorporate the use of biological criteria in their reporting and will require it by 1995. These methods, which involve counting the number of different aquatic species and the abundance of each, cost more than the chemical approach and require a staff of biologists— which most states don't have.

"The use of bio-criteria will result in the re-direction of money we currently spend on wastewater treatment plants to pay for the restoration of aquatic ecosystems," says James Karr, director of the Institute for Environmental Studies at the University of Washington. "Historically, states have concentrated on waters they suspected had toxic concentrations. Now, with bio-criteria, we're going to find problems in rivers the states thought were okay."

Pointless Pollution

When there is a blanket of dead fish covering a river just downstream from an industrial treatment plant, it's not hard for state and local authorities to find the source of the pollution and slap the guilty party with a fine. Non-point source pollution, however, is much more evasive — everyone is to blame, and no one is to blame. But there's no arguing with the fact that the agricultural industry is the primary culprit in polluting our nation's rivers. The EPA reports that 50 to 70 percent of impaired or threatened surface

---

"I wouldn't take my child swimming in the Willamette. You can see toilet paper floating by after a rain."

---

waters are affected by non-point source pollution from agricultural activities. According to the Conservation Foundation, nearly five tons of soil erode off of one acre of farmland each year in the U.S., carrying sediment (the number one pollutant), fertilizers, herbicides and insecticides into rivers and streams.

"The industrialization of agriculture over the last two decades has had a tremendous effect on rivers," says Kevin Coyle. "The U.S. decided it would maximize farm production and be the bread basket for the world. Farmers started spraying their crops with all kinds of chemicals, and spraying weeds with herbicides instead of tilling the land. There are more chemicals being used by farmers today than ever before. The EPA approves these substances, but they don't really monitor their potentially hazardous effects. And people wonder why the cancer rate is going up in Iowa?"

The 1990 Farm Bill encourages farmers to voluntarily participate in programs that minimize the use of chemicals and control agricultural runoff, but the financial incen-

tives for these programs are not nearly as enticing as other U.S. Department of Agriculture (USDA) programs that encourage exploitation of the land. For example, more than two-thirds of U.S. cropland is enrolled in a program where farmers receive price supports based on the average crop yield over a five-year period. This discourages the farmer from rotating crops or letting the land lay dormant, which would reduce runoff into rivers as well as the need for pesticides.

An obvious step would be to place stricter land use regulations on the agricultural industry. But the notion of making farmers comply with federal requirements is politically taboo, and downright un-American in much of the country. "Voluntary action through education is more effective than regulation in addressing our environmental issues," says James Moseley, USDA assistant secretary for agriculture for natural resources and environment. "Prohibiting the use of certain chemicals and policing and fining polluters is not the best way to deal with water quality concerns, particularly in a diversified industry such as agriculture. Regulations undermine agriculture's flexibility in determining production options."

The other major cause of non-point source pollution is urban run-off. Poisonous substances wash down city streets and storm drains from everything and everyone—from residential landscaping to road construction to motor oil. But finding ways to control the conglomeration of wastes that merge into urban runoff is a new and evolving science. "There is a lot of emphasis being placed on figuring out how to deal with non-point source pollution," says Curtis Dalpra of the Interstate Commission on the Potomac River Basin. "A lot of methods for abating non-point in urban areas are not turning out to be as effective as people first thought they would be."

Accidentally Toxic

For all of the talk about non-point source pollution, toxic spills are still a serious threat to both water quality and public health. In 1990, 31 states reported concentrations of toxic contaminants in fish tissue that exceeded public health standards. Deadly substances like DDT, chlordane, PCBs and dioxin live for decades in a river's aquatic food chain. "Toxic spills and the legacy of toxic substances along riverbanks are a real problem," says Richard Sparks. "Industries have to meet requirements in order to get discharge permits, but accidents happen—and unless there's a fish kill, small spills often go undetected."

In fact, much to the surprise of unsuspecting river recreationists, accidents happen quite often. A U.S. Public Interest Research Group (U.S. PIRG) study has found that 21 percent of the nation's 7,185 major industrial and public waste treatment facilities are in serious or chronic violation of their discharge permits, and another 19 percent of the facilities report at least occasional violations. What's more, the study says, many of these facilities violate their permits on purpose because the cost of paying minimal fines to the EPA is less than that of investing in the equipment needed to comply with water quality standards. An occasional toxic spill may seem like a negligible crime, but the long-term cost to public health can be high. Sparks says that it takes many of these toxic substances decades to degrade, and as time passes the toxin "bio-magnifies" as it moves up the food chain. "A fish can tolerate a certain amount of mercury," he says, "but a human that eats several of those fish could get too big a dose." EPA studies show that dioxin bioaccumulates at very high rates, and that even immeasurably small quantities cause a range of health problems in humans from immunological disorders to cancer.

Since states want to promote clean looking rivers for recreational users, it's not uncommon to find waterways where the fish are known to contain toxins but there are no signs warning people of the danger. On a stretch of Virginia's scenic Shenandoah River there's an advisory against eating fish tainted with mercury from an industrial spill 17 years ago, but there are no signs. "People tore the signs down a long time ago and we haven't put them back up," says a Virginia Department of Health official.

"Citizens have a right to know when significant threats to their health or environment are present in their communities," says U.S. PIRG staff attorney Carolyn Hartman. "We want the Clean Water Act to require public postings on waterways that don't meet standards, or have a fishing or shellfish ban." U.S. PIRG and the Clean Water Network also support legislation to require dischargers to post a sign on the entrance of their facility that details what is going into the river and who people can contact for more information.

Congress added a section to the Clean Water Act in 1987 to accelerate efforts to control and eventually eliminate toxic point-source discharges into rivers. It requires states to identify and put on three

---

YOUNG MAN RIVER

BY JESSICA SPEART

The 25-foot boat darted toward the Exxon tanker under cover of night, stealthily taking water samples of the liquid being discharged. For a half year, the tiny boat had trailed tankers along New York's Hudson River as they rinsed their tanks clean of sea water tainted with jet fuel—an illegal activity Exxon would live to regret. But back in 1983 few people knew that the Hudson River had recently acquired its very own watchdog: John Cronin, better known as the Riverkeeper.

As a boy, Cronin never paid much attention to the river. "To me, the good Lord put the Hudson there to separate New York from New Jersey," quips the 43-year old. Tall and lanky with blue eyes and a mop of curly blond hair, Cronin was working as a roofer in 1972 when he first heard of an environmental sloop named the Clearwater that would be sailing into Beacon, New York. There he met Pete Seeger, folk singer and activist, who was using the sloop in his fight to clean up the 315-mile river. The Clearwater crew still sails the Hudson today, giving people a firsthand look at the river's problems. Cronin soon joined Clearwater's "Pipewatch Project"; his subsequent investigative work, taking pipe discharge samples from an adhesive tape manufacturing company, resulted in the company being fined $50,000 for violating the 1972 Clean Water Act. Cronin was hooked. For the next 10 years he worked as a lobbyist, then a legislative aide on environmental issues in Albany. But Cronin missed the river. Quitting his job, he headed down to Nyack, New York to work as a commercial fisherman where he came face-to-face with the run-off, waste and sewage that fishermen on the Hudson were dealing with.

At about the same time, writer Robert Boyle coined an idea to protect the river, derived from the tradition of the old English riverkeeper who watched over private trout streams and protected them from poachers. Cronin seemed the perfect choice for the job. Within his first three months as Riverkeeper, Cronin nabbed several offending Exxon tankers. Settling out of court, Exxon contributed $250,000 to the Riverkeeper Fund, along with an additional $1.5 million to New York State to help clean up the river. The Hudson River Improvement Fund was established by Governor Cuomo with this money, and is still in existence today.

Cronin has been busy ever since, taking on landfills, nuclear facilities and sewage treatment plants. In doing so, he's made his share of enemies. "He's overcome by his own importance," claims Robert Kirkpatrick, former town supervisor of Newburgh, New York. Cronin had sued the town for illegally dumping alum (a toxic aluminum compound) into the river in the middle of the night. Cronin agrees he has little tolerance for those who pollute: "I'm uncompromising because somebody has to be. The area we've carved out for ourselves is to be unwavering in our enforcement of environmental law because the government is not doing it," responds Cronin.

Together with the Fund's full-time attorney, Robert F. Kennedy Jr., Cronin has won 60 cases in federal court since 1983. His focus is now on the power plants along the Hudson. "The Indian Point plant uses one million gallons of water a minute for cooling purposes and has been killing millions of fish a year," explains Cronin. Fish had been sucked in and crushed against the plant's intake screens. Cronin's suit forced Con Edison, an electric company, to spend $20 million on fish-saving equipment. Says Cronin, thinking aquatically as usual, "We may be a small fish in a big pond, but when it comes to the river, we mean business."

JESSICA SPEART is an environmental writer living in New York City.

EPA lists all sources of "priority" pollutants, and then stringently regulate those sites. A 1991 report by the U.S. General Accounting Office found, however, that the process of who gets listed is often politicized at the state and local levels. "Nationwide, EPA deleted a total of 309 facilities from the lists," says the report, primarily because "the fear of the negative image associated with being listed as a toxic pollutant discharger prompted certain industries to pressure states to make their water quality standards less stringent. In Alabama, this action resulted in nine out of 10 paper mills being deleted from the discharger list because they were no longer in violation of the new, less stringent dioxin standard."

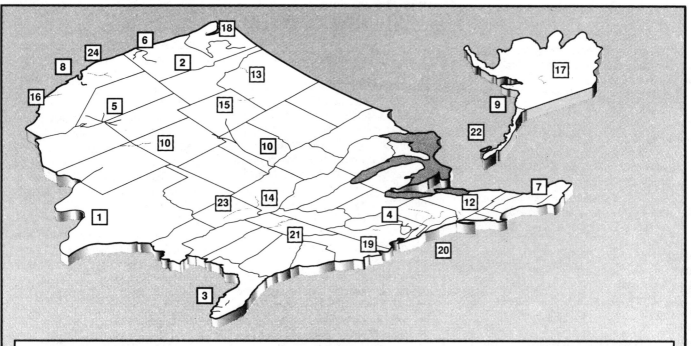

Endangered Rivers

1. Rio Grande and Rio Conchos River System, Colorado, New Mexico, Texas, Mexico
2. Columbia and Snake River System, Northwest U.S., Canada
3. Everglades, Florida
4. Anacostia River, Washington D.C., Maryland
5. Virgin River, Utah, Arizona, Nevada
6. Rogue and Illinois River System, Oregon

7. Penobscot River, Maine
8. Clavey River, California
9. Alsek and Tatshenshini River System, Alaska and British Columbia
10. Platte River, Nebraska

Threatened Rivers

11. Animas River, Colorado
12. Beaverkilt (lower) and Willowemec River System, New York
13. Blackfoot River, Montana

14. Eleven Point River, Missouri
15. Little Big Horn River, Wyoming
16. Los Angeles River, California
17. Moose Creek, Alaska
18. Skokomish River, Washington
19. St. Mary's River, Virginia
20. Susquehanna River, Pennsylvania
21. Tennessee River, Kentucky
22. Thorne River, Alaska
23. White River, Arkansas
24. Yuba South, California

Source: American Rivers

Bruce Kerr

"I'm in favor of more federal control when it comes to toxic discharges," says Nina Bell of NEA. "States are undercutting human and wildlife health in order to appeal to industry. There needs to be a provision in the Clean Water Act that recognizes some pollutants are so dangerous that there is no safe level for discharging them into our rivers. Why are we haggling over how much substances like dioxin should be diluted? Congress had to ban DDT, but it should have been eliminated through the Clean Water Act."

But focusing only on the worst pollution may create problems of its own. "Our national fixation on cleaning up the nation's most polluted water resources to minimally acceptable standards has blinded us to the imminent decline of existing pristine water bodies," says a 1992 report from the National Wildlife Federation. "If the current course remains unchanged, the consequence of continued neglect of outstanding water resources in the United

States will be equally mediocre water quality everywhere."

Fulfilling the goal of the Clean Water Act, "to restore and maintain the chemical, physical and biological integrity of the nation's water," remains a daunting task, but many environmental leaders say it is still possible. It will require federal leadership and plenty of funding, and most important, a renewed commitment to the original idea of the Clean Water Act rather than to the bureaucratic quagmire that it now represents. It will also require compromise on the part of business, government and environmental groups, but the one thing that can't be compromised is the river itself. "I hate the word 'mitigate,'" says Neal Emerald, grassroots coordinator for Trout Unlimited. "You can't make up for pollution that's already gone into a river. It's much better not to mess it up in the first place. What it comes down to is having a commitment to protecting the environment instead of raping and pillaging our natural resources." And if we don't?

Beautiful, healthy rivers will go the way of the Blackfoot, and become the stuff of fiction rather than real life treasures.

For information on what's being done to improve rivers and streams and how you can get involved, contact:

• *American Rivers*, 801 Pennsylvania Avenue SE, Washington, DC 20003/(202) 547-6900.

• *Clean Water Network*, 1350 New York Avenue NW, Washington, DC 20005/(202) 624-9357.

• *EPA Office of Wetlands, Oceans, and Watersheds*, 401 M Street SW WH-556F, Washington, DC 20460/(202)260-7166.

• *National Wildlife Federation*, "Keeping Clean Waters Clean" Program, 1400 16th Street NW, Washington, DC 20036/(202) 797-6800.

• *Northwest Environmental Advocates*, 133 SW Second Avenue, Portland, OR 97204/(503)295-0490.

• *U.S. Public Interest Research Group*, 215 Pennsylvania Avenue SE, Washington, DC 20003/(202)546-9707.

A Current Catastrophe:

EL NIÑO

Floods, drought, famine, and disease all over our planet. Meet the cause, the Pacific Ocean's problem child, El Niño.

Pamela Naber Knox

Pamela Naber Knox serves as the Wisconsin State Climatologist. Her interest in climatic change stems in part from her experiences with the National Weather Service and from her role as an instructor in physics, astronomy and meteorology.

Imagine planning the trip of a lifetime, your destination perhaps someplace distant and exotic, someplace that requires careful planning. Certainly you would want to know about the local weather. You might even call your library so you would know whether to pack sandals or a parka. But climatologists know that descriptions of "average" weather, or climate, are not always accurate. The atmosphere swings continually from one state to another. And these swings, manifested as floods, heat waves, and droughts, may have enormous impact on your travel plans.

Scientists now know that climate changes many ways, as periodic oscillations, long-term trends, or unforeseeable short-term events. Unfortunately, climatologists and your television weather expert have not always been able to predict most of these fluctuations. But that is changing, too. Scientists are beginning to understand and even to predict one of the strongest climate swings, the phenomenon commonly called El Niño.

Once considered a local fluctuation affecting only the coasts of Peru and Ecuador, today we know that El Niño is part of a global climate shift that affects weather worldwide. It influences not only the weather but the economy, health, and well-being of communities under its sway. How can this apparently local event have such widespread effects? The answer lies in the pulse of the tropical climate: fluctuating trade winds that interact with the ocean surfaces beneath them.

Originally, the name El Niño was given to a warm current that appeared each December in the eastern Pacific Ocean and flowed southward along the coasts of Peru and Ecuador. Local Peruvian fishermen named it El Niño ("the child") after the Christ Child, because it appeared shortly after Christmas. Eventually the name was reserved for years when this current, flowing counter to its normal direction, was unusually warm and strong. These anomalous currents, which occur roughly every three to seven years, mark the onset of abnormal weather patterns that drastically affect the economies of Ecuador and Peru.

First, an unusually warm current amasses along the coast in a deep pool of surface water. This pool then blocks nutrient-filled, colder waters rising from deeper levels in the ocean. Millions of fish offshore starve from lack of food. Their deaths not only devastate the local fishing industries but decimate the populations of guano-producing birds that feed on them. Yet at the same time, adjacent inland areas are blessed by an uncommon abundance of rain. Pastures and cotton fields grow far beyond their usual ranges, and families in these regions may flourish as

2. LAND-HUMAN RELATIONSHIPS

a result of the atypical weather. Although these climatic fluctuations were known to scientists around the world, they considered them strictly local phenomena little worth studying.

Not until the 1960s did oceanographers and atmospheric scientists begin to make the connection between local El Niño events and global variations in oceans and atmosphere. Following the International Geophysical Year (IGY 1957-58), scientists throughout the world had access to an unprecedented set of ocean data. The IGY also fortuitously coincided with the strongest El Niño episode since 1941.

With new data, scientists finally discovered that the anomalously warm waters of El Niño were not confined to coastal Peru, but extended westward halfway across the Pacific Ocean. Other observations showed that at the same time, trade winds lighter than normal and rainfall heavier than normal developed in the central and eastern Pacific. These discoveries marked the beginning of our modern understanding of El Niño.

Scientists gradually recognized these air and water patterns to be El Niño's trademark. In the strong El Niño of 1982, for example, a vast pool of warm water extended over the entire breadth of the tropical Pacific Ocean, diminishing the effects of the cold Humboldt current that flows northward along South America's western coast toward the equator. Thunderstorms, strong up- and down-drafts, and light trade winds accompanied this warm pool.

By late 1983, however, the warm water's expanse contracted and retreated to the western Pacific Ocean, along with the stronger air currents. The Humboldt current was then free to extend abnormally far into the tropical Pacific, where it brought huge volumes of cold polar water toward the equator and suppressed clouds and rain in the air above it. This opposite weather pattern is often called La Niña.

The expanded perspective of El Niño that scientists developed from these new observations led to the rediscovery of some old wisdom—the works of Sir Gilbert Walker, written early in the twentieth century. The Director-General of Observatories in India, Walker originally began his studies of tropical climate fluctuations after a disastrous failure of the monsoons. His goal was to predict monsoon variations and, from 1905 to 1937, he discovered statistically strong links between many climatic variables in the tropics and mid-latitudes. An irregular oscillation between atmospheric pressures at Tahiti in the central equatorial Pacific and Darwin in northern Australia appeared particularly significant.

Tahiti normally has higher atmospheric pressure than Darwin. But both pressures fluctuate irregularly around their average values. Walker linked this periodic pressure seesaw, which he

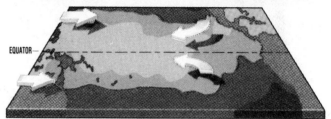

EARTH: All illustrations by Steven Davis

Normally, the strong trade winds blowing from east to west keep the warm equatorial waters of the Pacific Ocean well to the west (top). But in an El Niño event, the trades weaken, allowing warm equatorial waters to move east (bottom).

called the Southern Oscillation, to changes in many climatic variables across the world (including temperatures in southeastern Africa and the northwestern United States and Canada) and precipitation in the southeastern United States. Unfortunately, he was never able to use these relationships to accurately predict monsoon failures.

As scientists studied atmospheric oscillation more thoroughly, oceanographers began to form a more complete picture of the tropical ocean. Oceanographers determined that the trade winds, which blow from northeast and southeast toward the west, create a surface ocean current from east to west along the equator. This warm surface water pools in a deep layer in the western Pacific Ocean along Indonesia and the eastern coast of Australia. The water piles up to such an extent that sea levels near Indonesia are about 16 inches (40 centimeters) higher than sea levels near South America.

In addition, the pooling drives the thermocline, an ocean layer marking the boundary between warm surface water and cold subsurface water, to much greater depths in the western Pacific than off the coast of South America. The warmth of the water intensifies convective rainfall in the western Pacific, pumping moisture and energy high into the atmosphere. This fountain of heat and moisture connects the tropics to the global atmosphere.

Walker's early insights, plus investigations in the 1980s and 1990s, led to a more comprehensive knowledge of what causes El Niño. In most years, the trade winds follow a seasonal cycle that creates variations in surface ocean currents. In some years, however, the surface trade winds slow down more

Pacific Ocean Air and Water Currents

Normal Conditions

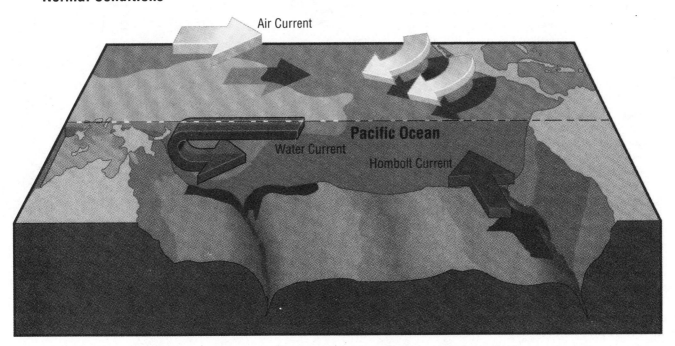

In a normal year (above), the winds across the equatorial Pacific Ocean drive a west-flowing current of warm surface water. When the water reaches Indonesia, it sinks and creates a deep pool of warm water. Under El Niño conditions (below) however, the prevailing westerlies overpower the trade winds and a large current of warm water surges eastward toward the coast of South America.

El Niño Conditions

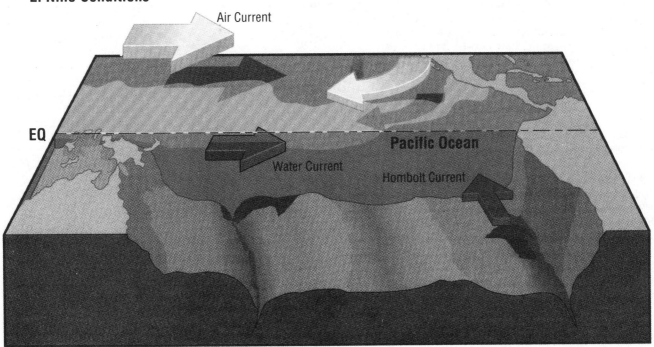

than usual, which reduces wind stress on the ocean surface. The lessened force of the winds no longer holds back the pool of warm water in the western Pacific, and it begins to drift eastward.

The influx of warm water in the eastern Pacific Ocean also creates a blanket of warm water over the cold, nutrient-filled water that normally lies near the surface. This cuts off the flow of nutrients to plankton at the base of the food chain and begins the cycle of biological devastation. At the same time, a warm countercurrent appears off the coast of South America. This countercurrent is the phenomenon that Peruvian fishermen christened El Niño. The warm water surging eastward from Indonesia and the warm surface waters in the eastern Pacific then merge, creating a band of warm water that may stretch one-third of the way around the Earth.

The eastward shift of the warmest surface water sets up changes in atmospheric circulation that affect areas far outside the tropics. Warm surface waters heat the overlying atmosphere, which enhances thunderstorms and encourages unusually high atmospheric pressure to develop along the equator. (See graph this page.) The above-normal pressure sets up a planetary chain of low- and high-pressure areas that radiate from the air overlying the pool of warm water. The presence of these atmospheric "hills and valleys" at mid-latitudes may divert storms from their normal paths and lead to unusual weather conditions in some locations. These are the phenomena that Walker examined.

Shifting weather patterns may be accompanied by stronger-than-normal upper-level winds called jet streams. These winds, capable of moving weather systems hundreds of miles across, may also literally blow the tops off tropical storms before they develop into hurricanes. Scientists compared Atlantic hurricane paths in the years before, during, and after 14 El Niño events. They found that El Niño years have noticeably fewer hurricanes than non-El Niño years. They also found that the El Niño storms lack the recurved paths typically seen in most hurricanes, probably because of a weaker Atlantic high-pressure area that oscillates in conjunction with the Southern Oscillation. These critical connections between the tropics and extra-tropical regions that scientists call "teleconnections" are the reason the media watch El Niños so carefully. Knowing how an El Niño evolves may provide clues to weather trends in other parts of the world and improve scientists' ability to make accurate long-range forecasts.

Once the countercurrent appears off the coast of South America, we consider El Niño events mature. Their impact on the atmosphere is powerful. And although the location of the warmest water varies considerably from event to event, development of El Niños is fairly well defined after the countercurrent appears. Sea surface temperatures in the eastern Pacific increase as warm water moves eastward from Indonesia and the thermocline is suppressed. About nine months after the onset of the mature El Niño, a

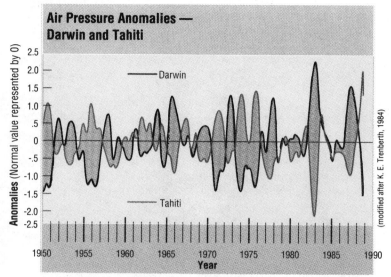

The black line depicts changes in air pressure over Darwin, Australia, with values above and below zero being abnormal; the gray line depicts changes over Tahiti, French Polynesia. Scientists have linked these anomalies, measured from 1950 to 1990, to El Niños.

broad band of warm water stretches over most of the equatorial Pacific. This warm pool continues to grow in the central Pacific through the next few months, while conditions along Peru and Ecuador swing back toward normal. By the summer of the following year, cold water replaces the warm pool in the eastern Pacific, winds resume their normal force, and La Niña, the reverse of El Niño, usually develops.

Though scientists now believe they know how an El Niño evolves, they face uncertainty about what triggers El Niño events. Some intriguing possibilities exist — although few researchers declare the puzzle solved. Mark Cane and Stephen Zebiak, of Columbia University's Lamont-Doherty Geological Observatory, developed a computer model that simulates the physics of the tropical atmosphere/ocean system. They believe their model explains that variabilities in El Niño's strength appear in internal oscillations within the system. Other scientists believe the trigger may lie outside the tropics. Cold air surges from higher latitudes and bursts of westerly winds in the western Pacific may initiate the eastward flow of warm surface water that leads to El Niño. Variations in the Indian monsoon or rainfall deficits in Southeast Asia, associated with increased snow cover over Eurasia, could also weaken trade winds, which might initiate an El Niño.

Paul Handler, a physicist at the University of Illinois, believes El Niños are launched by the effects of particles erupted from tropical volcanoes in the months preceding the onset of an El Niño episode. His controversial theory suggests that the haze produced by the particles blocks sunlight and cools the tropics, which may affect atmospheric pressures and surface winds enough to initiate an El Niño. No one has identified a proven mechanism for this connection, although Handler has made statistical correlations between volcanic activity and El Niño onset. Other scientists believe changes in Antarctic pressure patterns may alter the atmospheric circulation near New Zealand, also influencing El Niño's development. Another theory claims hot water spewed from underwater vents leads to warming in the eastern Pacific on a more or less regular schedule.

The effects of El Niño on world climate are widespread and highly variable, depending on the exact pattern of ocean temperatures in each episode and how the event evolves. The position of the warmest water may vary from one event to another and lead to different effects in separate El Niños, possibly causing floods one time and drought the next. Other areas, however, are affected more consistently. Some global effects most frequently observed during an El Niño include wet conditions in the southeast United States, drought in eastern Java and southern Peru, and warmer-than-normal winters in the northern United States and Canada.

In general, areas most sensitive to climate variability are those likely to be damaged by El Niño effects. These include lands prone to flooding or prolonged droughts, those lands often least able to deal with the climatic costs of El Niño and La Niña. Extremes of heat and moisture at some locations during an El Niño can affect insect and plant growth and the productivity of fisheries, and can cause heat stress to animals. The same extremes affect human health and comfort, often with life-threatening results.

One local economic effect of climatic fluctuation is the devastation of the Peruvian fishing and fertilizer industries. Anchoveta, small fish considered the primary source of fishing income, are also the major source of food for millions of seabirds in Peru. These birds excrete rich guano harvested for fertilizer, another source of income for the local population. The catch of anchoveta by Peru's rapidly growing fishing fleet sharply increased in the 1950s and 1960s but precipitated a concurrent decrease in the population of the birds that ate the fish. Anchoveta populations, however, experienced significant declines in El Niño years, when the warm surface waters cut off the flow of the nutrients needed to sustain the fish population. Because of heavy overfishing of coastal waters and the impact of several El Niños on the already stressed fish population, anchoveta levels declined to record low amounts, seriously affecting the local fishing industry.

El Niño-derived droughts cause enormous problems with health and food supplies. Droughts in Australia almost always occur in El Niño years, again affecting the economy. Losses in the millions of dollars occur in some years; in others, late planting or insect infestations reduce agricultural yields. These problems may relate to changes in rainfall and wind patterns during El Niño years. Outbreaks of insect-borne diseases such as encephalitis may occur as a result of El Niño- or La Niña-related climate shifts, especially in wet years when insect populations may boom. Past El Niños have reduced monsoons in India, causing enormous devastation. During the 1888 event, approximately 1.5 million people died because of drought-induced famine. In more recent El Niños, countries still suffered immense economic setbacks, although fewer lives have been lost because of expanded grain storage and increased international aid.

In addition to the large toll on life and health, floods and drought shatter the infrastructure of many countries. Loss of housing, hospitals, and schools further weakens the vitality of affected communities. Destruction of factories and other commercial enterprises impedes the production of goods, limiting what's available to consumers.

While El Niño has positive effects on some parts of the world, the pluses are a mixed blessing. A warm winter in the northern United States reduces heating and snowplowing bills but brings lighter snowfalls, endangering businesses dependent on winter recreation. Ample rain in normally dry regions may increase crop yields and replenish reservoirs, but too much rain, like that occurring in early 1992 in Texas, leads to flooding, forcing people from their homes, damaging crops, and killing livestock. Decreases in Atlantic hurricanes may reduce widespread destruction on the U.S. East Coast, but in other areas of the world, such as Tahiti, the number of hurricanes increases with El Niño, which increases storm damage. Unfortunately, the positive impacts on the biosphere and human life seem to be much harder to quantify than the negative.

For the past two years, scientists have carefully watched the onset and evolution of an El Niño, hoping to predict where the major impacts will occur. The recent El Niño's development appears typical, being slightly stronger than the previous episode in 1986-87. Floods in southeast Texas, an unusually warm winter in the northern United States and Canada, and February avalanches in Turkey may be symptoms of the present El Niño. Other climatic influences may be greenhouse warming and volcanic aerosols from Mount Pinatubo. El Niño may have contributed to the current Ethiopian famine. Fortunately, modern advances in medicine and agriculture reduce the enormous tolls taken by climatic changes induced by El Niño. Future advances in predicting El Niño events may help global leaders plan effectively for the next event.

To understand how long El Niños will affect the world's weather, we must look at past climatic records. Glacial ice cores taken high in the Andes tell us El Niños have occurred for at least the last 1,500 years. Data from ocean sediment cores and coral reefs hint at a much longer time span. So we suspect that El Niños will recur at their usual irregular pace for the foreseeable future.

Another major question asks how the projected greenhouse warming will affect future El Niños. The answer depends on El Niño's trigger. If an internal oscillation causes an El Niño, as Cane and Zebiak suggest, El Niño events are not likely to change, as tropical temperatures aren't expected to rise much above their current values. But if events outside the tropics trip an El Niño, then the answer critically de-

pends on how climatic factors change in the mid- and high latitudes. For example, warmer temperatures in mid-latitudes might reduce temperature gradients from the tropics to the poles, resulting in a less dynamic atmosphere and weakened jet streams. We must also take into account synergistic effects, whereby changes in one trigger mechanism may increase or decrease the effect of other potential triggers, creating highly complex scenarios that spell trouble for the scientists who are attempting to construct models of future El Niños.

We are a long way from completely understanding the dynamics of our planet. But by understanding more completely the connections between atmosphere, ocean, biosphere, and geosphere, in the future scientists can forecast long-term climatic shifts more accurately. Our expanding knowledge of El Niño is merely the promise of things to come.

The Deforestation
D · E · B · A · T · E

Estimates vary widely over the extent of forest loss

RICHARD MONASTERSKY

As tales of burning forests captured headlines in the late 1980s, a string of rock stars, movie actors, and even ice cream makers joined the fight to save tropical woodlands, helping to transform the awkward term "deforestation" into a household word. But recent studies have produced markedly different estimates of the pace of clearing, raising questions about the accuracy of deforestation figures that have floated around policy circles in recent years.

While tropical forests are certainly vanishing at a disturbing rate, the widespread disagreement over deforestation estimates makes it difficult for government officials and scientists to assess the problem. That, in turn, hampers efforts to gauge the threat of related issues, such as habitat destruction and global warming.

Concerns about previous deforestation estimates emerged in the last few years as researchers from a number of countries looked into the problem, often using more reliable methods than before. Most recently, a study published in the June 25 SCIENCE confirmed suspicions that several earlier assessments had drastically overestimated the rate of forest destruction in the Brazilian Amazon basin, thereby inflating some global estimates.

The Brazilian case provides a dramatic example of how different researchers can arrive at markedly divergent conclusions concerning the extent of deforestation. In 1988, Alberto Setzer of Brazil's National Space Research Institute (INPE) used data collected by infrared sensors on a U.S. weather satellite to gauge the number and extent of fires within the legally defined Brazilian Amazon — an area that includes only part of Brazil's tropical forests. Assuming that 40 percent of the fires occurred on recently cleared forest,

Setzer's team calculated that 8 million hectares of forest were cleared during 1987 within the legal Amazon — an almost unfathomable amount equal to 2.2 percent of the forest.

Although contested by other researchers, that alarming number found its way into several global deforestation estimates at the time. In particular, the Washington, D.C.-based World Resources Institute (WRI) included Setzer's Amazon figure in a 1990 worldwide assessment. The high number for Brazil drove up WRI's global estimate for tropical forest loss, which was calculated at 16.4 to 20.4 million hectares per year.

Despite the controversy over the Brazilian estimate, WRI's global total seemed to agree with a provisional number issued by the United Nations Food and Agriculture Organization (FAO), which put tropical deforestation at 17 million hectares per year for the period 1981 to 1990 (SN: 7/21/90, p.40).

Brazil emerged from the WRI study and others looking like the ultimate forest destroyer, responsible for roughly one-third to one-half of the global deforestation total. That triggered a round of international finger-pointing, focusing criticism on Brazil for allowing such rapid clearing of the Amazon. Brazil, however, complained that the estimates were inaccurate and that deforestation rates had never reached such heights, says Jayant A. Sathaye, an energy and forestry analyst at the Lawrence Berkeley (Calif.) Laboratory.

More recent studies have backed up Brazil's claims. In the last few years, researchers at INPE and the National Institute for Research on Amazonia, based in Manaus, Brazil, challenged Setzer's fire-counting technique and began gauging deforestation by mapping cleared areas on images taken by Landsat satellites. Studies that relied partly on this technique indicated that deforesta-

tion within the Brazilian Amazon averaged 2.1 million hectares per year between 1978 and 1989 and 1.4 million hectares from 1989 to 1990.

The newest estimate for Brazil goes even lower. David Skole of the University of New Hampshire in Durham and Compton Tucker of NASA's Goddard Space Flight Center in Greenbelt, Md., studied some 200 Landsat images covering the entire Brazilian Amazon for 1978 and 1988, allowing them to map the extent of forest and cleared land for those two years. The images showed that deforested areas covered 7.8 million hectares in 1978 and 23 million hectares in 1988, implying an average annual loss rate of 1.5 million hectares, Skole and Tucker report in their recent SCIENCE paper.

Conventional wisdom holds that Brazil's deforestation slackened dramatically after peaking in 1987, in part because the country's economy slowed, reducing the land speculation that had motivated people to clear forest for new farms or rangeland. If true, that standard theory could explain why Setzer and others found so much more deforestation going on at the peak period than others have seen in the last four years.

But Skole dismisses that explanation. "A lot of people are using it as a convenient excuse for being wrong," he contends. Although he, too, finds that deforestation rates have dropped in Brazil, he says the average for the late 1980s did not range much above 2 million hectares per year — not high enough to explain the earlier estimates.

If Brazil's actual rate of forest loss is so much lower than studies had previously suggested, how accurate are the various global estimates? Skole, for one, has little faith in the sea of numbers. "All of the published global studies do not use a systematic approach," he says. "They use secondary and tertiary sources, anecdotal reports, different time periods, different methodologies, different terminology. It's the state of affairs right now in deforestation monitoring. It's kind of a sad state of affairs, but people are using whatever resources they have, which are not adequate."

Before deforestation gained widespread attention in the late 1980s, most researchers relied on an estimate by FAO, which concluded in 1981 that tropical deforestation worldwide averaged 11.3 million hectares per year during the late 1970s. That number stood uncontested because it was the only figure available.

Since 1989, FAO has been working to update that global estimate. In 1990, it released a provisional figure of 17 million hectares of deforestation per year; but this March it lowered the global total to 15.5 million hectares per year for the period 1981 through 1990. K.D. Singh, leader of FAO's forestry assessment team in Rome, explains the change by saying that his organization's 1981 study overestimated the amount of forest remaining in the tropics at that time, a mistake that had inflated its 1990 estimate.

By all accounts, the latest FAO assessment represents the most comprehensive global study to date, having collected scattered data from individual countries or provinces and woven them together using a mathematical model based on information about forest conditions, population density, and ecological zones. Singh says his team has striven to find the best data available, although the quality and type of data vary from one region to another. Some local and national governments have conducted forest surveys using satellite images, while others have used ground-based approaches for estimating deforestation.

The success of the FAO effort remains uncertain, however. Forestry researchers cannot yet check the new global assessment against other studies because FAO released only regional deforestation estimates and has not yet issued tallies for individual countries.

For instance, the FAO numbers reported in March include 6.2 million hectares of forest lost annually in tropical South America from 1981 through 1990. That region includes Brazil and six other nations. To some researchers, the South American total appears unrealistically high. If FAO chose the best data available at the time (the Brazilian estimates of 2 million hectares lost annually in the legal Amazon during the 1980s), that would leave more than 4 million hectares cleared each year outside the Brazilian Amazon.

Where, then, is all that missing South American deforestation? That is precisely the issue raised by an international

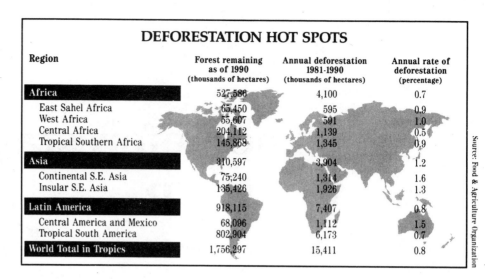

DEFORESTATION HOT SPOTS

Region	Forest remaining as of 1990 (thousands of hectares)	Annual deforestation 1981-1990 (thousands of hectares)	Annual rate of deforestation (percentage)
Africa	527,586	4,100	0.7
East Sahel Africa	65,450	595	0.9
West Africa	55,607	591	1.0
Central Africa	204,112	1,139	0.5
Tropical Southern Africa	145,868	1,345	0.9
Asia	310,597	3,904	1.2
Continental S.E. Asia	75,240	1,314	1.6
Insular S.E. Asia	135,426	1,926	1.3
Latin America	918,115	7,407	0.8
Central America and Mexico	68,096	1,112	1.5
Tropical South America	802,904	6,173	0.7
World Total in Tropics	1,756,297	15,411	0.8

Source: Food & Agriculture Organization

committee of researchers called the Intergovernmental Panel on Climate Change. In a 1992 report, the panel questioned FAO's South American estimate (which has since decreased slightly), saying, "This seems to ascribe a very high proportion (about 70 percent) of the total deforestation in South America in the late 1980s to the region outside the Brazilian Amazonia, even though this region accounts for very little of the total amount of forest on the continent."

But FAO's Singh says researchers have not appreciated the extent of South American deforestation in the open forest outside of the rainforest of the Brazilian Amazon. "This has been one of our main findings — that the deforestation outside the legal Amazon basin is quite high," he told SCIENCE NEWS. The open forest faces a greater threat because it is more accessible and because more people live there than in the rainforest, Singh says.

Beyond those hints, Singh says he cannot respond fully to criticisms of the new deforestation assessment until FAO publishes its more complete report, which will include a breakdown of figures for individual countries. The report should come out in late summer, he says. Singh also hopes to submit the study to a panel of scientists for peer review.

Some researchers, however, wonder whether the FAO will delay releasing the long-awaited numbers, perhaps indefinitely, in hopes of avoiding the intense scrutiny that will descend on this study.

Ken Andrasko, chief of a forestry section in the Environmental Protection Agency's climate change division, suggests that deforestation studies have acquired political significance in the wake of last year's Earth Summit in Rio de Janeiro, where countries signed treaties on biodiversity and climate change. The climate change convention requires all countries to provide an inventory of their greenhouse gas emissions, including

those caused by deforestation. Since future treaties may call for all countries to limit their emissions, the deforestation estimates will partly determine how much a country needs to cut back.

"It's going to be increasingly difficult to publish those individual country [deforestation] estimates. If the number is higher or lower, it has very real political consequences now," Andrasko says.

That lesson was not lost on Brazil, which released its own deforestation data on the eve of the Rio summit. Such studies, as well as the one by Skole and Tucker, show that while Brazil may still be number one, it no longer sits in a category of its own, far and above every other nation on the list of forest destroyers.

Although the FAO did not provide estimates for individual countries, the data released earlier this year clearly indicate that tropical forests are shrinking worldwide. Indeed, parts of other continents are losing a greater percentage of their forests than is South America, which has much more intact forest than other parts of the globe. According to FAO, the highest percentages of deforestation during the 1980s occurred in southeast Asia, Central America and Mexico, and West Africa.

Looking beyond the problem of absolute deforestation, experts say fragmentation and degradation of remaining forests also present substantial threats, especially to the diversity of plant and animal life in some of the most biologically rich habitats on Earth. Indeed, Skole and Tucker found that the area of disturbed habitat surrounding cleared areas grew by more than 4 million hectares per year in Brazil's Amazon, much faster than the pace of deforestation there. "Even though the rate of deforestation was much lower than previous estimates, the effect on biological diversity was much greater," says Tucker.

To address such concerns, the second phase of the FAO study will examine the extent to which people have broken up forest or removed trees without stripping an area bare. As part of this work, Singh's team has purchased high-resolution satellite images for 117 randomly selected regions corresponding to about 10 percent of the forest-covered land in the tropics. For quality control, his group also used these satellite images in the first phase of the study to improve its deforestation estimates.

Ideally, FAO would purchase satellite images for the entire tropical belt. But Singh says the organization cannot afford the thousands of images needed to cover the tropics.

An ongoing NASA project may help complete the picture. In a program called Landsat Pathfinder, NASA is funding the purchase of satellite photos covering some three-quarters of the world's tropical rainforest. According to Skole, a participant in the project, it will provide the most accurate assessment of deforestation to date. As with the Brazilian case, the research done with satellite data could show that past estimates have been inflated. But Skole says the new study may also uncover much more deforestation than had been suspected.

The NASA effort does have some critical limitations. Because it focuses mostly on tropical rainforest, it will not include all open forests and dry forests, which cover just as much territory as do rainforests. Moreover, the project is a one-shot deal, not to be repeated. Experts say periodic updates are needed, not only to assess the changing deforestation threat but also to gauge the resulting increase in concentrations of carbon dioxide.

Despite its drawbacks, however, the NASA project should finally provide a means of checking the global estimates. "I feel confident," says Skole, "that in the next two to three years, as our work comes forward, we will have a much better idea because we are applying a consistent method."

The greenhouse effect could wreak havoc on human health, allowing tropical diseases to affect nontropical populations, causing heat-related stress and illness, and worsening air pollution.

Climate Change

The Threat to Human Health

George F. Sanderson

George F. Sanderson was until recently deputy director of information and public affairs, United Nations Environmental Programme, Nairobi, Kenya. He is currently with the Department of External Affairs and International Trade Canada in Ottawa. His address is 483 Fraser Avenue, Ottawa, Ontario K2A 2R1, Canada.

We have come a long way in recent years toward realizing how extensively global warming will jeopardize both planetary and human health.

Carbon-dioxide buildup—mainly from combustion of fossil fuels such as oil, gas, and coal, and from clearing and burning of forests—is believed responsible for about half of this worldwide warming, while chlorofluorocarbons (CFCs), methane, ground-level ozone, and nitrous-oxide emissions account for the rest. Together, these gases act like glass in a greenhouse: They allow passage of incoming solar radiation but trap some of the outbound heat radiation from the earth.

One very significant aspect of global warming is the variety of effects it will have on human health—some will be subtle and indirect, others dramatic and direct. Many of the very factors contributing to global warming are themselves harmful to humans, such as the burning of fossil fuels: A typical automobile emits carbon monoxide, sulfur and nitrogen oxides, hydrocarbons, low-level ozone, and lead, all of which are hazardous to health. According to a World Health Organization (WHO) task group on the potential health effects of climate change and the Intergovernmental Panel on Climate Change (IPCC), climate change will likely worsen air pollution—especially in heavily populated urban areas—by altering the composition, concentration, and duration of chemical pollutants in the atmosphere.

Chlorine released by CFCs and bromine released by halons (used in fire extinguishers) both deplete stratospheric ozone. The use of CFCs and halons thus escalates the risk of skin cancer, eye cataracts, snow blindness, and weakened immunity to a host of other illnesses by exposing humans to increased ultraviolet B radiation from the sun. "Skin cancer risks are expected to rise most among fair-skinned Caucasians in high-latitude zones," according to the IPCC. The WHO task group reached a similar conclusion, noting that "the incidence of non-melanoma skin cancer could increase between 6% and 35% after the year 2050. These increases may be larger in the Southern Hemisphere, where total ozone depletions have been larger."

Health Inside the Greenhouse

On the basis of present trends, scientists predict that greenhouse gases will warm the earth further by about 0.3° C in each decade of the next century. This rise, faster than any experienced over the past 10,000 years, could increase the planet's mean temperature by 3° C before the year 2100, making it warmer on average than it has been for 100,000 years.

This may not sound especially ominous, but left unchecked, global warming could alter rainfall patterns, flood vast areas of low-lying land as warmed seas rise (possibly by as much as a meter), and drive countless species to extinction as fragile ecosystems collapse. A warmed planet will affect human health by disrupting food and fresh water supplies, displacing millions of people, and altering disease patterns in dangerous and unpredictable ways.

The populations most vulnerable to the negative impacts of the greenhouse effect are in developing countries, in the lower-income groups, residents of coastal lowlands and islands, those living in semiarid grasslands, and those in the squatter set-

From *The Futurist,* March/April 1992, pp. 34-38. Reprinted with permission from *The Futurist,* published by the World Future Society, 7910 Woodmont Avenue, Suite 450, Bethesda, Maryland 20814.

tlements, slums, and shanty-towns of large cities.

Present strategies for immunization, coping with disease vectors or carriers, providing safe drinking water, and improving nutrition are all based on existing climate regimes, ecosystems, and sea and solar-radiation levels. These are all expected to change, but exactly how much cannot be predicted with any certainty, making it virtually impossible to adjust health and nutritional strategies now to take possible climate changes into account.

Humans adapt well to moderate changes in temperature and to occasional extremes. But this adaptive capacity—developed over many thousands of years—is relatively low in infants and the elderly; it rises through childhood and adolescence to reach a maximum that can be maintained up to about 30 years of age, then begins to decline.

When the Temperature Rises

Heat-related illness is one problem that will likely proliferate. Currently, the temperature in Washington, D.C., exceeds 38° C (100° F) on an average of one day per year; it rises above 32° C (90° F) about 35 days every year. "But by the middle of the next century, these figures could rise to 12 and 85 days respectively per year," according to the World Meteorological Organization. "The effect of such temperature rises on human health in Washington and similar cities throughout the world is difficult to predict. But there is no question that increased urban heat stress could come to claim many lives."

The same conclusion was reached by IPCC, which warned in June 1990 that the increase in deaths caused by a greater number of summer heat waves "would be likely to exceed the number of deaths avoided by reduced severe cold in winter."

A changing climate will also probably shift the range of conditions favoring certain pests and diseases, according to the final scientific statement issued by the Second World Climate Conference in November 1990.

As temperatures rise, the boundaries of the tropics may extend into the present subtropics, and parts of temperate areas may become subtropical. This will allow the insects and animals that carry or cause many tropical diseases (e.g., mosquitoes, snails, etc.) to move poleward in both the Northern and Southern hemispheres. Some communicable illnesses, including those transmitted through air, water, and food, could therefore become common in regions that once rarely knew them, with a possible rise in death rates.

Diseases such as malaria, hepatitis, meningitis, polio, yellow fever, dengue fever, tetanus, cholera, and dysentery, which flourish in hot, humid weather, could increase, while those associated with cold weather would be expected to diminish.

In a warmer climate, malarial mosquitoes and other disease carriers also may migrate vertically, up into formerly inhospitable highlands. This may be particularly hazardous in tropical highland areas where there is no natural resistance to malaria. Researchers in Kenya have already found malaria-carrying

---

Fighting the Greenhouse Effect

Scientists and policy makers disagree even among themselves on the exact scenario that the greenhouse effect will follow in the years ahead—whether we will see only a mild warming or a major disruption of Earth's climate or something in between.

Because there is no single "silver bullet" that will mitigate the effects of global warming, a variety of options are under evaluation by the Committee on Science, Engineering, and Public Policy, administered by the National Academies of Sciences and Engineering and the Institute of Medicine. A report by the committee's mitigation panel concludes that current knowledge about the pollutants and other factors leading to global warming warrants only the lowest-cost (or no-cost) mitigation options.

The panel analyzed a wide range of options and rated them on the basis of their cost effectiveness. Those rated "best practice" are those that could reduce U.S. greenhouse-gas-equivalent emissions by 25% from 1990 levels at a relatively low cost or even a savings.

Among the no-cost and low-cost options the panel recommends are:

• Implement conservation and energy-efficiency improvements in residential, commercial, and industrial buildings. Examples of strategies include using fluorescent lamps and superefficient appliances, planting more vegetation around buildings, and painting roofs and road surfaces white.

• Reduce emissions through improved motor-vehicle efficiency by, for instance, reducing vehicle weight, improving aerodynamics, and reducing engine warm-up time.

• Continue aggressive phaseout of chlorofluorocarbons.

• Reduce global deforestation and reforest marginal lands.

• Consider greenhouse warming in design and operation of electricity-generating systems. Existing industrial energy systems could be replaced with cogeneration plants that produce heat and energy simultaneously.

The panel points out that, if global climate change proves to be more severe than now predicted, higher-cost "backstop" options could be implemented, such as increasing phytoplankton growth in the oceans.

Source: *Policy Implications of Greenhouse Warming: Report of the Mitigation Panel*, Committee on Science, Engineering, and Public Policy, National Academy of Sciences, National Academy of Engineering, Institute of Medicine. 1991. 500 pages. Available from the National Academy Press, 2101 Constitution Avenue, N.W., Washington, D.C. 20418, for $35 (plus $3 shipping) prepaid.

mosquitoes in areas where they were previously unknown.

Changes in temperature, rainfall, humidity, and storm patterns may affect insect- and animal-borne diseases in two ways. First, they may directly affect the carrier's range, longevity, reproduction rate, biting rate, and the duration and frequency of human exposure. Second, they may modify agricultural systems or plant species, thus changing the relationship between carrier and host.

Development rates of mosquitoes, for example, would increase with warmer temperatures, provided these pests have wet areas in which to breed, and snail-borne diseases are likely to spread if global warming forces increased irrigation or causes people to migrate toward irrigation projects. Changed human migration patterns, along with increased temperature and rainfall, may extend the geographic range of hookworms, too.

Moreover, "warmer, humid conditions may enhance the growth of bacteria and molds and their toxic products, such as aflatoxins," cautioned the WHO task group. "This would probably result in increased amounts of contaminated and spoilt food."

Such changes would not be limited to developing countries. For example, in the United States, tick-borne diseases such as Rocky Mountain spotted fever and Lyme disease could spread northward. Americans could face the risk of five separate mosquito-borne diseases that have at present been virtually eradicated, according to Andrew Haines, a professor at University College and Middlesex School of Medicine in London.

Higher, Warmer Waters

As ocean temperatures rise and nutrients from agricultural fertilizers leach into rivers and coastal waters, toxic "red tides" may become more frequent, disrupting marine food stocks. This proliferation of minute marine organisms called dinoflagellates sets off a toxic chain reaction up the food chain: Incidences of food poisoning would increase when people eat tropical fish or shellfish that have eaten organisms that have eaten dinoflagellates.

Sea-level rise could spread infec-

Malaria on the Rise

Malaria is raging at unprecedented levels and will continue to do so unless governments support research to prevent the disease, according to an Institute of Medicine committee. Malaria is a tropical disease caused by mosquito-borne parasites; it kills some 1.5 million people and infects as many as 300 million worldwide each year.

The mosquitoes carrying the parasites that cause malaria have become resistant to insecticides, and the disease itself has become resistant to the drugs traditionally used to treat it. Thus, vaccination offers the greatest hope. A malaria vaccine is still years away but could offer enormous benefits, according to committee chairman Charles C.J. Carpenter, professor of medicine at Brown University.

Source: *Malaria: Obstacles and Opportunities*, Institute of Medicine, National Academy of Sciences. 1991. Available for $42.95 (postpaid) from the National Academy Press, 2101 Constitution Avenue, N.W., Washington, D.C. 20418.

---

tious disease by flooding sewerage and sanitation systems in coastal cities, and increase the incidence of diarrhea in children. The flooding of hazardous waste dumps and sanitation systems could result in long-term contamination of croplands.

Rising, warmer seas may also disrupt marine habitats and aquatic food chains. Since fish constitute 40% of all animal protein consumed by the people of Asia, such a disruption of the marine ecosystem would affect the food supplies of many millions of people and dramatically increase protein deficiency and malnutrition.

Food shortages, reaching "famine proportions in some regions," could also follow the inundation of fertile coastal land by rising seas, the WHO task group noted. And the potential scarcity in some developing countries of food, cooking fuel, and safe drinking water because of drought may further increase the extent of malnutrition, with "enormous consequences for human health and survival," according to the IPCC. The most-serious implications are for Indonesia, Pakistan, Thailand, the Ganges Delta in Bangladesh, and the Nile Delta in Egypt, all low lying and densely populated.

Human Disruptions

Finally, changes in the availability of food and water as well as radical shifts in disease patterns could initiate large migrations of people. An increased number of "environmental refugees" would lead to overcrowd-

ing, social stress, and instability, all of which may impair human health and increase health inequality between peoples of developed and developing countries.

Much more emphasis must be placed on research into how people contribute to and cope with climate change and on public awareness and education programs. "Not only do we need more information about environmental conditions . . . we also need information about health conditions if we are to target our efforts and use our ever-limited resources to best serve health needs," notes Wilfried Kreisel, director of WHO's Division of Environmental Health. "Sad to say, environmental health globally suffers from informational malnutrition."

Equally important in Kreisel's view is the global need to generate more and better human resources for environmental health, to develop more-coherent environmental health policies, and to influence not only the leaders of business and industry but also people in all walks of life to be more sensitive to the health implications of their choices and decisions.

"Day by day, the image of the world as the 'global village' becomes more of a reality," Kreisel points out. And as all people are affected by environmental degradation, including that caused by global warming, "communication and sharing of resources among peoples is essential for the survival of the planet and our species."

Exploring the links between

Desertification and Climate Change

Mike Hulme and Mick Kelly

Mike Hulme is a research climatologist in the Climatic Research Unit at the University of East Anglia in Norwich, England. He specializes in African climate, global climate change, and climate remodeling. Mick Kelly is an atmospheric scientist in the Climatic Research Unit. He is also research director of the Climate and Development Programme for the Centre for Social and Economic Research on the Global Environment at the University of East Anglia and director of the Climate Programme at the International Institute for Environment and Development in London.

M ore than 100 countries are suffering the consequences of desertification, or land degradation in dryland areas.[1] Loss of productivity and other social, economic, and environmental impacts are directly affecting the perhaps 900 million inhabitants of these nations. There is also concern that the environmental impact of dryland degradation may be felt further afield. Some have suggested that this impact might even be felt worldwide.

The first international effort to address desertification occurred in 1977, when the United Nations Conference on Desertification (UNCOD) recognized that desertification was a major environmental problem with high human, social, and economic costs. The conference adopted the Plan of Action to Combat Desertification (PACD), a 20-year, worldwide program to arrest further dryland degradation. Sixteen years later, after several reviews, PACD has achieved little success.[2] A second phase in the international response to desertification began at the United Nations Conference on Environment and Development (UNCED) in June 1992. It was agreed then that a Convention to Combat Desertification should be ready for signing and ratification by June 1994 and that an Intergovernmental Negotiating Committee on Desertification(INC-D) should be established to guide this process. It has also been decided that projects to mitigate land degradation in drylands will qualify for allocations from the Global Environ-

ment Facility (GEF),[3] but only insofar as the projects pertain to the GEF goals of protecting the global environment by reducing greenhouse-gas emissions, preserving biodiversity, and protecting international waters.

Desertification should be ready for signing and ratification by June 1994 and that an Intergovernmental Negotiating Committee on Desertification (INC-D) should be established to guide this process. It has also been decided that projects to mitigate land degradation in drylands will qualify for allocations from the Global Environment Facility (GEF),[3] but only insofar as the projects pertain to the GEF goals of protecting the global environment by reducing greenhouse-gas emissions, preserving biodiversity, and protecting international waters.

A significant obstacle to the work of INC-D is that desertification is a difficult word to define. In 1991, the UN Environment Programme (UNEP) defined desertification as "land degradation in arid, semi-arid and dry sub-humid areas resulting mainly from adverse human impact."[4] Just

From *Environment*, July/August 1993, pp. 4-11, 39-45. Reprinted with permission of the Helen Dwight Reid Educational Foundation. Published by Heldref Publications, 1319 Eighteenth St., NW, Washington, DC 20036-1802. © 1993.

one year later, UNCED adopted the definition of "land degradation in arid, semi-arid and dry sub-humid areas resulting from various factors including climatic variations and human activities."[5] The different emphasis placed on climate variation in these two definitions is indicative of the disagreement that exists concerning the relative importance of the various causes of dryland degradation. This disagreement may appear to be an impractical, academic issue for the large numbers of people dependent on drylands for their livelihood, but misconceptions and arguments must be resolved for any adequate response to desertification to be made.

The first step in discussing such issues must be to define the most important terms and thereby avoid confusion. The terms *climate change* and *climate variation* are used here to indicate climate variability and trends arising from both natural and anthropogenic causes. The term *global-mean warming* indicates climate change resulting from greenhouse-gas emissions. *Desertification* is taken here to mean land degradation in dryland regions, or the permanent decline in the potential of the land to support biological activity and, hence, human welfare. Desertification should not be confused with drought or desiccation. *Drought* refers to a period of two years or more with below-average rainfall, and *desiccation* is aridification resulting from a dry period lasting a decade or more.[6]

Climate change undoubtedly alters the frequency and severity of drought and can cause desiccation in various regions of the world. It does not necessarily follow, however, that drought and desiccation will, by themselves, induce, or even contribute to, desertification. Whether or not desertification occurs depends upon the nature of resource management in these dryland regions. Identifying the contribution of climate variation to desertification is not a simple matter, and the difficulties are compounded by the possibility that desertification itself may generate climate change.

Does Climate Change Cause Desertification?

The definitions of desertification adopted by UNEP in 1991 and by UNCED in 1992 both implicitly link climate change and the assessment of the extent of desertification. Because arid, semi-arid, and dry subhumid areas are climatically defined,[7] any change in climate that results in an expansion or contraction of these areas is likely to change the formal, measured extent of the problem. For example, when an arid area becomes extremely dry or hyperarid, because of climate change, the area defined as being prone to desertification decreases because hyperarid areas are not included in the accepted definition. Conversely, when a humid area converts to subhumid, the defined area within which desertification is considered possible increases.

That climates do change over the decades has now been established beyond dispute. In the African Sahel, for example, annual rainfall during the most recent three decades has been between 20 and 40 percent less than it was from 1931 to 1960. Table 1 on this page illustrates this change in a different way. Within contiguous Africa, there has been a net shift of land area toward aridity, especially toward hyperaridity, and a consequent net loss of semi-arid and dry subhumid land. Overall, areas prone to desertification have decreased from 52.4 percent of mainland Africa to 51.5 percent between these two 30-year periods—a reduction of 25.3 million hectares. The amount of hy-

A single rain storm like this one in the northern Sahel may constitute a significant percentage of the area's annual rainfall.

TABLE 1 ▰▰▰▰▰▰▰
CHANGE IN AREAS OF MOISTURE ZONES IN CONTIGUOUS AFRICA

Moisture zone	Mean land area from 1931 to 1960		Mean land area from 1961 to 1990		Net change between two periods	
	(millions of hectares)	(percentage of total)[a]	(millions of hectares)	(percentage of total)	(millions of of hectares)	(percent)
Hyperarid	450.8	15.1	501.5	16.8	+ 50.7	+ 1.7
Arid	676.9	22.7	680.0	22.8	+ 3.1	+ 0.1
Semi-arid	620.9	20.8	606.9	20.3	− 14.0	− 0.5
Dry subhumid	264.4	8.9	250.0	8.4	− 14.4	− 0.5
Humid	972.4	32.6	947.0	31.7	− 25.4	− 0.9

[a]Percentages total to 100.1 because of rounding.

SOURCE: M. Hulme, R. Marsh, and P. D. Jones, "Global Changes in a Humidity Index Between 1931–60 and 1961–90," *Climate Research* 2 (1992):1–22.

perarid land, however, has increased by more than 50 million hectares.

Determining the precise contribution of climate change to the problem of desertification is not an easy matter. When resource management failure has occurred, there is no doubt that climate variation can aggravate the problem. But separating out the interrelated impacts of climatic and human factors is extremely difficult. Some progress has, however, been made. To cite one example, Compton Tucker and his colleagues at the U.S. National Aeronautics and Space Administration's Space Flight Center in Maryland have used a satellite index of active vegetative cover to determine the extent of the Sahara Desert between 1980 and 1989.[8] Their analysis shows that very substantial interannual variations exist in the extent and quality of surface vegetation in dryland regions. Because much of the vegetation response detected by the satellite is caused by changes in rainfall, much (but not all) of the variability in the extent of the Sahara is due to interannual rainfall variations (see Figure 1). Thus, the index can be used to discriminate between the degradation of vegetation cover caused by rainfall and that which is due to other factors, most notably failures in resource management. Figure 1 relates the estimates of change in the extent of the Sahara to independently derived annual rainfall data for the years from 1980 to 1989. Linear regression analysis indicates that a considerable amount of the year-to-year variation in areal extent—83

percent—can be explained by the rainfall data. This is a statistically significant relationship. The relationship does, however, leave some residual variability in the extent of the Sahara unexplained, as shown by the lowest curve in Figure 1. This residual component of the variability has tended to increase over the past decade, and statistical analysis suggests that it amounts to an average annual increase in the extent of the Sahara of about 41,000 square kilometers per year. This is equivalent to an average annual areal increase not directly related to annual rainfall variations of almost 0.5 percent from 1980 onwards, which would amount to almost a 5-percent increase in total area over the decade. This trend could be the result of the cumulative impact of a series of dry years on vegetation recovery. For example, a particular rainfall amount in 1989 may generate less vegetation than the same amount in 1980 because of the drought years preceding 1989. Alternatively, the increase in extent may well be due to a deterioration of vegetative cover caused by human activity.

Clearly, the relative contributions of human activity and climate change to desertification will vary from region to region and from time to time. Separating out the relative roles of these factors to identify the most appropriate response in any particular situation—and to accord each factor due weight in the desertification convention—is a pressing challenge.

Can Desertification Change Climate?

Separating cause and effect is rendered more difficult by the fact that desertification may, in turn, affect both local climate and climates further afield. Recently, Bob Balling, Jr., of Arizona State University has suggested that surface air temperature has increased significantly in desertified regions owing to changes in land cover and that this effect has substantially affected global-mean temperature.[9] Desertification is likely to lead to reductions in surface soil moisture, which result in more energy available to heat the air (sensible energy) because less goes to evaporate water (latent energy). While it is conceivable that the warming of desertified areas may have been great enough to produce a measurable increase in global-mean temperature (see the box on page 79), the influence would have been small compared to the potential impact of an enhanced greenhouse effect. The question of whether desertification has had or will have a detectable effect on global climate is, nevertheless, a critical one. If a clear influence could be established, it might be argued that dryland degradation should be classed as a global environmental problem in its own right. At present, though, the evidence of a substantial effect must be considered extremely weak.

There is a better-established, if less direct, link, however, between dryland degradation and global-mean warming through the influence of desertification on the sources and sinks of greenhouse gases. Progressive desertification of drylands in the tropics and elsewhere is likely to reduce a potential carbon sink by reducing the carbon sequestered or stored in these ecosystems. As vegetation dies and soil is disturbed, emissions of carbon dioxide will increase. Desertification may also affect emissions of other greenhouse gases. For example, nitrous oxide emissions might increase because of greater fertilizer use. Methane production may increase in poorly fed cattle. On the other hand, because dry soils are methane sinks, desertification might reduce the gas's atmospheric concentration. There remains a large measure of uncertainty about the relative magnitudes of the various sources and sinks for carbon dioxide and the other greenhouse gases,[10] and such uncertainty adds to the difficulty of quantifying the precise effect of desertification on global-mean warming.

The problem of the "missing" carbon sink illustrates this difficulty: It is impossible to balance the global carbon budget on the basis of current understanding of the major carbon sources and sinks; a certain amount of carbon released into the atmosphere cannot be accounted for. Several possibilities might account for the discrepancy, including a substantial carbon dioxide and/or nitrogen fertilization effect on plant growth; a larger uptake of carbon by the oceans than has previously been thought likely; greater carbon sequestering as a result of recent reforestation programs in northern midlatitudes; and a larger carbon-storing capacity of annual grasses in tropical and subtropical regions. Desertification is clearly relevant to this last possibility as it would alter the effectiveness of a sink that may prove more important than is now estimated.

Although the importance of the net carbon flux associated with desertification is impossible to quantify at this time, a rough order of magnitude can be estimated. Data from the 1980s indicate that atmospheric carbon dioxide accounts for about 55 percent of

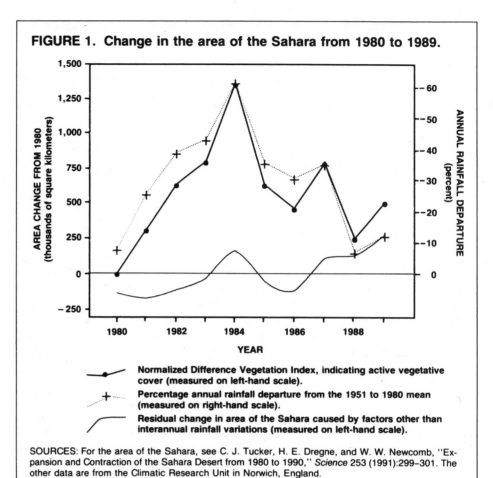

FIGURE 1. Change in the area of the Sahara from 1980 to 1989.

● Normalized Difference Vegetation Index, indicating active vegetative cover (measured on left-hand scale).

+ Percentage annual rainfall departure from the 1951 to 1980 mean (measured on right-hand scale).

Residual change in area of the Sahara caused by factors other than interannual rainfall variations (measured on left-hand scale).

SOURCES: For the area of the Sahara, see C. J. Tucker, H. E. Dregne, and W. W. Newcomb, "Expansion and Contraction of the Sahara Desert from 1980 to 1990," *Science* 253 (1991):299–301. The other data are from the Climatic Research Unit in Norwich, England.

all greenhouse-gas forcing.[11] Of the global carbon emissions, the net biospheric contribution (that is, the nonindustrial component, largely resulting from land-use changes) is variously estimated at between 10 percent and 30 percent (or between 5 percent and 15 percent of total greenhouse forcing). The bulk of these biospheric emissions results from tropical deforestation; land conversion in dryland areas is a minor contributor.[12] When considering the net contribution of dryland regions to greenhouse-gas sources and sinks, a distinction should be made between the contribution arising from nondegrading changes in land use (almost certainly the primary contribution) and that from desertification (the secondary contribution). (An example of a sustainable land-use change is the conversion of a dryland from shrubs to grassland with no subsequent degradation in soil quality.) Therefore, desertification's contribution of carbon

to greenhouse forcing and, hence, to global-mean warming is almost certainly less than a few percent of the global total. It is not possible at this time to estimate net emissions of other greenhouse gases resulting from dryland degradation.

Despite the relative unimportance of desertification as a direct contributor to global warming, there is a clear need to improve understanding of the desertification-related sources and sinks of greenhouse gases. Greater knowledge is important for both scientific and political reasons. The scientific reason is that improved projections of future atmospheric carbon dioxide concentrations and estimates of future rates of global-mean warming depend on better quantification of where and how fast atmospheric carbon is sequestered. The same is true for all the major greenhouse gases. The political reason is that national inventories of greenhouse-gas sources and sinks will be an important element in future negotiations connected with the UN Framework Convention on Climate Change.

By restoring a terrestrial carbon sink and reducing direct emissions of carbon, a reversal of desertification could measurably contribute to reducing global greenhouse forcing but would not by itself lead to a significant reduction in future global-mean warming. The contribution of dryland degradation is too small to have a substantial impact. Nevertheless, because most dryland countries have relatively low industrial carbon emissions, desertification could be a major element of such nations' individual net carbon budgets. In such cases, arresting desertification should be considered a priority action. Only a modest stimulation or protection of carbon sinks in such countries would offset a significant proportion of their other emissions of carbon.

Desiccation in the Sahel

No region has been at the center of the debate over the causal links between climate change, desiccation, and desertification more than the African Sahel. Over the past 25 years, the Sahel has undergone severe desiccation and increasing deterioration of the soil quality and vegetative cover (see box on next page). More than in any other dryland region in the world, it is the simultaneous occurrence of these phenomena in the Sahel that raises questions about the links between climate change and desertification and, in particular, about the cause of the sustained decline in rainfall. Arguments over these questions date back to the mid 1970s, when the meteorology behind the Sahelian crisis of 1972 and 1973 was first discussed.[13] Current ideas about the causes of the desiccation have crystalized around two central themes: internal biogeophysical feedback mechanisms within Africa associated with land-cover changes, such as desertification; and global circulation changes associated with particular patterns of heat distribution in the oceans (ocean temperature departures from the his-

torical mean). With regard to the ultimate cause of the oceanic changes, two possibilities present themselves: They may be a manifestation of quasi-periodic natural fluctuations in ocean circulation, the result of natural climate variability; and/or they may be a response of the ocean system to anthropogenic greenhouse-gas and sulfate aerosol forcing of the climate system. There is some evidence for each of these three causal agents, and each of them has very different impli-

cations in terms of both the appropriate remedial actions and the political repercussions for the INC-D negotiations.

Land-Cover Changes in Africa

The idea that modification of land-cover characteristics in dryland regions might affect regional rainfall was first proposed by Joseph Otterman, an environmental scientist at Tel Aviv University, in 1974 and arose from his empirical work in the

---

HAS DESERTIFICATION "CONTAMINATED" THE GLOBAL-MEAN TEMPERATURE RECORD?

Bob Balling, Jr., of Arizona State University recently suggested that, during the 20th century, desertified areas have warmed by about 0.5° C relative to nondesertified areas.[1] When averaged globally, he argues, this warming significantly "contaminates" the global-mean temperature record and so complicates the search for a warming signal, or component, caused by enhancement of the greenhouse effect.[2] However, Balling's identification of areas that are severely desertified and nondesertified is derived from a map of desertification prepared for the UN Conference on Desertification back in 1977. That map has been superseded by more accurate data collected for the recent assessment by the UN Environment Programme (UNEP).[3] Moreover, Balling's calculated desertification warming signal of 0.5° C per 100 years is based on only a small subset of these "desertified" areas. He extrapolates from this subset to the global scale by suggesting that more than 30 percent of all land (including 90 percent of all drylands) is prone to this desertification warming signal.

UNEP's 1992 global assessment of desertified areas estimated, however, that only 20 percent of dryland regions were seriously degraded. This figure suggests that only 6 percent of all land may exhibit a desertification warming signal, rather than the 30 percent assumed by Balling. The potential bias is, therefore, comparable in magnitude to other biases that may affect the global temperature record. Finally, it should be noted that the differential warming between desertified and neighboring nondesertified areas found by Balling at a resolution

of 5° latitude/longitude may simply indicate that these areas differ in their sensitivity to climate variability and may not be evidence of warming caused by desertification per se. Thus, although Balling may be correct in principle, he has greatly overstated his case, and his analysis provides no convincing evidence that warming caused by desertification has substantially affected the global-mean temperature record.

---

1. R. C. Balling, Jr., "Impact of Desertification on Regional and Global Warming," *Bulletin of the American Meteorological Society* 72 (1991):232-34. The ideas expressed in this article have received a moderate amount of attention in certain circles skeptical of the influence of greenhouse-gas emissions on global climate and have been introduced into discussions of the links between climate change and desertification. See, for example, UN Sudano-Sahelian Office and UN Development Programme, *GEF and Desertification: UNSO/UNDP Workshop, Nairobi, 28–30 October 1992* (New York: UNSO/UNDP, 1992).

2. T. M. L. Wigley, G. I. Pearman, and P. M. Kelly, "Indices and Indicators of Climate Change: Issues of Detection, Validation and Climate Sensitivity," in I. M. Mintzer, ed., *Confronting Climate Change: Risks, Implications and Responses* (Cambridge, England: Cambridge University Press, 1992), 85-96. For a discussion of other potential sources of bias in the global-mean temperature record, see C. K. Folland, T. Karl, and K. Ya. Vinnikov, "Observed Climate Variations and Change," in J. T. Houghton, G. J. Jenkins, and J. J. Ephraums, eds., *Climate Change: The IPCC Scientific Assessment* (Cambridge, England: Cambridge University Press, 1990), 195-238; and C. K. Folland et al., "Observed Climate Variability and Change," in J. T. Houghton, B. A. Callander, and S. K. Varney, eds., *Climate Change 1992: The Supplementary Report to the IPCC Scientific Assessment* (Cambridge, England: Cambridge University Press, 1992), 135-70.

3. UN Environment Programme, *World Atlas of Desertification* (Sevenoaks, U.K.: Edward Arnold, 1992).

2. LAND-HUMAN RELATIONSHIPS

Negev Desert.[14] His initial contention was that bared, high-reflecting soils would increase surface albedo (reflectivity), reduce convective processes, and thus decrease rainfall. Around the same time, Jule Charney, a meteorologist at the Massachusetts Institute of Technology, was developing his biogeophysical hypothesis that land-cover changes, primarily around the Sahara, could enhance aridity.[15] Charney's proposed mechanism involved a desertification-induced change in the vertical energy flux in the atmosphere over dryland regions. Charney's mechanism was subsequently criticized because of his omission of the role of soil moisture and the absence of any discussion of latent/sensible heat ratios.[16]

Charney's hypothesis received considerable attention because it, or some variant, would provide an apparent explanation for self-reinforcing drought (that is, desiccation) in dryland regions. According to this hypothesis, an initial change in land-cover characteristics occurs in association with desertification. The initial change may involve a change or removal of vegetation and/or a deterioration in soil quality and moisture-holding capacity. The land-cover change is then amplified as land surface-atmosphere interaction suppresses rainfall, either by reducing surface moisture or by increasing atmospheric subsidence. Lower rainfall, in turn, increases moisture stress on vegetation, lowers soil moisture levels, and further reduces rainfall amounts, thereby closing the feedback loop. The significance of this hypothesis for the present discussion is that, if such land-cover changes can account for the rainfall decline in the Sahel or even for a significant proportion of that decline, then it is the complex matrix of processes leading to desertification in recent decades that is responsible for the Sahel's desiccation.

A substantial amount of effort has been directed over the last 15 years to refining Charney's basic hypothesis and to examining the sensitivity of regional rainfall to large-scale changes

THE DESICCATION OF THE SAHEL

Within recent years, the climate of the Sahel has exhibited a continuing trend toward desiccation. For about 25 years, rainfall has been substantially lower than it was during the first seven decades of the century (see the figure below). This desiccation represents the most substantial and sustained change in rainfall for any region in the world ever recorded by meteorological instruments. Individual years, such as 1984 and 1990, have seen rainfall totals drop to less than 50 percent of those received during the 1930s, 1940s, and 1950s. Contrasting two successive 30-year periods (from 1931 to 1960 and from 1961 to 1990), the rainfall decline over this region has been between 20 and 40 percent. Although this magnitude of desiccation is unprecedented in the instrumental record, it is harder to assess how unusual it is for the longer-term history of the Sahel. Using a combination of lake levels, landscape descriptions, and historical accounts, Sharon Nicholson at Florida State University has shown that recurrent droughts enduring from one to two decades have been a feature of the Sahelian climate over the last few centuries.[1] Quantifying the severity and precise duration of such droughts, however, is impossible. Examining the historical levels of Lake Chad, one of the great inland lakes of Africa, which lies toward the south of the Sahel, may provide some clues. A comparison of the current decline in the level of Lake Chad to its historical variations suggests that the present desiccation in the Sahel is at least as severe as anything experienced during the last millennium.

---

1. S. E. Nicholson, "Climatic Variations in the Sahel and Other African Regions During the Past Five Centuries," *Journal of Arid Environments* 1 (1978):3-24.

Annual rainfall departure index for the Sahel from 1901 to 1992, expressed as a percent departure from the mean annual rainfall between 1951 and 1980.

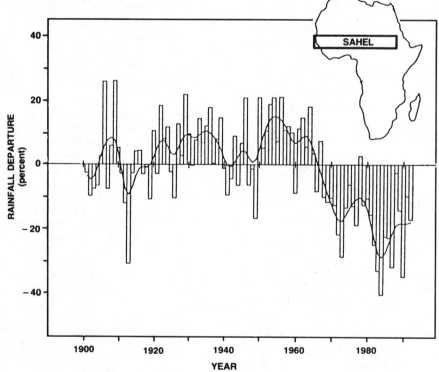

Note: Up to 60 rainfall stations contributed data to this regional series. The smooth curve results from applying a filter that emphasizes variations on a time-scale longer than 10 years.

FIGURE 2. Global field of annual surface air temperature departures associated with drought in the Sahel.

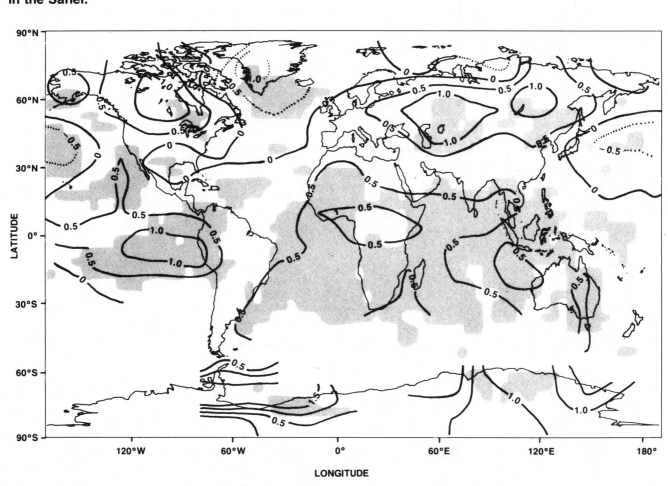

Note: The numbers represent the temperature departures in degrees Celsius, or the differences in temperature between a set of five dry years and a set of five wet years. Shaded areas indicate where the differences are statistically significant.

in land cover through climate modeling experiments. These experiments have been performed for various regions, including the Amazon, the Sahara, and tropical Africa.[17] Such experiments have also addressed a wide range of physical mechanisms for desertification-induced desiccation by modeling interactions among surface albedo, soil moisture and evaporation, and changes in surface roughness and vegetation.[18] These experiments clearly show that large-scale conversion of land-cover characteristics can generate climate change on local and regional scales.

There appears, however, to be a fundamental difficulty in attributing the recent desiccation in the Sahel to land-cover changes on the basis of these model experiments. Observational evidence of the marked, large-scale, sustained changes in surface albedo in dryland regions that are introduced into most model experiments remains weak.[19] (Surface albedo is the measure of land-cover characteristics used in these experiments.) The albedo increases caused by desertification that have been observed are on the order of 25 to 50 percent,[20] and yet a doubling of albedo is used in many model experiments. Moreover, the observed changes have been localized in extent and often short-term, rather than widespread and sustained as assumed in the modeling studies. All of the modeling experiments that have displayed substantial regional rainfall reduction as a response to

land-cover changes have been "sensitivity" rather than "simulation" experiments; rather than imposing observed perturbations to surface vegetation, soil moisture, and so on, they have imposed arbitrarily determined changes, which in all cases have been much larger than those that have actually been observed. Although these experiments are important to understanding how the various physical systems are linked, it is dangerous to draw the conclusion from their results that observed land-cover changes have accounted for observed rainfall changes in the recent past.

A surer way to proceed is to conduct simulation experiments, which impose known perturbations on the model (for example, the observed soil moisture conditions in a given year) and then to examine whether the model reproduces the observed rainfall anomaly of that year. The most impressive set of such simulation experiments has been completed at the British Meteorological Office. The investigators simultaneously perturbed both ocean temperatures and initial soil moisture conditions in a manner consistent with observations.[21] They concluded that ocean temperature forcing appears to dominate the effects of the land surface moisture feedback. Although this work confirms that land surface feedback can play a part in generating self-sustaining drought, the role of this mechanism is secondary to that of variability within the wider climate system.

In light of current empirical and modeling evidence, then, it appears that desertification is not, in itself, a primary cause of the recent desiccation in the Sahel. The degradation of both soil and vegetative cover in dryland regions could well have contributed to the rainfall decline, but this contribution cannot have accounted for anything more than a small fraction of the observed trend. If there is severe and sustained degradation of a substantial dryland area over the next few decades, however, the significance of this internal feedback mechanism may well increase. Over the next 50 years, though, it is more likely that land-cover changes in the humid and subhumid regions of the tropics will lead to substantial changes in regional climate than will those occurring in dryland areas.

Natural Changes in Ocean Circulation

The British Meteorological Office's experiments confirmed the importance of a set of natural mechanisms of climate change that appear to be responsible for the Sahelian desiccation. These mechanisms involve links between Sahelian drought and sea surface temperature (SST) anomalies in the neighboring Atlantic Ocean and other oceans. Research has shown that there is a significant correlation on the

---

WHY IS THE SAHEL DRY?

The Sahel of Africa possesses a monsoonal climate—that is, the climate of the region exhibits a very strong seasonality with a nearly 180° reversal of the prevailing surface wind direction between the wet and dry halves of the year. The winter, or dry, monsoon lasts from October through April and is characterized by northerly or northeasterly surface winds circulating clockwise around the Saharan anticyclone. These winds are extremely dry and lead to no rainfall during these months. The summer, or wet, monsoon commences sometime between April and June and arrives progressively from the south. The moisture in these rain-bearing southerly or southwesterly surface winds originates mostly over the Atlantic Ocean and, to a lesser extent, the Indian Ocean. The sea surface temperatures of these oceans, therefore, exert important control over the rainfall in the Sahel by altering both the moisture characteristics and the vigor of the wet monsoon flow into northern tropical Africa. The wet monsoon varies in duration from five or six months in the southern Sahel (about 10°N) to only a month or two in the far north (about 16°N). This variation in duration creates a very tight gradient in annual rainfall—from less than 100 millimeters north of 16°N to more than 800 millimeters south of about 10°N. Because of this tight gradient, relatively small interannual variations either in the northward penetration or moisture load of the wet monsoon or in the strength of the atmospheric disturbances that lead to the rain outfall result in relatively large variations in the total volume of rainfall received by a locality. Consequently, the rainfall of the Sahel is highly variable from year to year.

---

interannual time-scale between higher-than-normal SSTs south of West Africa and reduced Sahelian rainfall.[22] In the early 1980s, it was argued that a change in atmospheric circulation was affecting both the SST pattern and rainfall over the Sahel. In other words, the SST pattern was just an indicator of the processes affecting Sahelian rainfall rather than the primary cause. It is now thought, however, that the SST pattern may well be the direct cause of the shift in the atmospheric circulation that subsequently affects Sahelian rainfall.

During the mid 1980s, Chris Folland and his colleagues at the Meteorological Office confirmed this statistical correlation between local variations in ocean temperatures and African rainfall on a time-scale of years to decades and found evidence of a broader relationship between Sahelian rainfall and worldwide ocean temperatures.[23] They demonstrated that differences in SST anomalies between the Northern and Southern Hemispheres, most marked in the Atlantic sector, were related to Sahelian rainfall. Higher temperatures south of the equator and lower temperatures north of the equator

(see Figure 2) were associated with lower rainfall over much of northern Tropical Africa. Modeled simulations of the effects of these observed SST anomaly patterns confirmed this association. The success of these model experiments in simulating the observed Sahelian rainfall anomalies suggested that the SST anomaly pattern was the direct cause of the rainfall anomalies. This work has since been extended, confirming the original empirical and model results, and the relationship has provided the basis of an experimental seasonal forecasting scheme.[24]

The physical basis of this relationship appears to lie in a disturbance to the meridional, Hadley circulation of the atmosphere over the Atlantic/Africa sector that is induced by the pattern of contrasting hemispheric ocean temperature anomalies. The Hadley circulation exerts a controlling influence on African rainfall patterns. It determines, in part, the position of the Intertropical Convergence Zone—specifically, the extent of its annual north-south migration, which, in turn, affects the strength of the southwesterly airflow originating in the tropical Atlantic that brings the Sahel much of its rain (see the box on

FIGURE 3. Temperature fluctuations from 1861 to 1991.

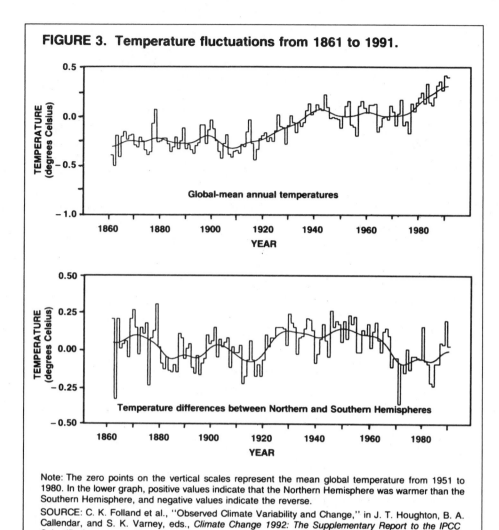

Global-mean annual temperatures

Temperature differences between Northern and Southern Hemispheres

Note: The zero points on the vertical scales represent the mean global temperature from 1951 to 1980. In the lower graph, positive values indicate that the Northern Hemisphere was warmer than the Southern Hemisphere, and negative values indicate the reverse.
SOURCE: C. K. Folland et al., "Observed Climate Variability and Change," in J. T. Houghton, B. A. Callendar, and S. K. Varney, eds., *Climate Change 1992: The Supplementary Report to the IPCC Scientific Assessment* (Cambridge, England: Cambridge University Press, 1992), 135–70.

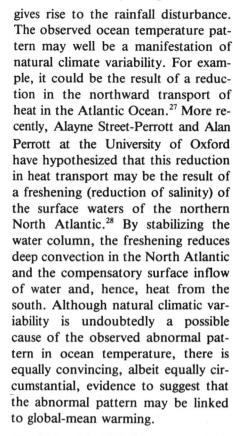

gives rise to the rainfall disturbance. The observed ocean temperature pattern may well be a manifestation of natural climate variability. For example, it could be the result of a reduction in the northward transport of heat in the Atlantic Ocean.[27] More recently, Alayne Street-Perrott and Alan Perrott at the University of Oxford have hypothesized that this reduction in heat transport may be the result of a freshening (reduction of salinity) of the surface waters of the northern North Atlantic.[28] By stabilizing the water column, the freshening reduces deep convection in the North Atlantic and the compensatory surface inflow of water and, hence, heat from the south. Although natural climatic variability is undoubtedly a possible cause of the observed abnormal pattern in ocean temperature, there is equally convincing, albeit equally circumstantial, evidence to suggest that the abnormal pattern may be linked to global-mean warming.

The Link with Global-Mean Warming

Is there a greenhouse-related mechanism that could account for the interhemispheric temperature contrast associated with Sahelian desiccation? Recent research has indicated that there are two alternative—or, more likely, complementary—mechanisms that could induce a temperature difference between the hemispheres.

First, the emission of sulfur compounds as a result of human activity (specifically, fossil fuel combustion) increases the amount of sulfate aerosols in the atmosphere. These aerosols reflect solar radiation, both directly and by altering cloud albedo. Any increase in their concentration in the atmosphere is, therefore, likely to have a cooling effect, offsetting greenhouse warming. Estimates of the scale of this effect vary, but it is considered possible that sulfur dioxide emissions may have reduced the level of warming that rising greenhouse-gas concentrations might have effected by a significant amount.[29] As most sulfur emissions come from the Northern Hemisphere and the sulfate aerosols have a short residence time in the at-

previous page).[25] Thus, the link between lower rainfall in the Sahel and a particular pattern of SST anomalies in the oceans has been well established. The initial cause of this SST pattern, however, has yet to be determined. The spatial scale of the phenomenon may provide some evidence of the cause. Although temperatures in the Atlantic Ocean are probably the dominant influence on Sahelian rainfall, the recently observed pattern of SST anomalies in the Atlantic sector is part of a much larger trend in surface air temperatures. Large-scale warming has affected both hemispheres since the late 19th century and has resulted in a net global-mean warming of about 0.5° C (see the top graph in Figure 3). There has been a clear difference, however, in the warming rates of the two hemispheres during recent decades, with the Northern Hemisphere

warming more slowly than the Southern Hemisphere (see the second graph in Figure 3).

This relationship between conditions in the Atlantic/Africa sector and worldwide climatic trends is not confined to temperature; a link also exists between the desiccation of the Sahel in recent decades and global rainfall fluctuations. Although the rainfall deficit in the Sahel is the most striking rainfall change of recent decades, rainfall has been lower in many parts of the northern tropics and subtropics but has increased at higher latitudes (see Figure 4).[26] The Sahel's desiccation could be considered a regional manifestation of the global shift in the climate system that has occurred since the 1950s.

The outstanding question in this line of reasoning concerns the initial cause of the temperature change that

FIGURE 4. Precipitation fluctuations in the Northern Hemisphere.

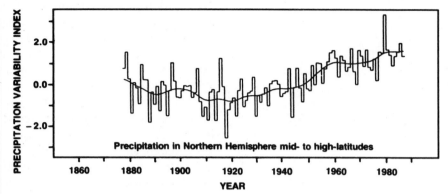

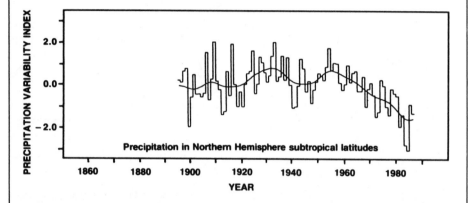

Note: The zero point on the vertical scale of the upper graph represents the mean precipitation from 1877 to 1986. The zero point on the lower graph represents the mean precipitation from 1895 to 1986. The scale is an index of large-scale variations in precipitation, which takes into account the marked changes in mean rainfall that occur from place to place and from month to month.

SOURCE: R. S. Bradley et al., "Precipitation Fluctuations over Northern Hemisphere Land Areas Since the Mid-19th Century," *Science* 237 (1987):171–75.

mosphere, this cooling effect would be largely confined to the Northern Hemisphere. As a result, greenhouse-gas-induced warming would be offset in the Northern Hemisphere but relatively unaffected in the Southern Hemisphere, thereby inducing a differential warming rate and a temperature contrast between the two hemispheres. Estimates of the scale of this effect are not inconsistent with the observed differential between the warming trends in the two hemispheres.[30]

A second greenhouse-related factor exists, particularly in the Atlantic Ocean, that may have caused different rates of warming in the hemispheres. Recent time-dependent ("transient") experiments of enhanced greenhouse-gas forcing using general circulation models (GCMs) that incorporate a dynamic ocean model have suggested that the rate of warming may be retarded in the northern North Atlantic sector and over the Southern Ocean around Antarctica.[31] In these areas, the sinking of dense, saline water masses results in a localized increase in the effective heat capacity and, therefore, in the thermal inertia of the ocean. The increase in thermal inertia slows the warming of the overlying air as heat is drawn down into the ocean. This process induces a meridional gradient in temperature in the Atlantic Ocean north of 60°S, which is not unlike the temperature pattern associated with lower rainfall in the Sahel (see Figure 2). This mechanism may amplify, or even be triggered by, changes in the temperature field and atmospheric circulation that are induced by the effects of sulfate aerosols.

Thus, a physically plausible argument can be advanced linking the recent desiccation in the Sahel to global-mean warming. However, just as it is impossible to ascribe with any certainty the observed global-mean warming to enhancement of the greenhouse effect,[32] neither can the interhemispheric temperature contrast be attributed with confidence to greenhouse-gas-plus-sulfate forcing or to any other greenhouse-related mechanism. For now, the evidence must be considered circumstantial.

The Future of the Sahel

GCM experiments have recently been conducted to determine whether significant rainfall changes over northern tropical Africa may result from future greenhouse-gas forcing. (Unfortunately, none of these experiments incorporates the sulfate aerosol effect, and only one allows the ocean circulation to respond realistically to greenhouse-gas forcing.) The experiments indicate that global-mean warming should lead to an overall increase in global-mean precipitation because evaporation over the warmer oceans increases the moisture content of the atmosphere.[33] The distribution of rain and snowfall, however, will also be determined by changes in the atmospheric circulation and by other climatic factors.

Figure 5 shows a composite estimate of the percentage change in mean annual rainfall that may accompany each 1° C rise in global-mean temperature induced by greenhouse-gas forcing. The composite is based on a set of seven GCM experiments.[34] An increase in rainfall is apparent in most areas, but annual rainfall decreases over the Mediterranean, North Africa, and a large part of the Sahel, especially the Western Sahel. The effect is most marked over the southwestern margins of the Sahara, in Mauritania and northern Mali and Niger. In the latter two areas, annual rainfall decreases by more than 6 percent for every 1° C of global-mean warming. If global-mean warming follows the 1992 projections of the Intergovernmental Panel on Climate Change,[35] rainfall in these areas decreases by between 6 and 30 percent by 2100. This is, however, a slow rate of decline, as long as it occurs progressively, in comparison with the 20 to 40 percent decline in rainfall experienced by the Sahel during recent decades.

The current performance of climate models in estimating regional rainfall patterns is considered quite weak. Model-to-model differences in predicted rainfall changes over northern Africa are large, and individual models predict a complex spatial pattern of change. Nevertheless, the model composite does show a well-defined reduction in rainfall over much of the Sahel. Compounded by increased temperatures (which can be predicted with greater confidence), lower rainfall would inevitably cause substantial reductions in soil moisture availability.

Negotiating for Survival

Negotiations for the Convention to Combat Desertification will be complicated by the technical and scientific uncertainties underlying many aspects of the desertification issue. It is to be hoped that the negotiators will be assisted by the kind of technical support that was provided by the Intergovernmental Panel on Climate Change during the negotiations for the Framework Convention on Climate Change. Although varying degrees of uncertainty surround the links between climate change, desiccation in dryland regions, and desertification, this brief assessment has shown that there are intrinsic links that should not be ignored (see Figure 6). Even the rather speculative link between global-mean warming and desiccation in the Sahel warrants serious consideration on a precautionary basis because of the serious implications for those living in this dryland area.

The area prone to desertification is, by definition, determined by climatic conditions and, hence, by climate change. It is more difficult to determine the precise balance between the human and climatic factors that lead to desertification. There is, however, no doubt that, against a background of resource management failure in dryland regions, climate change will aggravate the problem. It is also clear that, by modifying surface characteristics, desertification can induce significant changes in local temperature. But it is far less likely that the global-mean temperature has been affected to any significant extent by dryland degradation.

Progressive desertification of the dryland tropics and other areas is likely to reduce a potential carbon sink by reducing the carbon stored in these ecosystems. Moreover, as vegetation dies and soil is disturbed, desertification increases emissions of carbon dioxide. Desertification may also increase or decrease emissions of other greenhouse gases. Although desertification contributes only a small percentage of all greenhouse-gas emissions, understanding of desertification-related sources and sinks of greenhouse gases should still be improved and desertification rates reduced to enable dryland countries to offset growth in their emissions of other greenhouse gases.

The hypothesized link between the

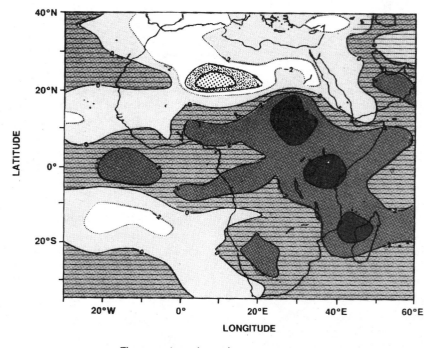

FIGURE 5. Percentage change in mean annual rainfall over Africa per degree Celsius of global-mean warming resulting from increased greenhouse-gas forcing.

The percentage change is:

- ■ greater than +4
- ▨ greater than +2 but less than or equal to +4
- ▤ greater than zero but less than or equal to +2
- □ greater than −2 but less than or equal to zero
- □ greater than −4 but less than or equal to −2
- ▨ greater than −6 but less than or equal to −4
- ▨ less than or equal to −6

Note: The percentage change in mean annual rainfall is standardized by global-mean warming. The increase in greenhouse-gas forcing is as predicted by an ensemble of seven global climate modeling experiments.

SOURCE: P. M. Kelly and M. Hulme, *Climate Scenarios for the SADCC Region* (Norwich, England: Climatic Research Unit/London and International Institute for Environment and Development, 1992).

2. LAND-HUMAN RELATIONSHIPS

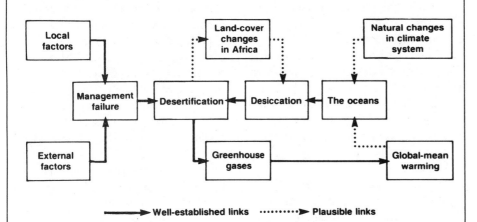

FIGURE 6. The matrix of cause and effect surrounding desertification and the role of climate change.

NOTE: Desertification is the result of resource management failure, the product of both local factors, such as population pressure and inequity, and external factors, such as the state of the global economy, commodity prices, and the burden of debt. Desertification is aggravated by climate change—desiccation—which may be the result of natural mechanisms with the climate system, such as ocean-atmosphere feedback; by desertification itself through, for example, surface-atmosphere interaction; or possibly by global-mean warming. Finally, desertification contributes to global-mean warming through its effect on the sources and sinks of greenhouse gases, such as carbon dioxide. Many uncertainties affect assessment of the relative role of these various factors.

recent desiccation of the Sahel and an interhemispheric contrast in ocean temperature is supported by empirical studies, model simulations, and theoretical argument. The initial cause of the interhemispheric temperature contrast has yet to be determined. It may be the result of natural climate variability or a manifestation of global-mean warming resulting from greenhouse-gas and other anthropogenic emissions. Both possibilities are physically plausible, and neither possibility is contradicted by the available data.

Finally, on the basis of current evidence, it appears likely that the role of desertification in causing (via climate change and land-cover changes) the recent desiccation in the Sahel is very much secondary to that of forcing by ocean temperature anomalies. The rainfall decline estimated to have resulted from past degradation of both soil and vegetative cover in Africa is only sufficient to account for a small fraction of the observed desiccation in the Sahel.

In the future, however, the relative importance of these various factors may change. For example, a sustained degradation of a substantial

dryland area may well increase the significance of desertification as a causative agent. On the other hand, climate model experiments suggest that rainfall over the Sahel may decrease even more as global-mean warming develops. In this case, the role of greenhouse-gas emissions in reducing rainfall in the Sahel would become more prominent. Whatever the complexity of these mechanisms, their investigation cannot be considered to be solely of academic interest; as the inhabitants of the Sahel well know, the trend of reduced rainfall continues.

ACKNOWLEDGMENTS

The authors wish to thank Sarah Granich for her contribution to this article and Tim O'Riordan for his helpful comments. The article is based on a technical report prepared for the Overseas Development Administration in London, but it does not necessarily reflect the administration's views.

NOTES

1. UN Environment Programme, *World Atlas of Desertification* (Sevenoaks, U.K.: Edward Arnold, 1992).

2. R. S. Odingo, "Implementation of the Plan of Action to Combat Desertification (PACD) 1978–1991,"

Desertification Control Bulletin 21 (1992):6–14.

3. The Global Environment Facility was established in 1990 to help developing nations respond to global environmental change insofar as this response will reduce the global impact of the problems.

4. UN Environment Programme, *Status of Desertification and Implementation of the UN Plan of Action to Combat Desertification*, UNEP/GCSS.III/3 (Nairobi: UN Environment Programme, 1991).

5. United Nations, "Managing Fragile Ecosystems: Combating Desertification and Drought," chapt. 12 of *Agenda 21* (New York: United Nations, 1992), pt. 2.

6. A. Warren and M. M. Khogali, *Assessment of Desertification and Drought in the Sudano-Sahelian Region: 1985 to 1991* (New York: UN Development Programme and UN Sudano-Sahelian Office, 1992).

7. In the 1992 UNEP desertification assessment, a simple moisture index (the ratio of precipitation to potential evapotranspiration) was used to define these boundaries.

8. The index, known as the Normalized Difference Vegetation Index (NDVI), is derived from the visible and near infrared sensors on polar-orbiting satellites and is an indicator of the photosynthetic vigor of surface biomass. See C. J. Tucker, H. E. Dregne, and W. W. Newcomb, "Expansion and Contraction of the Sahara Desert from 1980 to 1990," *Science* 253 (1991):299–301.

9. R. C. Balling, Jr., "Impact of Desertification on Regional and Global Warming," *Bulletin of the American Meteorological Society* 72 (1991):232–34.

10. R. T. Watson, H. Rodhe, H. Oeschger, and U. Siegenthaler, "Greenhouse Gases and Aerosols," in J. T. Houghton, G. J. Jenkins, and J. J. Ephraums, eds., *Climate Change: The IPCC Scientific Assessment* (Cambridge, England: Cambridge University Press, 1990), 1–40; and R. T. Watson, L. G. Meira Filho, E. Sanhueza, and A. Janetos, "Greenhouse Gases: Sources and Sinks," in J. T. Houghton, B. A. Callander, and S. K. Varney, eds., *Climate Change 1992: The Supplementary Report to the IPCC Scientific Assessment* (Cambridge, England: Cambridge University Press, 1992), 25–46.

11. K. P. Shine, R. G. Derwent, D. J. Wuebbles, and J.-J. Morcrette, "Radiative Forcing of Climate," in Houghton, Jenkins, and Ephraums, eds., note 10 above, pages 41–68. The results of more recent research suggest that the carbon dioxide contribution to the total anthropogenic climate forcing may be higher when the effects of emissions of sulfur compounds and ozone depletion are considered. See Watson, Meira Filho, Sanhueza, and Janetos, note 10 above; and T. M. L. Wigley and S. C. B. Raper, "Implications of Revised IPCC Emissions Scenarios," *Nature* 357 (1992):293–300.

12. A. F. Bouwman, "Land Use Related Sources of Greenhouse Gases," *Land Use Policy*, April 1990, 154–64.

13. M. H. Glantz, "The Value of a Long-Range Weather Forecast for the Sahel," *Bulletin of the American Meteorological Society* 58 (1977):150–58; P. J. Lamb, "Large-Scale Tropical Atlantic Surface Circulation Patterns Associated with Sub-Saharan Weather Anomalies," *Tellus* 30 (1978):240–51; and S. E. Nicholson, "The Nature of Rainfall Fluctuations in Subtropical West Africa," *Monthly Weather Review* 108 (1980):473–87.

14. J. Otterman, "Baring High-Albedo Soils by Overgrazing: A Hypothesised Desertification Mechanism," *Science* 186 (1974):531–33.

15. J. G. Charney, "Dynamics of Deserts and Drought in the Sahel," *Quarterly Journal of the Royal Meteorological Society* 101 (1975):193–202.

16. See, for example, S. B. Idso, "A Note on Some Recently Proposed Mechanisms of Genesis of Deserts," *Quarterly Journal of the Royal Meteorological Society* 103 (1977):369–70.

17. R. E. Dickinson and A. Henderson-Sellers, "Modelling Tropical Deforestation: A Study of GCM Land-Surface Parameterisations," *Quarterly Journal of the Royal Meteorological Society* 114 (1988):439–62; W. M. Cunnington and P. R. Rowntree, "Simulations

of the Saharan Atmosphere: Dependence on Moisture and Albedo," *Quarterly Journal of the Royal Meteorological Society* 112 (1986):971–99; M. F. Mylne and P. R. Rowntree, "Modelling the Effects of Albedo Change Associated with Tropical Deforestation," *Climate Change* 21 (1992):317–43; and Y. C. Sud and M. J. Fenessey, "A Study of the Influence of Surface Albedo on July Circulation in Semi-Arid Regions Using the GLAS GCM," *Journal of Climatology* 2 (1982):105–25.

18. Y. C. Sud and M. J. Fenessey, "Influence of Evaporation in Semi-Arid Regions on the July Circulation: A Numerical Study," *Journal of Climatology* 4 (1984):393–98; and J. Lean and D. A. Warrilow, "Simulation of the Regional Climatic Impact of Amazon Deforestation," *Nature* 342 (1989):411–13.

19. M. Hulme, "Is Environmental Degradation Causing Drought in the Sahel?" *Geography* 74 (1989):38–46.

20. S. I. Rasool, "On Dynamics of Deserts and Climate," in J. T. Houghton, ed., *The Global Climate* (Cambridge, England: Cambridge University Press, 1984), 107–20.

21. D. P. Rowell et al., "Causes and Predictability of Sahel Rainfall Variability," *Geophysical Research Letters* 19 (1992):905–08.

22. Lamb, note 13 above; and J. M. Lough "Atlantic Sea Surface Temperatures and Weather in Africa" (Ph.D. diss., University of East Anglia, 1980).

23. C. K. Folland, D. E. Parker, and F. E. Kates, "Worldwide Marine Temperature Fluctuations, 1856–1981," *Nature* 310 (1984):670–73; and C. K. Folland, T. N. Palmer, and D. E. Parker, "Sahel Rainfall and Worldwide Sea Temperatures, 1901–1985," *Nature* 320 (1986):602–07.

24. C. K. Folland, J. A. Owen, M. N. Ward, and A. W. Colman, "Prediction of Seasonal Rainfall in the Sahel Region Using Empirical and Dynamical Methods," *Journal of Forecasting* 10 (1991):21–56.

25. For a more detailed discussion of the proposed mechanism, see C. K. Folland and J. A. Owen, "GCM Simulation and Prediction of Sahel Rainfall Using Global and Regional Sea Surface Temperatures," in *Modelling the Sensitivity and Variations of the Ocean-Atmosphere System*, WMO/TD no. 254 (Geneva: World Meteorological Organization, 1988), 107–15.

26. R. S. Bradley et al., "Precipitation Fluctuations over Northern Hemisphere Land Areas Since the Mid-19th Century," *Science* 237 (1987):171–75.

27. R. E. Newell and J. Hsiung, "Factors Controlling Free Air and Ocean Temperature of the Last 30 Years and Extrapolation to the Past," in W. H. Berger and L. D. Labeyrie, eds., *Abrupt Climatic Change: Evidence and Implications* (Dordrecht, the Netherlands: Reidel, 1987), 67–87.

28. F. A. Street-Perrott and R. A. Perrott, "Abrupt Climate Fluctuations in the Tropics: The Influence of Atlantic Ocean Circulation," *Nature* 343 (1990):607–12.

29. I. Isaksen, V. Ramaswamy, H. Rodhe, and T. M. L. Wigley, "Radiative Forcing of Climate," in Houghton, Callander, and Varney, eds., note 10 above, pages 47–68; and Wigley and Raper, note 11 above.

30. See, for example, Wigley and Raper, note 11 above.

31. W. L. Gates, P. R. Rowntree, and Q-C. Zeng, "Validation of Climate Models," in Houghton, Jenkins, and Ephraums, eds., note 10 above, pages 93–130; and W. L. Gates et al., "Climate Modelling, Climate Prediction and Model Validation," in Houghton, Callander, and Varney, eds., note 10 above, pages 97–134.

32. T. M. L. Wigley and T. P. Barnett, "Detection of the Greenhouse Effect," in Houghton, Jenkins, and Ephraums, eds., note 10 above, pages 243–55.

33. J. F. B. Mitchell, S. Manabe, V. Meleshko, and T. Tokioka, "Equilibrium Climate Change and Its Implications for the Future," in Houghton, Jenkins, and Ephraums, eds., note 10 above, pages 137–64.

34. Climatic Research Unit, *A Scientific Description of the ESCAPE Model* (Norwich, England: Climatic Research Unit, 1992); and P. M. Kelly and M. Hulme, *Climate Scenarios for the SADCC Region* (Norwich, England: Climatic Research Unit/London and International Institute for Environment and Development, 1992).

35. Houghton, Callander, and Varney, eds., note 10 above; and Wigley and Raper, note 11 above.

WATER TIGHT

Four cities' innovative conservation efforts prove that saving water makes economic and environmental sense.

SANDRA POSTEL

Sandra Postel is vice president for research at the Worldwatch Institute and author of Last Oasis: Facing Water Scarcity, *from which this article is excerpted*

Mexico City's historic plaza offers a strange sight. The imposing Metropolitan Cathedral, built soon after the 16th-century Spanish conquest, droops rather dramatically on its right side, and slightly less so on the left. Inside, an array of tension wires and green metallic beams support the weakening edifice. The capital's revered cathedral is sinking, and it has little to do with the engineering skills of the early Spaniards. Large parcels of land are slumping as the city siphons its underground water supply, and the resulting structural damage is just the most visible of many consequences.

Mexico City is an extreme case, but there are many cities around the world that have overstepped the limits of their water supply. Homes, apartments, offices, stores, restaurants, and government buildings account for less than a tenth of all the water used in the world today—the remainder goes to farmers and industries—but their demands are concentrated in relatively small geographic areas. And, in many cases, those demands are escalating. As cities expand, they strain the capacity of local water supplies and force engineers to reach out to ever more distant sources.

Beyond the challenge of finding enough water, it costs a lot to build and maintain the reservoirs, canals, pumping stations, pipes, sewers, and treatment plants that make up modern water and wastewater systems. Collecting and treating water and wastewater also takes a great deal of energy and chemicals, adding to environmental pollution and the operating costs of a community's water system. Under this financial strain, many cities are having difficulty meeting the water needs of their residents, and many low-income residents in developing countries get no service at all.

Conservation, once viewed as just an emergency response to drought, has been transformed in recent years into a sophisticated package of measures that offers one of the most cost-effective and environmentally sound ways of balancing urban water budgets. Just as energy planners have discovered that it is often cheaper to save energy up front—for instance, by investing in home insulation and compact fluorescent lights—than to build more power plants, water planners are realizing that water efficiency measures can yield permanent water savings and thereby delay or avert the need for expensive new dams and reservoirs, groundwater wells, and treatment plants.

Slowly the idea is catching on that *managing* demand rather than continuously striving to meet it is a surer path to a secure water supply—and will save money and protect the environment at the same time.

Water Balance

Many urban areas simply have no feasible way to balance supply and demand without conservation and more efficient water use. Mexico City is a prime example. This sprawl-

---

The idea is catching on that managing the demand for water rather than continuously striving to meet it will save money and protect the environment at the same time.

---

ing metropolis of 18 million people relies on groundwater for more than 80 percent of its supplies. Pumping exceeds natural replenishment by 50 to 80 percent, which has caused water tables to drop and aquifers—water-holding geologic formations—to compress. This, in turn, has led to sinking of the land and damage to surface structures, including the cathedral.

Mexico City, which sits in a cramped, high mountain bowl, has outstripped its supply of groundwater and is forced to search out water on the other side of the mountains that hem it in. The city now meets 17 percent of its demand by bringing water from the Cutzamala River system 80 miles away and lifting it 3,900 vertical feet—all at enormous cost. With the metropolitan area expanding by more than half a million people each year, officials are racing against time to achieve some degree of water stability.

Faced with such an intractable problem, the Mexican government and city officials are orchestrating an aggressive water conservation effort. In 1989, the federal government took a bold step in adopting a strict set of nationwide efficiency standards for household plumbing fixtures and appliances.

These require toilets—the biggest water guzzlers in the home—to use no more than 1.6 gallons of water per flush. Maximum limits have been set for showers, faucets, dishwashers, and washing machines as well.

Mexico City has launched an ambitious program to replace conventional toilets using about 4.2 gallons with the 1.6-gallon models in public places, commercial buildings, and private residences. By late 1991, more than 350,000 toilets had already been upgraded, saving nearly 7.4 billion gallons of water each year—enough to meet the household needs of more than 250,000 residents. Officials also hiked the city's water rates in 1990, while encouraging residents to install a package of low-flow shower heads, toilet dams, faucet aerators, and other home water-saving devices, and to be more thrifty overall. To bolster the whole effort, a large-scale public information campaign—including educating schoolchildren and airing radio and television spots—is under way to raise awareness of the city's water plight and tell people how they can conserve.

It is too early to judge the program's effectiveness, but officials are projecting that water use will fall from the current level of 79 gallons per person per day, to 66 gallons by 1996. Unfortunately, without a slowdown in birthrates and migration to the capital, population growth will negate these savings, and total water use in Mexico City will continue to climb, albeit at a slower pace.

Although not in as dire straits as the Mexican capital, the Canadian city of Waterloo, in Ontario, had similar reasons for shifting from the traditional approach of expanding supplies to managing demand. The Waterloo program has made conservation an effective part of the city's long-term water strategy through raising the price of water, educating the public to the need and ease of conserving water, and distributing water-saving devices to make home plumbing fixtures more efficient. Volunteer groups have distributed retrofit kits—including toilet dams, faucet aerators, and low-flow showerheads—to nearly 50,000 homes, and homeowners have been urged to conserve water outdoors as well. Waterloo's per-capita water use fell 10 percent in just three years.

As in Mexico City, Waterloo's efforts will be bolstered by province-wide efficiency standards for new plumbing fixtures that will take effect in 1993. By 1996, new toilets throughout Ontario must meet the 1.6-gallon standard, the strictest required today. Ontario has also set an ambitious goal of zero growth in water use for the next 20 years, which will be helped along by the conservation initiatives Waterloo already has under way. "If we achieve zero growth, we will reduce stress on the environment, lessen the likelihood of water shortages and reduce energy costs," says Bud Wildman, Ontario's natural resources minister.

Waste Not, Want Not

The only motivation some cities need for conservation is a look at the money they shell out to treat wastewater. In the mid-1980s, the sewage treatment plant in San Jose, California, was nearing capacity, and the city was faced with the prospect of building a new one at a cost of $180 million. The city decided to pursue a different option. Since less water used indoors translates into less wastewater released to the sewer system, the city initiated a large-scale program to reduce residential and industrial water use quickly. Officials hoped to delay the need for this huge capital investment and save the city and its residents money.

San Jose set a goal in 1986 of cutting wastewater flows to the treatment plant by 10 percent by 1996. The centerpiece of the conservation program was a massive retrofit campaign, in this case involving the distribution of water-saving devices door-to-door to some 220,000 households. Diligent canvassers made at least three attempts to talk with residents about the importance of installing the devices. As a result, 90 percent of the targeted households cooperated—an unmatched success rate. Water use in participating homes dropped 10 to 17 percent. When industry conservation was factored in, the program had cut wastewater flows as of 1991 by an estimated 1.4 billion gallons a year, about the amount 20,000 U.S. households would send to a treatment plant.

Low Flow

In the greater Boston metropolitan area, the environmental consequences of expanding the water supply forced local officials to take a hard look at conservation. That soul-searching resulted in one of the most comprehensive and successful water-effi-

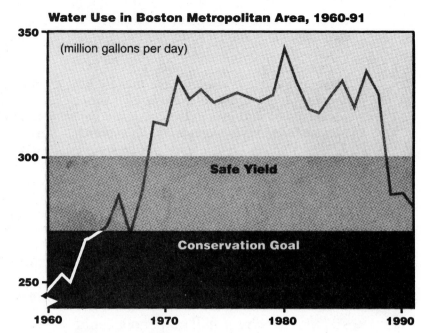

Water Use in Boston Metropolitan Area, 1960-91

(million gallons per day)

Source: Massachusetts Water Resources Authority

ciency programs in the United States. When demand in the metropolitan region rose above the dependable yield of the water supply system in the early 1970s, water planners did what most of their colleagues do—they looked for another river to dam or divert. The Boston engineers proposed diverting some of the flow of the Connecticut and Merrimack rivers eastward by tunnel to the metropolitan area.

Environmental groups asserted that diverting the rivers would contaminate the city's otherwise relatively pure water supply, which did not require filtration, and added that the increased concentration of pollutants resulting from diminished river flows would damage salmon restoration efforts. They organized opposition to the projects in some 48 towns across Massachusetts, which, along with the engineering scheme's high cost, led the city to consider seriously ways to curb water demand instead.

As a result, in March 1987, the Massachusetts Water Resources Authority (MWRA) launched an aggressive strategy of conservation and increased efficiency throughout its service area, which includes some 2.5 million people. Water-saving devices were installed in about 100,000 homes, leaks in old pipes were found and repaired, more than a million pieces of conservation literature were distributed to schoolchildren, and advice on water-saving measures was given to hundreds of businesses and industries.

The results were impressive. Total annual water demand fell from 122 billion gallons in 1987, when the program began, to 102 billion in 1991, a 16-percent drop (see Figure). Boston's water use is now below the system's safe yield. The MWRA plans to install water-saving devices in an additional 330,000 households during the next few years, at no direct cost to the customer, and to extend other aspects of the program as well. The expected savings have prompted MWRA officials to recommend postponing an expansion of the water supply until at least 1995. Moreover, the program has been a bargain, with the conservation measures costing a third to half as much as supply-side options. "For the first time in 20 years, we are living within our means," says Paul Levy, former MWRA executive director.

Water Break

As the experiences of Mexico City, Waterloo, San Jose, and Boston illustrate, conservation makes sense for many reasons, and a different mix of measures may be appropriate in each situation. In almost every case, however, successful efforts to curb domestic water use include some combination of economic incentives, regulations, and public outreach to promote the use of water-saving technologies and behaviors. These measures are mutually reinforcing, and together they constitute a water supply option as reliable and predictable as a new dam and reservoir. They also are often less expensive and better for the environment.

Raising the price of water to better reflect its true cost is one of the most important steps any city can take to propel conservation efforts. Proper pricing gives consumers an accurate signal of just how costly water use is, and it allows them to change their behavior accordingly. Studies in a number of countries, including Australia, Canada, Israel, and the United States, suggest that household water use drops 3 to 7 percent with a 10-percent increase in water prices.

Water is consistently undervalued, and as a result is chronically overused. Not only do current water prices typically fail to promote efficiency, the water rate structures of many utilities actually reward waste by charging less per gallon when more is consumed. Seven out of 10 residents in Manitoba, Canada, for instance, are charged according to this perverse "declining block" pricing policy, as are one out of three in Alberta and

Ontario. Amazingly, water charges for most British households are linked to the value of the home, not to actual consumption.

Many residences in industrial and Third World cities are not equipped with water meters, which precludes even the possibility of charging people appropriately for their water use. Metering encourages savings by tying the bill for water to the amount used. The city of Edmonton, Alberta, meters all residential users, and its per-capita water use is half that of Calgary, which is only partially metered. The areas of Calgary that are metered, however, register water use rates much like Edmonton's. Trials in the United Kingdom have shown that metering can cut household use by 10 to 15 percent.

Raising water prices is not a popular political move, but if accompanied by a public outreach campaign that explains the need for a price hike and the steps consumers can take to keep their water bills down, it can be both doable and beneficial. For example, when officials in Tucson, Arizona, were faced with dire water supply conditions in the mid-1970s, they raised water rates sharply to better reflect

---

*W**ater is consistently undervalued and, as a result, is chronically overused.*

---

the true cost of water services. At about the same time, the city ran a public education campaign called "Beat the Peak" with a goal of curbing water use on hot summer afternoons, when there was the greatest danger of demand outrunning supply. The result was a 16-percent drop in per-capita use within a few years, which, along with the lowered peak demand, allowed the Tucson water utility to cut its planned expansion costs by $75 million.

Pricing was the main tool of a conservation strategy adopted by the water utility serving Bogor, Indonesia, as well. With a proposed water project estimated to cost twice as much per gallon as existing supplies, the utility opted to try reducing demand through more effective pricing. It tripled or quadrupled water prices, depending on the amount used, to encourage households to conserve. Between June 1988 and April 1989, average monthly residential water use dropped nearly 30 percent, enough to allow the utility to connect more households to the urban water system at a lower cost.

Built-In Efficiency
Since economic incentives and public outreach won't motivate everyone to conserve, water-efficiency standards for common fixtures—toilets, showerheads, and faucets—are a necessary component of a dependable conservation strategy. Standards establish technological norms that ensure a certain level of efficiency is built into new products and services. As already noted, Mexico has established nationwide standards, and Ontario is including standards in its conservation strategy as well.

In the United States, there has been a growing movement at the state level to mandate the use of water-efficient plumbing fixtures. In 1988, Massachusetts became the first state to require that all new toilets installed use no more than 1.6 gallons. Since then, 14 other states have followed suit, most of them adopting efficiency standards for showerheads and faucets as well.

Legislation setting national standards was signed into law in October 1992 as part of a broad energy bill. It requires that showerheads and faucets manufactured in the United States after January 1, 1994, use no more than 2.5 gallons per minute; domestic toilets manufactured after that date are to use no more than 1.6 gallons per flush. As all new homes and major remodeling nationwide incorporate these more-efficient fixtures, average U.S. indoor residential water use should fall gradually from 77 gallons per person a day to 54 gallons, a 30-percent reduction, according to estimates by Boston-based water conservation consultant Amy Vickers.

Effective pricing, regulations, and public outreach can also help curb the use of water outdoors. In many dry regions, lawn sprinkling accounts for anywhere from a third to *half* of residential water demand. This water has a particularly high economic and environmental price, since it is used most during hot summer days when utilities experience their highest level of use. Meeting this peak demand requires planners to develop more water sources and treatment capacity than is typically needed during the rest of the year.

Many U.S. communities have turned to "xeriscape landscaping" on public and private land in recent years to save water. From the Greek word *xeros*, meaning dry, xeriscape designs draw on a wide variety of attractive indigenous and drought-tolerant plants, shrubs, and ground cover to replace the thirsty green lawns found in most suburbs. A study in Novato, California, found that xeriscape landscaping cut water use by 54 percent, fertilizer use by 61 percent, and herbicide use by 22 percent.

Just a decade old, the xeriscape concept has spread rapidly through parts of the United States. Programs in at least eight states—including several in the more humid East—actively support this form of landscaping as a way to conserve water and improve the urban environment. Tucson, Arizona, gave the approach a boost in early 1991 by forbidding new developments from having more than 10 percent of their landscape area planted in grass. Xeriscape landscaping is making inroads in a handful of other countries, including Australia, Canada, and Mexico.

Patching Leaks
Water conservation can't ignore waste at the source—leaks in the water distribution system itself. Finding and repairing leaks usually yields a big payoff, especially in older cities. As urban water systems deteriorate because of age or lax maintenance, a lot of water can be lost through broken pipes and faults in the distribution network. More than half the urban water supply in Cairo, Jakarta, Lagos, Lima, and Mexico City simply disappears. Although some of this water is probably siphoned off by poor residents not served by the system, much of it is wasted. This makes for big financial losses since cities still pay to have this unaccounted-for water collected, stored, treated, and distributed, even though it never reaches a billable customer.

In most cases, finding and fixing leaks rewards a city not only with water savings, but with a quick payback on its investment as well. At a cost of $2.1 million, the Massachusetts Water Resources Authority's leak detection program cut system-wide demand in greater Boston by about 10 percent, making it one of the most cost-effective measures in the city's conservation strategy. Leak detection and repair can be especially beneficial in Third World cities that suffer extremely large losses. If Jakarta, Indonesia, could cut its unaccounted-for water from 51 percent to 31 percent of output, for example, it could save 11.9 billion gallons annually—enough to supply 800,000 people.

With some notable exceptions, such as Mexico City and Bogor, Indonesia, few Third World cities are trying to conserve water. Most are preoccupied with the daunting challenge of providing reliable water services to the large number of people now lacking them. Since average household use in most developing countries is a fraction of that in industrial countries, conservation and efficiency are often viewed as irrelevant or, at best, as options to pursue later.

But, in fact, conservation is an integral part of any practical solution to poorer nations' water supply problems. The Third World's population is growing by 90 million each year, and widespread migration from rural areas to cities guarantees explosive urban growth. Water-efficient plumbing, pricing policies, and other measures offer these cities an opportunity to build conservation into their water plans early on, allowing more needs to be met with fewer resources and curbing water costs overall.

It would be a costly mistake for the developing world to adopt the water-intensive ways of industrial countries, many of which find their water practices unsustainable. It costs between $450 and $700 per resident to build water distribution networks, connect each individual household to water and sewer pipes, and construct centralized water and wastewater treatment plants. Efficient plumbing fixtures and other conservation measures can help shrink these costs by lowering each household's water demand. They allow cities to scale down the size of expensive new treatment plants and distribution pipes, cutting both capital and operating expenditures.

There are signs that the idea of incorporating conservation into developing countries' long-term water supply planning may be catching on. The World Bank and the United Nations Development Program have started working with a number of countries—including Chile, China, India, and South Korea—to identify cities that could serve as good demonstration sites for urban water conservation.

In rich and poor countries alike, conservation and efficiency are the wave of the future in meeting urban water needs in a cost-effective and environmentally sound manner. Astute entrepreneurs with new conservation technologies and services already are finding new markets to pursue. And efficiency measures show clearly how economic gains and environmental protection can go hand-in-hand. Conservation's potential is vast—and it has barely been tapped.

The Region

The region is one of the most important concepts in geography. The term has special significance for the geographer, and it has been used as a kind of areal classification system in the discipline.

Two of the regional types most used in geography are "uniform" and "nodal." A uniform region is one in which a distinct set of features is present. The distinctiveness of the combination of features marks the region as being different from others. Examples of these features include climate type, soil type, prominent languages, resource deposits, and virtually any other identifiable phenomenon having a spatial dimension.

The nodal region reflects the zone of influence of a city or other nodal place. Imagine a rural town in which a farm implement service center is located. Now imagine lines drawn on a map linking this service center with every farm within the area that uses it. Finally, imagine a single line enclosing the entire area in which the individual farms are located. The enclosed area is defined as a nodal region. The nodal region implies interaction; regions of this type are defined on the basis of banking linkages, newspaper circulation, and telephone traffic, among other things.

The articles in this unit present examples of a number of regional themes. These selections can provide only a hint of the scope and diversity of the region in geography. There is no limit to the number of regions; there are as many as the researcher sets out to define.

"The Rise of the Region State" suggests that the nation-state is an unnatural and even dysfunctional unit for organizing human activity. "Africa's Geomosaic Under Stress" reviews the independence movement in Africa and draws attention to core-periphery problems in South Africa, a nation torn by cultural strife for decades. Saul Cohen's article on the Middle East highlights the chang-ing geopolitical position of this region in the post–cold war era. An article from *American Demographics* draws attention to the nearly eight million people aligned along the U.S.–Mexican border and the maquiladoras (U.S.–owned factories) located there.

"Reclaiming Cities for People" addresses the attention that is being focused on making cities more physically humane and attractive. The next article points out that a major environmental catastrophe is occurring with the literal demise of the Aral Sea, as water is diverted for irrigation agriculture. The following article deals with "China's Sorrow," the great river of the north, the Huang He or Yellow River, and its tremendous impact on the Chinese through more than four centuries. "Low Water in the American High Plains" focuses on mismanagement of a valuable water resource in the United States.

Finally, in "The Key to Understanding the Former Soviet Union," author Joel Garreau suggests that there are a set of cultural parallels existing between North America and the former Soviet Union.

Looking Ahead: Challenge Questions

To what regions do you belong?

Why are maps and atlases so important in discussing and studying regions?

What major regions in the world are experiencing change? Which ones seem not to change at all? What are some reasons for the differences?

What regions in the world are experiencing tensions? What are the reasons behind these tensions? How can the tensions be eased?

Why are regions in Africa suffering so greatly?

Is the nation-state system an anachronism?

Unit 3

THE RISE OF THE REGION STATE

Kenichi Ohmae

Kenichi Ohmae is Chairman of the offices of McKinsey & Company in Japan.

The Nation State Is Dysfunctional

THE NATION STATE has become an unnatural, even dysfunctional, unit for organizing human activity and managing economic endeavor in a borderless world. It represents no genuine, shared community of economic interests; it defines no meaningful flows of economic activity. In fact, it overlooks the true linkages and synergies that exist among often disparate populations by combining important measures of human activity at the wrong level of analysis.

For example, to think of Italy as a single economic entity ignores the reality of an industrial north and a rural south, each vastly different in its ability to contribute and in its need to receive. Treating Italy as a single economic unit forces one—as a private sector manager or a public sector official—to operate on the basis of false, implausible and nonexistent averages. Italy is a country with great disparities in industry and income across regions.

On the global economic map the lines that now matter are those defining what may be called "region states." The boundaries of the region state are not imposed by political fiat. They are drawn by the deft but invisible hand of the global market for goods and services. They follow, rather than precede, real flows of human activity, creating nothing new but ratifying existing patterns manifest in countless individual decisions. They represent no threat to the political borders of any nation, and they have no call on any taxpayer's money to finance military forces to defend such borders.

Region states are natural economic zones. They may or may not fall within the geographic limits of a particular nation—whether they do is an accident of history. Sometimes these distinct economic units are formed by parts of states, such as those in northern Italy, Wales, Catalonia, Alsace-Lorraine or Baden-Württemberg. At other times they may be formed by economic patterns that overlap existing national boundaries, such as those between San Diego and Tijuana, Hong Kong and southern China, or the "growth triangle" of Singapore and its neighboring Indonesian islands. In today's borderless world these are natural economic zones and what matters is that each possesses, in one or another combination, the key ingredients for successful participation in the global economy.

Look, for example, at what is happening in Southeast Asia. The Hong Kong economy has gradually extended its influence throughout the Pearl River Delta. The radiating effect of these linkages has made Hong Kong, where GNP per capita is $12,000, the driving force of economic life in Shenzhen, boosting the per capita GNP of that city's residents to $5,695, as compared to $317 for China as a whole. These links extend to Zhuhai, Amoy and Guangzhou as well. By the year 2000 this cross-border region state will have raised the living standard of more than 11 million people over the $5,000 level. Meanwhile, Guangdong province, with a population of more than 65 million and its capital at Hong Kong, will emerge as a newly industrialized economy in its own right, even though China's per capita GNP may still hover at about $1,000. Unlike in Eastern Europe, where nations try to convert entire socialist economies over to the market, the Asian model is first to convert limited economic zones—the region states—into free enterprise havens. So far the results have been reassuring.

These developments and others like them are coming just in time for Asia. As Europe perfects its single market and as the United States, Canada and Mexico begin to explore the benefits of the North American Free Trade Agreement (NAFTA), the combined economies of Asia and Japan lag behind those of the other parts of the globe's economic triad by about $2 trillion—roughly the aggregate size of some 20 additional region states. In other words, for Asia to keep pace existing regions must continue to grow at current rates throughout the next decade, giving birth to 20 additional Singapores.

Many of these new region states are already beginning to emerge. China has expanded to 14 other areas—many of them inland—the special economic zones that have worked so well for Shenzhen and Shanghai. One such project at Yunnan will become a cross-border economic zone encompassing parts of Laos and Vietnam. In Vietnam itself Ho Chi Minh City (Saigon) has launched a similar "sepzone" to attract foreign capital. Inspired in part by Singapore's "growth triangle," the governments of Indonesia, Malaysia and Thailand in 1992 unveiled a larger triangle across the Strait of Malacca to link Medan, Penang and Phuket. These developments are not, of course, limited to the developing economies in Asia. In economic terms the United States has never been a single nation. It is a collection of region states: northern and southern California, the "power corridor" along the East Coast between Boston and Washington, the Northeast, the Midwest, the Sun Belt, and so on.

Reprinted by permission of *Foreign Affairs*, Vol. 72, No. 2, Spring 1993, pp. 78-87. © 1993 by the Council on Foreign Relations, Inc.

What Makes a Region State

THE PRIMARY linkages of region states tend to be with the global economy and not with their host nations. Region states make such effective points of entry into the global economy because the very characteristics that define them are shaped by the demands of that economy. Region states tend to have between five million and 20 million people. The range is broad, but the extremes are clear: not half a million, not 50 or 100 million. A region state must be small enough for its citizens to share certain economic and consumer interests but of adequate size to justify the infrastructure—communication and transportation links and quality professional services—necessary to participate economically on a global scale.

It must, for example, have at least one international airport and, more than likely, one good harbor with international-class freight-handling facilities. A region state must also be large enough to provide an attractive market for the brand development of leading consumer products. In other words, region states are not defined by their economies of scale in production (which, after all, can be leveraged from a base of any size through exports to the rest of the world) but rather by their having reached efficient economies of scale in their consumption, infrastructure and professional services.

For example, as the reach of television networks expands, advertising becomes more efficient. Although trying to introduce a consumer brand throughout all of Japan or Indonesia may still prove prohibitively expensive, establishing it firmly in the Osaka or Jakarta region is far more affordable—and far more likely to generate handsome returns. Much the same is true with sales and service networks, customer satisfaction programs, market surveys and management information systems: efficient scale is at the regional, not national, level. This fact matters because, on balance, modern marketing techniques and technologies shape the economies of region states.

Where true economies of service exist, religious, ethnic and racial distinctions are not important—or, at least, only as important as human nature requires. Singapore is 70 percent ethnic Chinese, but its 30 percent minority is not much of a problem because commercial prosperity creates sufficient affluence for all. Nor are ethnic differences a source of concern for potential investors looking for consumers.

Indonesia—an archipelago with 500 or so different tribal groups, 18,000 islands and 170 million people—would logically seem to defy effective organization within a single mode of political government. Yet Jakarta has traditionally attempted to impose just such a central control by applying fictional averages to the entire nation. They do not work. If, however, economies of service allowed two or three Singapore-sized region states to be created within Indonesia, they could be managed. And they would ameliorate, rather than exacerbate, the country's internal social divisions. This holds as well for India and Brazil.

The New Multinational Corporation

WHEN VIEWING the globe through the lens of the region state, senior corporate managers think differently about the geographical expansion of their businesses. In the past the primary aspiration of multinational corporations was to create, in effect, clones of the parent organization in each of the dozens of countries in which they operated. The goal of this system was to stick yet another pin in the global map to mark an increasing number of subsidiaries around the world.

More recently, however, when Nestlé and Procter & Gamble wanted to expand their business in Japan from an already strong position, they did not view the effort as just another pin-sticking exercise. Nor did they treat the country as a single coherent market to be gained at once, or try as most Western companies do to establish a foothold first in the Tokyo area, Japan's most tumultuous and overcrowded market. Instead, they wisely focused on the Kansai region around Osaka and Kobe, whose 22 million residents are nearly as affluent as those in Tokyo but where competition is far less intense. Once they had on-the-ground experience on how best to reach the Japanese consumer, they branched out into other regions of the country.

Much of the difficulty Western companies face in trying to enter Japan stems directly from trying to shoulder their way in through Tokyo. This instinct often proves difficult and costly. Even if it works, it may also prove a trap; it is hard to "see" Japan once one is bottled up in the particular dynamics of the Tokyo marketplace. Moreover, entering the country through a different regional doorway has great economic appeal. Measured by aggregate GNP the Kansai region is the seventh-largest economy in the world, just behind the United Kingdom.

Given the variations among local markets and the value of learning through real-world experimentation, an incremental region-based approach to market entry makes excellent sense. And not just in Japan. Building an effective presence across a landmass the size of China is of course a daunting prospect. Serving the people in and around Nagoya City, however, is not.

If one wants a presence in Thailand, why start by building a network over the entire extended landmass? Instead focus, at least initially, on the region around Bangkok, which represents the lion's share of the total potential market. The same strategy applies to the United States. To introduce a new top-of-the-line car into the U.S. market, why replicate up front an exhaustive coast-to-coast dealership network? Of the country's 3,000 statistical metropolitan areas, 80 percent of luxury car buyers can be reached by establishing a presence in only 125 of these.

The Challenges for Government

TRADITIONAL ISSUES of foreign policy, security and defense remain the province of nation states. So, too, are macroeconomic and monetary policies—the taxation and public investment needed to provide the necessary infrastructure and incentives for region-based activities. The government will also remain responsible for the broad requirements of educating and training citizens so that they can participate fully in the global economy.

Governments are likely to resist giving up the power to intervene in the economic realm or to relinquish their impulses for protectionism. The illusion of control is soothing. Yet hard evidence proves the contrary. No manipulation of exchange rates by central bankers or political appointees has ever "corrected" the trade imbalances between the United States and Japan. Nor has any trade talk between the two governments. Whatever cosmetic actions these negotiations may have prompted, they rescued no industry and revived no economic sector. Textiles, semiconductors, autos, consumer electronics—the competitive situation in these industries did not develop according to the whims of policymakers but only in response to the deeper logic of the competitive marketplace. If U.S. market share has dwindled, it is not because government policy failed but because individual consumers decided to buy elsewhere. If U.S. capacity has migrated to Mexico or Asia, it is only because individual managers made decisions about cost and efficiency.

The implications of region states are not welcome news to established seats of political power, be they politicians or lobbyists. Nation states by definition require a domestic political focus, while region states are ensconced in the global economy. Region states that sit within the frontiers of a particular nation share its political goals and aspirations. However, region states welcome foreign investment and ownership—whatever allows them to employ people productively or to improve the quality of life. They want their people to have access to the best and cheapest products. And they want whatever surplus accrues from these activities to ratchet up the local quality of life still further and not to support distant regions or to prop up distressed industries elsewhere in the name of national interest or sovereignty.

When a region prospers, that prosperity spills over into the adjacent regions within the same political confederation. Industry in the area immediately in and around Bangkok has prompted investors to explore options elsewhere in Thailand. Much the same is true of Kuala Lumpur in Malaysia, Jakarta in Indonesia, or Singapore, which is rapidly becoming the unofficial capital of the Association of Southeast Asian Nations. São Paulo, too, could well emerge as a genuine region state, someday entering the ranks of the Organization of Economic Cooperation and Development. Yet if Brazil's central government does not allow the São Paulo region state finally to enter the global economy, the country as a whole may soon fall off the roster of the newly industrialized economies.

Unlike those at the political center, the leaders of region states—interested chief executive officers, heads of local unions, politicians at city and state levels—often welcome and encourage foreign capital investment. They do not go abroad to attract new plants and factories only to appear back home on television vowing to protect local companies at any cost. These leaders tend to possess an international outlook that can help defuse many of the usual kinds of social tensions arising over issues of "foreign" versus "domestic" inputs to production.

In the United States, for example, the Japanese have already established about 120 "transplant" auto factories throughout the Mississippi Valley. More are on the way. As their share of the U.S. auto industry's production grows, people in that region who look to these plants for their livelihoods and for the tax revenues needed to support local communities will stop caring whether the plants belong to U.S.- or Japanese-based companies. All they will care about are the regional economic benefits of having them there. In effect, as members of the Mississippi Valley region state, they will have leveraged the contribution of these plants to help their region become an active participant in the global economy.

Region states need not be the enemies of central governments. Handled gently, region states can provide the opportunity for eventual prosperity for all areas within a nation's traditional political control. When political and industrial leaders accept and act on these realities, they help build prosperity. When they do not—falling back under the spell of the nationalist economic illusion—they may actually destroy it.

Consider the fate of Silicon Valley, that great early engine of much of America's microelectronics industry. In the beginning it was an extremely open and entrepreneurial environment. Of late, however, it has become notably protectionist—creating industry associations, establishing a polished lobbying presence in Washington and turning to "competitiveness" studies as a way to get more federal funding for research and development. It has also begun to discourage, and even to bar, foreign investment, let alone foreign takeovers. The result is that Boise and Denver now prosper in electronics; Japan is developing a Silicon Island on Kyushu; Taiwan is trying to create a Silicon Island of its own; and Korea is nurturing a Silicon Peninsula. This is the worst of all possible worlds: no new money in California and a host of newly energized and well-funded competitors.

Elsewhere in California, not far from Silicon Valley, the story is quite different. When Hollywood recognized that it faced a severe capital shortage, it did not throw up protectionist barriers against foreign money. Instead, it invited Rupert Murdoch into 20th Century Fox, C. Itoh and Toshiba into Time-Warner, Sony into Columbia, and Matsushita into MCA. The result: a $10 billion infusion of new capital and, equally important, $10 billion less for Japan or anyone else to set up a new Hollywood of their own.

Political leaders, however reluctantly, must adjust to the reality of economic regional

> "Political leaders, however reluctantly, must adjust to the reality of economic regional entities if they are to nurture real economic flows."

entities if they are to nurture real economic flows. Resistant governments will be left to reign over traditional political territories as all meaningful participation in the global economy migrates beyond their well-preserved frontiers.

Canada, as an example, is wrongly focusing on Quebec and national language tensions as its core economic and even political issue. It does so to the point of still wrestling with the teaching of French and English in British Columbia, when that province's economic future is tied to Asia. Furthermore, as NAFTA takes shape the "vertical" relationships between Canadian and U.S. regions—Vancouver and Seattle (the Pacific Northwest region state); Toronto, Detroit and Cleveland (the Great Lakes region state)—will become increasingly important. How Canadian leaders deal with these new entities will be critical to the continuance of Canada as a political nation.

In developing economies, history suggests that when GNP per capita reaches about $5,000, discretionary income crosses an invisible threshold. Above that level people begin wondering whether they have reasonable access to the best and cheapest available products and whether they have an adequate quality of life. More troubling for those in political control, citizens also begin to consider whether their government is doing as well by them as it might.

Such a performance review is likely to be unpleasant. When governments control information—and in large measure because they do—it is all too easy for them to believe that they "own" their people. Governments begin restricting access to certain kinds of goods or services or pricing them far higher than pure economic logic would dictate. If market-driven levels of consumption conflict with a government's pet policy or general desire for control, the obvious response is to restrict consumption. So what if the people would choose otherwise if given the opportunity? Not only does the government withhold that opportunity but it also does not even let the people know that it is being withheld.

Regimes that exercise strong central control either fall on hard times or begin to decompose. In a borderless world the deck is stacked against them. The irony, of course, is that in the name of safeguarding the integrity and identity of the center, they often prove unwilling or unable to give up the illusion of power in order to seek a better quality of life for their people. There is at the center an understandable fear of letting go and losing control. As a result, the center often ends up protecting weak and unproductive industries and then passing along the high costs to its people—precisely the opposite of what a government should do.

The Goal is to Raise Living Standards

THE CLINTON administration faces a stark choice as it organizes itself to address the country's economic issues. It can develop policy within the framework of the badly dated assumption that success in the global economy means pitting one nation's industries against another's. Or it can define policy with the awareness that the economic dynamics of a borderless world do not flow from such contrived head-to-head confrontations, but rather from the participation of specific regions in a global nexus of information, skill, trade and investment.

If the goal is to raise living standards by promoting regional participation in the borderless economy, then the less Washington constrains these regions, the better off they will be. By contrast, the more Washington intervenes, the more citizens will pay for automobiles, steel, semiconductors, white wine, textiles or consumer electronics—all in the name of "protecting" America. Aggregating economic policy at the national level—or worse, at the continent-wide level as in Europe—inevitably results in special interest groups and vote-conscious governments putting their own interests first.

The less Washington interacts with specific regions, however, the less it perceives itself as "representing" them. It does not feel right. When learning to ski, one of the toughest and most counterintuitive principles to accept is that one gains better control by leaning down toward the valley, not back against the hill. Letting go is difficult. For governments region-based participation in the borderless economy is fine,

except where it threatens current jobs, industries or interests. In Japan, a nation with plenty of farmers, food is far more expensive than in Hong Kong or Singapore, where there are no farmers. That is because Hong Kong and Singapore are open to what Australia and China can produce far more cheaply than they could themselves. They have opened themselves to the global economy, thrown their weight forward, as it were, and their people have reaped the benefits.

For the Clinton administration, the irony is that Washington today finds itself in the same relation to those region states that lie entirely or partially within its borders as was London with its North American colonies centuries ago. Neither central power could genuinely understand the shape or magnitude of the new flows of information, people and economic activity in the regions nominally under its control. Nor could it understand how counterproductive it would be to try to arrest or distort these flows in the service of nation-defined interests. Now as then, only relaxed central control can allow the flexibility needed to maintain the links to regions gripped by an inexorable drive for prosperity.

Africa's Geomosaic Under Stress

H. J. de Blij

University of Miami, Department of Geography, Coral Gables, Florida, and Georgetown University, School of Foreign Service, Washington, DC.

In 35 years, the colonial map of Subsaharan Africa has been transformed into a patchwork of four dozen independent states, many subject to severe devolutionary stresses. Domino effects hastened the decolonization process; a short-lived but significant buffer zone slowed it in southern Africa. During the period of Africa's transition, polarization in South Africa (socially as well as spatially) intensified. Core-periphery contrasts in the republic deepened; whereas integration and modernization increasingly mark the South Africa's core areas, the prospect for the periphery is bleak and instability may develop. Insurgencies, refugee flows, and environmental stresses buffet a politico-geographical framework that may appear stable on maps, but which may collapse from within. **Key words: domino theory, buffer zone, insurgent state, devolution, boundaries, core-periphery, centripetal-centrifugal forces, front line states.**

In less than two generations, the political geography of Africa has been transformed. In the late 1940s, European powers, recovering from the effects of World War II, laid plans to reassert and consolidate their control over African colonial empires. In the early 1990s, Africa's last colony, South West Africa, became the independent state of Namibia, and South Africa faced the realities of the Africa of the future. Four decades of change saw what British Prime Minister Harold Macmillan called Africa's "Wind of Change" sweep southward from West Africa and the Sudan toward the Cape of Good Hope.

During the decade from the mid-1950s to the mid-1960s, developments in Africa rivaled cold-war competition for world attention. African leaders in the struggle for independence, such as Kwame Nkrumah, Patrice Lumumba, and Jomo Kenyatta, rose to international prominence in a manner not unlike Lech Walesa and Vaclav Havel today. President Kennedy appointed a secretary of state for African affairs, and his Peace Corps initiative, while a global venture, was directed principally toward Africa. Many academics were euphoric. I remember hearing a leading political scientist, addressing a convention of the APSA, describe political prospects for Ghana and Nigeria as "guaranteed" by the tested models on which their new constitutions were based. Ghana would become the embodiment of the Westminster model for Africa, and Nigeria would lead the continent with a federal system that was to be the vanguard for other states burdened with regional centrifugalism (Rothchild 1960). Shortly thereafter, I heard a future East African Union depicted as having "limitless opportunities for growth and development (once freed from the colonial yoke)...on a par with Western Europe." African leaders visited the United States and lectured at universities; African students in unprecedented numbers enrolled in American colleges. African studies centers, programs, and institutes arose at dozens of U.S. institutions. Funding for field research in Africa expanded, and scholarly journals in their contents reflected the new era.

In Africa, elation at the demise of colonialism and optimism about the future, fueled in substantial part by the sometimes careless predictions of non-African scholars, led to a revolution of rising expectations. The arts, especially literature, blossomed as perhaps never before. An outpouring of prose, poetry, and plays by African authors writing in English and French conveyed African feelings to European and American audiences. Leopold Senghor, a respected poet and author, became the first president of newly independent Senegal. The knowledge, judgment, and wisdom contained in works from this period seemed to confirm the favorable prospects and bright potentials for Africa's future (Nyerere 1961, 1963; Nkrumah 1957, Houphouet-Boigny 1957, Rooney 1988).

But even as the tide of African nationalism and independence swept eastward and southward from its West African sources, the salient problems that were to dash the dream of a peaceful, stable, prosperous Africa became manifest. Ghana achieved sovereignty in 1957 with a strong agricultural and industrial base, good educational and medical facilities, an adequate transport system, the highest per capita income of any tropical African country, substantial foreign currency reserves derived mainly from cocoa exports, a comparatively efficient civil service, and a dynamic political leadership. President Nkrumah spoke and wrote of pan-African unity and cooperation based on common interests and aspirations (Nkrumah 1963). Less than a decade later, the Ghanaian army had staged a coup, President Nkrumah was exiled, graft and corruption were rife, foreign reserves were depleted, the infrastructure was deteriorating, and living standards were in decline. Only part of this was due to the drop in cocoa prices

From *Journal of Geography*, Vol. 90, No. 1, January/February 1991, pp. 2-9. Reproduced with permission of the National Council for Geographic Education.

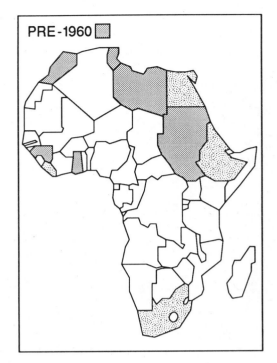

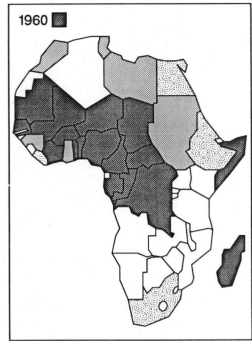

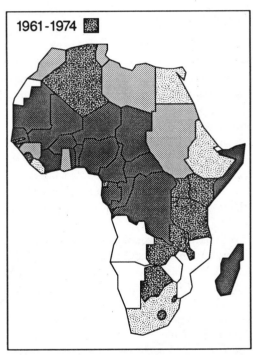

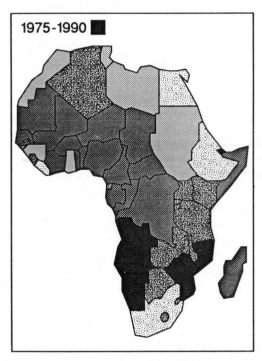

Figure 1. The progression of African independence in four stages. Before 1960, only Egypt, Ethiopia, Liberia, and South Africa were independent states not subject to decolonization.

on world markets. Economic mismanagement, political misjudgments, grandiose engineering and building projects, internal corruption worsened by dealings with corrupt multinational corporations, and inexperience all contributed to the collapse of black Africa's first great hope (Hadjor 1988).

Nigeria, Africa's most populous country and the regional cornerstone of West Africa, in 1960 became independent as a three-region federal state: North, West (really the southwest), and East (southeast). The federal parliament was regionally apportioned according to population, and the future depended on a delicate demographic balance (Azikiwe 1961). The first post-independence census, in 1962, showed the Northern Region to outnumber the other two regions combined, and a bitter controversy arose. As the 1965 elections approached, Nigeria plunged into anarchy that led to a military coup.

Following the military intervention, a wave of genocide occurred against Easterns living and working in the North. Eastern Nigeria responded by declaring itself independent as the state of Biafra. More than three years of devastating civil war ensued, setting Nigeria back severely despite the coinci-

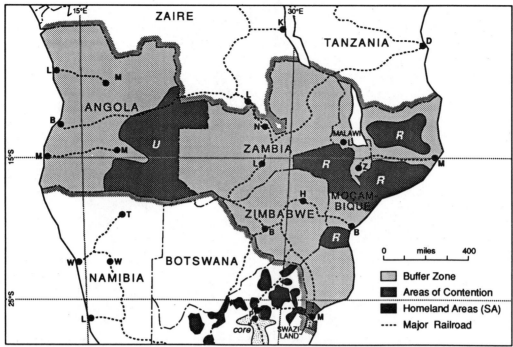

Figure 2. The buffer zone in its earliest stage. The stage of contention is indicated by the lettered areas: "U" for the UNITA movement in Angola, "R" for the RENAMO movement in Moçambique.

dental expansion of its oil industry. Thus, even before decolonization had reached southern Africa, disaster lay in its wake.

Domino Theory

The spatial sequence of decolonization in Subsaharan Africa suggests that the notion of a domino effect has merit despite fulminations to the contrary (O'Sullivan 1986). Simply defined, the domino theory holds that destabilization, or other significant transferable change, in one country or territory can affect the stability or normalcy of an adjacent state, and that this process can, in turn, affect other contiguous entities. The domino effect was much debated during the Indochina War, and it would be difficult to sustain the position that the conflict in Vietnam did not contribute to the disruption of neighboring Laos or Cambodia or, to a lesser degree, Thailand. More recently, observers have commented on the apparent domino effect in the Eastern Europe of the 1980s; Poland is often identified as the first "domino," and Albania as the last.

In Africa, Ghana was first in the 1950s, followed by East Africa in the 1960s and southern Africa in the 1970s and 1980s. British progress in West Africa made the French position untenable, and De Gaulle seized the initiative and moved not only Algeria, but also French West and French Equatorial Africa quickly toward independence in a "community" association with France (Houphouet-Boigny 1957). So rapidly was this process set in motion that Francophone Africa became independent in the same year long-prepared Nigeria did. Only Sekou Tourés Guinea chose to go it alone, asserting sovereignty in 1958 (Toure 1959). Zaire became independent following a violent upheaval in 1960; adjacent, landlocked,

Belgian-administered Rwanda and Burundi were decolonized in 1962. In East Africa, where the British had faced the Mau Mau uprising in Kenya, it was Tanzania that led the way (1961), followed by Uganda (1962) and Kenya (1963). To the north, Somalia had become independent in 1960; to the south, Malawi and Zambia waited until 1964. Kenya's early independence, despite substantial opposition among its white minority, is especially relevant in the domino context. Nearly encircled by independent states and in control of Uganda's outlet to the sea, Kenya's sovereignty could not be deferred (Fig. 1).

The British awarded nominal independence to Botswana and Lesotho in 1966 and to Swaziland in 1968, but the key southern African territories—the Portuguese dependencies of Angola and Moçambique as well as still-British Southern Rhodesia—slowed the Wind of Change nearly to a halt. Indeed, change in Angola and Moçambique, in further affirmation of the domino effect, did not accelerate until insurgents could secure safe haven in adjacent Zaire and Tanzania, respectively. For several years, the Moçambique nationalist movement, FRELIMO, had headquarters in Dar es Salaam, Tanzania's largest city. One of Angola's three insurgent movements, FNLA, received protection as well as fiscal support from Zaire across the border. The presence of insurgents in newly independent Zaire and Tanzania had severe impact on these countries and their economies. In the case of Zaire, the conflict in Angola disrupted the outflow of minerals from the reserves in the southeastern interior, which during colonial times had moved across the Portuguese territory to the port of Lobito. Tanzania suffered not only from the effects of war in Moçambique, but also from the disastrous events that occurred in its neighbor to the northwest, Uganda.

Eventually Tanzanian armed forces entered Uganda in the aftermath of Amin's murderous regime, placing a heavy burden on an already impoverished economy.

Independence in Moçambique and Angola came in 1975, and again the domino effect came into play. Landlocked Rhodesia, where the white minority had declared unilateral independence (UDI), and which was under international economic sanctions, lost security along its eastern flank when the Portuguese withdrew from Moçambique. Immediately, its internal conflict widened into Moçambique, where insurgents now found safer haven and where Rhodesian forces engaged in hot pursuit. Britain's last domino fell in 1980; even South African support could not prolong UDI.

Angola's independence exposed Africa's last colonial domino, Namibia, to the full force of the Wind of Change. Despite continuing civil war in Angola, Namibia's rebel movement SWAPO (South West African People's Organization) used Angola's territory in its struggle against South Africa. Independence was achieved.

Buffer Zones and Front Line States

The slowdown of Africa's drive for independence could be prophesied in the early 1960s, based on the dimensions and influence of white minorities in southern Africa and upon the capacity of South Africa to affect regional power balances (De Blij 1962). From the beginning, it was clear that the ultimate objective of black Africans would be the termination of white minority rule in what was then still the Union of South Africa. In less than ten years following Ghana's independence, the Wind of Change had penetrated southern Africa. But Portugal did not match British or French flexibility; and Angola at the time of Zaire's independence had the second largest European minority in all of Subsaharan Africa. Moçambique, on the opposite coast, had a white minority more than three times as large as Kenya's. And landlocked Rhodesia, situated between Angola and Moçambique, was in position to slow the advance of African nationalism by thwarting British plans for decolonization. Together, these territories, astride the northern tier of southern Africa, created a buffer zone that for more than a decade separated South Africa from independent black Africa (De Blij 1973).

The southern African buffer zone had several consequences, some of which are outlasting its existence. First, by keeping the so-called Front Line States territorially removed from South Africa, the republic was able to implement its grand design of "Separate Development" with an impunity it would have been denied under other circumstances. Second, South Africa was given time to plan and organize support not only for the minority colonial regimes in the buffer zone, but also for movements that would later emerge as opponents of newly installed black African governments. Third, South Africa could prepare for the eventual breakdown of the buffer zone by exporting its Separate Development program to Namibia, by intimidating Botswana through cross-border raids (and by other means), and by strengthening its presence in the northern Transvaal. South Africa's geopolitical action(s) during this period leave no doubt that the republic was intent on substituting an internal buffer zone for the external one once the latter collapsed. Northern Namibia, Botswana, the northern Transvaal's Bantu Homelands, and the Kruger National Park all were part of this preparation for the time when the so-called Front Line States would abut South Africa itself rather than its neighbors to the north.

The appellation "Front Line States" originated in the southward march of African independence alluded to previously. While the external buffer zone remained intact, Tanzania and Zaire were Front Line States, both supporting insurgent movements in the Portuguese colonies; and Zambia was on the front line opposite white-minority-ruled Rhodesia. Malawi, although adjoining Moçambique and certainly affected by the strife in this Portuguese dependency, never played a comparable role ideologically, a posture dictated by its conservative ruler, Hastings Banda. But all of the Front Line States paid a price for their relative location. Revolutionaries, refugees, and retribution spilled across boundaries into these countries, themselves still searching for post-independence stability. A prominent Moçambique leader, Eduardo Mondelane, was killed in a bomb attack in Dar es Salaam, Tanzania's capital. Zaire's Kasai province was seriously destabilized by cross-border warfare. The then-outlawed (South African) African National Congress set up headquarters in Zambia, where South African paramilitary forces repeatedly attacked it.

When Portugal yielded, and Angola and Moçambique became independent states (1975), these flanks of the buffer zone now became Front Line States themselves—confronting not only South Africa, but, in the case of Moçambique, revolution-ravaged Rhodesia as well. Immediately, both countries were drawn into regional as well as superpower conflicts. Even as three factions vied for control over Angola, the government in Luanda proclaimed the country a Marxist state; a similar position was taken by the new government in Maputo (the renamed capital of Moçambique). In Angola, the movement called UNITA, ostensibly an anticommunist, "democratic" organization, received support not only from South Africa and from the United States but also from China— because the Chinese wished to counter Moscow's influence in the region. The Luanda government received proxy support from Cuba as tens of thousands of Cuban soldiers rallied to the Marxist side. By the mid-1980s, the UNITA movement had created an insurgent state in eastern and southern Angola (McColl 1969). In Moçambique, the RENAMO movement challenged the Maputo government, inflicting terror and dislocation on a rural population already ravaged by famine and disruption. RENAMO, supported by South Africa, never achieved the territorial success of UNITA; it controlled several sections of the country including parts of the hinterlands of both Maputo and Beira (Fig. 2). The conflict in Rhodesia also damaged Moçambique severely. The port of Beira, the landlocked neighbor's nearest outlet, was in disrepair. The Beira corridor to the interior lay devastated: bridges, roads, dams, railroads, and power lines were destroyed. The cost of front-line location was incalculably high.

When the struggle in Rhodesia ended and the independent state of Zimbabwe appeared on Africa's map (1980), the

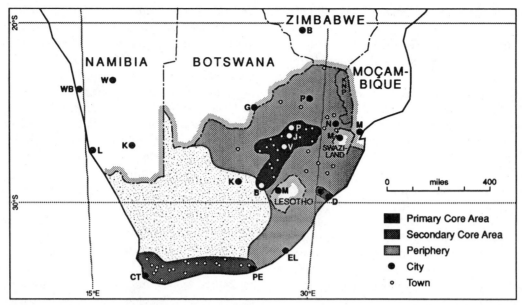

Figure 3. Core and periphery in South Africa. The lightly stippled area in the west is South Africa's emptiest quarter. Approximately 90 percent of the African population is located in the horseshoe-shaped zone enclosing the core area, and in the core area itself.

breakdown of the buffer zone was complete and South Africa confronted Front Line States across the entire length of its boundary except in the west, where Namibia still was under its control. At that time, the republic still pursued its campaign in Angola (in support of UNITA and in opposition to the Namibian independence movement, SWAPO) with enthusiasm. However, the cost of the war, both militarily in the field and politically at home, contributed to a search for settlement of the Angolan as well as the Namibian issue. The independence of Namibia in March 1990 with SWAPO members in the majority in the elected government, made Namibia a Front Line State and ended colonialism—not counting the domestic form within South Africa—on the continent.

South Africa's campaigns of support for UNITA and RENAMO in Angola and Moçambique, respectively, and of sabotage and disruption in Zimbabwe, were attempts to influence the geostrategic situation in the now-defunct buffer zone, intended to sustain conditions that had existed there previously. As long as the Front Line States were preoccupied with internal problems, they would be less able to cause or promote difficulties for South Africa. But in fact, the republic in the 1980s faced domestic challenges that could not have originated in the Front Line States. Urban insurrections, labor resistance, civil disobedience campaigns, and other forms of mass opposition to apartheid and Separate Development (and these systems' legal foundations) led to concessions that, in time, will be seen as the beginning of South Africa's political transformation.

Core, Periphery, and Front Line

The geopolitical situation in southern Africa in 1990 thus limited the roles of the final Front Line States in contributing to change within South Africa. Moçambique's primary objective was recovery. Zimbabwe, its economy the healthiest among regional states, was engaged in internal political reorganization. Botswana, many of its inhabitants dependent in one way or another on revenues earned in South Africa, was in no position to risk South African retribution. Namibia had become independent early in the year, its ties with (and dependence upon) the republic virtually unchanged as yet.

Should further change in South Africa take a revolutionary course, however, the relative location of the Front Line States would be likely to revive the role implied by their collective name. Cosmetic as well as consequential changes in the apartheid system have mainly affected the social geography of South Africa's heartland, its major cities, and their hinterlands. This has intensified core-periphery contrasts in a country where such regional disparities already are very strong (Fig. 3). The periphery was the scene of the establishment of the black homelands in the Separate Development scheme, and the horseshoe-shaped zone of ethnic "states" that partially envelops the republic's core area spatially defines much of the periphery. In much of this zone, South Africa resembles the poorer rural parts of the continent rather than the comparatively rich country that average-reporting statistics suggest it to be.

In Zimbabwe, the geographic situation was somewhat similar; there were no Bantustans in the South African sense, but core-periphery definitions and contrasts were sharp. It was in the periphery where the insurgencies of the 1970s arose and could not be suppressed. In South Africa, the periphery also will be vulnerable to insurgency—and the periphery marks the zone of contact between the republic and its front-line neighbors. Briefly during the 1980s, the vulnerability of this border area was underscored when land mines, planted by insurgents allegedly based in (or operating from) Zimbabwe, caused a number of casualties on Limpopo Valley farms. The South African government responded by compelling farmers to occupy their holdings for a substantial part of the year rather than leasing them to tenants. An increased military presence plus pressure on Zimbabwe suppressed the threat, but the potential for more serious problems along the republic's margins had been revealed.

The present Front Line States are strongly affected by their powerful neighbor—economically as well as politically. Their location relative to the geopolitical structure of South Africa will enmesh them in any revolutionary change the republic may undergo.

The Realm's Boundary Framework

Black Africa became independent partitioned by a boundary framework that was defined and delimited by colonial powers. Whatever the merits and demerits of this framework, it was decided by the member states of the Organization of African Unity that boundaries inherited at independence should be respected, and changed only by bilateral or multilateral consent (Brownlie 1979). Today, nearly four decades after the dawn of independence, the framework of African boundaries is substantially intact. Prescott (1987) attributes this to two causes: the colonial boundaries were defined with greater care and foresight than has generally been suggested, and African states have handled boundary conflicts with restraint.

In the early 1990s, the most serious boundary and territorial conflicts continue to afflict the Horn of Africa. The Ogaden question remains unresolved, and the Eritrean war appears on the verge of resurrecting the relic boundary between Eritrea and Ethiopia. Should the Eritreans' war for independence succeed, Ethiopia would become a landlocked country. This is undoubtedly the most consequential issue in black Africa's northeast today.

As a comparison between the maps of 1957 and 1990 shows, few boundary changes have been made during the period of decolonization. The boundary between former British and Italian Somaliland has been eliminated, and the Nigeria-Cameroon border was repositioned following a plebiscite in the borderland. Zanzibar was united with Tanganyika, eliminating a maritime boundary and resulting in the name Tanzania. In mid-1990, the issue of the South African exclave at Walvis Bay, on the coast of newly independent Namibia, remained unresolved. A United Nations resolution urged South Africa to yield the port and its immediate hinterland to Namibia, and there are indications that Pretoria will agree to a transfer. In that case, the exclave boundary will become a relic.

Given the intense fragmentation of Subsaharan Africa, the number of boundary disputes in the realm is small. Armed hostilities have occurred along the boundary between Mali and Burkina Faso, where a zone about ten miles wide and 100 miles long is contested. Nigeria and Chad have been in conflict over their boundary in Lake Chad, but this dispute has been overtaken by Chad's civil war involving Libya. In 1990, a boundary dispute erupted along the border between Senegal and Guinea-Bissau following the publication of reports that the area may contain fossil fuel reserves. And there is a

disagreement between Zaire and Zambia over their joint boundary where it reaches Lake Tanganyika. Finally, the position of the border between Malawi and Tanzania at the northern end of Lake Nyasa has given rise to intermittent negotiations.

Considering the origin of Africa's boundaries, the purposes they initially served, and the number of landlocked states they created, this is a low incidence of border conflict indeed. A larger number of functional disputes have arisen, such as that between Kenya and Tanzania, which closed this East African boundary for more than six crucial years (1977 to 1983); but none of these has led to positional conflict. These functional problems, moreover, have generally been of short duration.

There are, however, signs that Africa's boundary framework may yet change significantly—not by boundary repositioning, but as a result of devolutionary stresses within African states. The Ethiopian-Eritrean case is only one of a substantial number of internal problems of this kind. The possibility of a north-south fragmentation in the Sudan grows as Khartoum's capacity to control secessionism in the south diminishes. The war in Chad has underscored the fundamental contrasts between north and south in this landlocked state as well. Strong and intensifying centrifugal forces are affecting Nigeria, where Muslim fundamentalism is fueling Islamic-Christian discord. In 1987, the worst religious riots in memory ravaged the northern regions (Brooke 1987). Devolutionary forces also are strong, as noted earlier, in Angola and in Moçambique. And the possibility of Natal's secession from South Africa, now mentioned by Zulu conservatives, is nothing new: it was proposed as an option in the early 1950s, when the Nationalist government's apartheid policies were first imposed on that multiracial, mainly non-Afrikaner province.

Given Africa's strongly fragmented cultural mosaic, its ideological divisions, and the growing awareness of ethnic and historic identity among peoples ranging from the Zulu nation in Natal to the Muslim Hausa of northern Nigeria, the potential for the balkanization of postcolonial Africa exists—and not only in the larger, more populous countries. Rwanda and Brundi, the small but populous countries wedged between Tanzania and Zaire, have been the scene of such deathly tribal conflict between Tutsi and Hutu that the compartmentalization of these territories has been seriously advanced as a means to reduce the carnage. The African reaction to ethnic complexity of national populations has been the imposition of the one-party state; in early 1990, there were 18 one-party governments (and 17 military regimes). But the centrifugal forces inherent in Africa's anthropogeography will not be permanently accommodated by such centralization of authority. Many leaders in postcolonial Africa were much impressed by Soviet-style political control and economic policies (Ghana's Nkrumah was among the first), and the Soviet model is much in evidence in the African one-party state. The Soviet struggle to adjust to the realities of devolution will be equally relevant to African states trying to cope with similar stresses.

While boundary changes have been few, population shifts, mainly in the form of refugee movements, have been numerous and devastating. Both in terms of absolute numbers as well as per capita, Africa has the world's largest body of dislocated people. Again the northeast has been most severely afflicted. Ethiopia's long-term conflict with its ethnic Somali peoples has driven nearly a million Somali across the border into Somalia, where they live in squalid, inadequate camps. Another million Eritreans and Ethiopians, victims of the war for Eritrean independence, live, and starve, in the Sudan. The turmoil in Uganda produced 200,000 refugees, most of whom fled to the Sudan; many have returned to their home areas as Uganda stabilizes. Nearly a half million refugees are encamped in Zaire, where they seek safety from strife in Rwanda, Burundi, and Angola. By 1988, nearly a million inhabitants of Moçambique had fled into neighboring countries ill-equipped to accommodate them. Africa's refugee camps are places of misery, disease, malnutrition, and starvation. Because their occupants, in some areas, are perceived to support a political cause, governments have attacked and destroyed relief operations, condemning hundreds of thousands to death. Not even the most ruthless colonial rule (Portugal's in Moçambique probably was the most atrocious) caused such horrors.

And the numbers above do not begin to reflect the dimensions of dislocation in Africa. In Moçambique, for example, an estimated 1.2 million people have left their rural homes to seek protection from the civil war in towns and cities. Liberia, Chad, Uganda, Angola, and other strife-torn countries have seen their villages and towns swell from a rural exodus. The combination of deserted farmlands and overgrown urban centers spells economic disaster and political trouble. Thus, governments are impelled to take draconian action to protect the interests of their citizens. In 1983, the Nigerian government summarily expelled 2 million Ghanaian workers who had come to Nigeria from their dictatorial homeland and its collapsing economy, to find jobs. The hardship faced by Africa's dislocated people mocks the stability of the realm's international boundary framework. Whatever the OAU decided, changes are bound to come.

Geography, Demography, and Environment

The transformation of Africa's political geography has taken place against a background of population explosion and environmental deterioration. In 1957, the year Ghana became independent, Subsaharan Africa's population totaled just 200 million. In 1990, the realm counted 500 million inhabitants; the projected figure for 2000 is just under 700 million. No other world geographic realm even approached such a growth rate; Kenya during the 1970s and 1980s became the world's fastest-growing country at about four percent annually, a doubling time of approximately 17 years. The TFR (total fertility rate) in the early 1980s was an estimated eight children per woman in Kenya. In Kenya as in other African countries, hopes for economic progress and improvement were dashed by growing costs and burdens of larger numbers.

And yet, viewing the numbers in global perspective, Africa is not, in general, densely populated. India alone has a much larger population than Africa's nearly four dozen countries combined; India's area is one-seventh that of Subsaharan Africa. According to FAO statistics, Indians and Africans, in the early 1960s, consumed, on average, about the same number of calories (slightly over 2000) daily. By the late 1980s, the average intake in India had risen to 2150; in Africa it had declined to 1870. Dietary balances had improved in India, but deteriorated in Africa. The population explosion, plus the social dislocation discussed earlier, contributed to this. But more important was the failure to transfer the Indian "green revolution" to Africa. Western agricultural advisers' obsession with high yields and cash crops led to the introduction of farming systems that worked in India's fertile Punjab, but not in most of Africa's low-yield countryside. Irrigation schemes and fertilizer-dependent crop varieties could not be afforded—either by governments or by subsistence farmers. African farmers need more roads and market systems to move farm produce from the remote, sparsely settled, land-abundant areas to the food-deficient population clusters. Without an improving infrastructure to help traditional farmers, no massive dam-and-canal schemes could solve Africa's food problems. So the African farmer lost ground in every direction: the newly independent governments mostly had their eyes on other, more prestigious sectors of the economy, and Western-devised green-revolution techniques, often involving irrigation projects and labor-intensive cultivation systems, were inappropriate for the realm's needs. In the meantime, millions of Africans remain utterly dependent upon overseas food supplies.

Even as the political transformation proceeded and the population explosion gained momentum, Africa's regional environments also changed dramatically. There can be no doubt that human intervention (through the opening of marginal land and the herding of livestock into fragile ecosystems) played a significant role in this. But nature also dealt Africa a series of blows during the second half of the century. The Sahara's southward expansion created the Sahel crisis of the 1970s and its enormous cost in human and animal lives, but it was not all people's doing. The desert also spread northward, even where no human activity contributed to its advance. The evidence suggests that the Sahara during this period was embarked on one of its natural expansion phases. The Inter-Tropical Convergence (ITC), a bellwether for West African moisture conditions, made its northward move later and started its coastward retreat earlier, facilitating the Saharan advance in the interior and reducing the wet season in the coastal countries. The human (and livestock) presence exacerbated the environmental situation in the Sahel, but it was not the disaster's sole cause.

While attention was focused on the Sahel crisis, other areas of Africa also experienced droughts, including East and South Africa. Regional droughts in Africa have several consequences that last long after the dry spell has ended: people and livestock are induced to destroy a widening area of natural vegetation, and the loss of this vegetation appears to hinder the return of normal, average conditions. We have studied maps of climatic regions, rainfall distributions, and vegetative associations, perhaps taking insufficient account of the degree to which the mapped phenomena are interrelated. The evidence from the Amazon Basin suggests that, in areas where the forest has been cut down, the exposed surface reflects more sunlight back into space, creating layers of dry air close to the ground. Since as much as half of all rainfall comes from moisture evaporated from the land and plant surfaces, such a dry layer would absorb evaporated moisture without yielding beneficial "return" rains. This, in turn, inhibits the revival of vegetation. Break the ecological balance, the evidence suggests, and the maps may have to be redrawn. In Africa, the destruction of natural vegetation may worsen permanently an already severe water shortage. Rainforest and savanna destruction are widespread. The Sahel problem (which extends all the way across the north from Senegal to Ethiopia) is only part of a larger environmental threat.

Environmental problems can have politico-geographical consequences. While the Sahel drought affected primarily the states and peoples of the Sahara margin, coastal West African countries were involved as well. Bassett (1988) describes the impact of a group of Fulani herders who drove their cattle from the desiccating Sahel zone into greener Ivory Coast. The government of Ivory Coast welcomed these pastoralists, who would make beef supplies more dependable and cheaper than they had been previously. But the local farmers saw it rather differently: the Fulani's cattle trampled their crops of cotton, corn, rice, millet, and yams. They resented the inadequate protection from the capital, and a violent conflict erupted. When finally the government decided to allocate parcels of pastureland to the Fulani, these were inadequate and the Fulani left the country. In the meantime, much of northern Ivory Coast was in disarray, a victim of the Sahel crisis hundreds of miles away.

Directly and indirectly, environmental change has had political repercussions in other areas: northern Nigeria, northeastern Kenya, and highland Lesotho are examples. The long-term impact of environmental change on the African political framework has just begun to be felt.

Conclusion

The final chapter in the transformation of Africa's political geography is now being written. South Africans could see the dominoes fall and the buffer zone collapse, but until the mid-1980s the minority government pursued a course that was destined to make future accommodations more difficult. That future is now here, and the question is whether Africa will witness still another violent struggle for power (and not necessarily an exclusively black-white conflict) or if a negotiated transition can be achieved. Then, perhaps, South

Africa could become the regional force it should be, the provider of products and technologies, the example of a successful, multiracial African economy, the leader in finance and investment. Preoccupation with the South African situation has done much to screen the problems of the rest of Subsaharan Africa from world view and world opinion. Africa, the cradle of humankind, is now its sickbed, literally and figuratively. No geographic realm is so severely afflicted by disease (the total AIDS cases may number 3 million; some demographers are predicting an actual population decline in some areas as a result). No geographic realm has witnessed comparable economic and political failure; none is as hurt by falling commodity prices on world markets as is Africa.

The Wind of Change carried from Senegal to South Africa. Let us hope that a counterforce will soon emanate from the south, radiating the potential of a renewed republic, a prosperous Zimbabwe, and a stabilized Angola northward, eastward, and westward to redirect the destinies of a troubled realm.

References

Azikiwe, N. 1961. *Zik: a Selection from the Speeches of Nnamdi Azikiwe.* Cambridge: Cambridge University Press.

Bassett, T. J. 1988. The political ecology of peasant-herder conflicts in the Northern Ivory Coast. *Annals of the Association of American Geographers* 78:453-472.

Brooke, J. 1987. Moslem-Christian rioting tears Northern Nigeria. *New York Times*, 22 March, 25.

Brownlie, I. 1979. *African Boundaries: a Legal and Diplomatic Encyclopedia.* London: Hurst.

De Blij, H. J. 1962. *Africa South.* Evanston: Northwestern University Press.
_____. 1973. *Systematic Political Geography.* 2nd ed. New York: Wiley.

Hadjor, K. B. 1988. *Nkrumah and Ghana: the Dilemma of Postcolonial Power.* London: Kegan Paul International.

Houphouet-Boigny, F. 1957. Black Africa and the French Union. *Foreign Affairs* 35:579.

McColl, R. W. 1969. The insurgent state: territorial bases of revolution. *Annals of the Association of American Geographers* 59:613.

Nkrumah, K. 1957. *Ghana.* New York: Thomas Nelson.
_____. 1963. *Africa Must Unite.* London: Heinemann.

Nyerere, J. K. 1961. The African and democracy. In *Africa Speaks*, eds. J. Duffy and R. A. Manners. Princeton: Van Nostrand.
_____. 1963. A United States of Africa. *Journal of Modern African Studies* I:9

O'Sullivan, P. 1986. *Geopolitics.* New York: St. Martin's.

Prescott, J.R.V. *Political Frontiers and Boundaries.* London: Allen and Unwin.

Rooney, D. 1988. *Kwame Nkrumah: the Political Kingdom of the Third World.* London: Taurus.

Rothchild, D. S. 1960. On the application of the Westminster model to Ghana. *Centennial Review* 4:465.

Touré, S. 1959. *Toward Full Re-Africanization.* Paris: Présence Africaine.

Middle East Geopolitical Transformation: The Disappearance of a Shatterbelt

by Saul B. Cohen

DEPARTMENT OF GEOLOGY AND GEOGRAPHY, HUNTER COLLEGE, NEW YORK, NEW YORK.

The cold war's end has brought about major global geopolitical restructuring. It also is cause for regional geopolitical reordering. The gulf war and its aftermath are but one expression of Middle Eastern disequilibrium. This shatterbelt region has been caught up in both intraregional tensions and the post-World War II history of competition between the Maritime and Eurasian Continental realms. Now the Middle East is becoming strategically reoriented to the West. While powerful centrifugal forces still prevail, the reduction of external competitive pressures permits centripetal forces to become more salient. In addition to Arabism and Islam, these include migration and capital flows and water and oil transportation lines. A new balance among the Middle East's six regional powers can be fostered but not dictated by the outside world. Equilibrium can best be promoted, not by a Pax Americana, but by the United States and the European Community acting as two competitive but allied stabilizers. **Key words:** *capital flows, geopolitical equilibrium, Gulf War, migration, oil pipelines, shatterbelt.*

Introduction

The Persian/Arab Gulf War was heralded by many as a "defining" war, whose aftermath would be a new Middle Eastern order. The allied victory over Iraq was overwhelming and unprecedented. It was the first genuinely high-technology conflict that the world has known. Its devastating consequences will long be felt by Iraq, and the lessons learned will surely become a textbook for strategy and weaponry in future regional wars.

While the immediate allied objectives, freeing Kuwait and eliminating Iraq as a military threat to its neighbors, were speedily realized, the "new order" has yet to be achieved. We are learning, once again, that military victories do not translate into immediate and enduring political success. A decisive war is a one-time event. A decisive peace, however, can come only from a succession of reinforcing events.

Thus, the early aftermath of the 44-day war was the failed attempt by the Shia and Kurds of Iraq to overthrow Saddam Hussein. This was followed by the savage repression that triggered a massive wave of refugees fleeing the Iraqi troops—up to 1.75 million Kurds into the Turkish and Iranian borderlands and 100,000 Shia into Iran and the allied occupied South.

The allied effort to provide relief for the Kurds in a northern "safe-haven" zone, and the positioning of a United Nations' observer force to replace U.S. soldiers in the south, is part of the longer-term process of restoring stability to Iraq. It is not clear whether Saddam Hussein will retain his dictatorial powers or will be toppled, and whether a federated form of government that offers greater autonomy to the northern and southern provinces will emerge. As forceful as was United States leadership in "demonizing" Saddam during the war, it has been far more ambiguous in pursuing measures that would now drive him from office. What is clear, however, is that Iraq is in flux.

The same uncertainty holds true for an environmentally devastated Kuwait. Increasing pressures on the ruling Sabah family to open the country to political and economic reform could well lead to democratization of the kingdom. Moreover, the war may compel the Kuwaitis to forge a new kind of national identity in which they take responsibility for operating their own state, rather than depending on outside workers, as in the past.

Other questions about the post-war era have to do with whether the American initiative to bring the Arabs and Israel to the negotiating table will bear fruit and whether the Syrian-backed Lebanese government will be able to reunify the country. Finally, there is the danger that a new United States-sparked arms race in the Middle East will snatch defeat from the jaws of victory, making a mockery of the calls for a new Middle Eastern order.

If, then, the immediate aftermath of the Persian/Arab Gulf War is doubt and uncertainty, need one be pessimistic about long-term prospects for stability? A geopolitical assessment of the region offers an encouraging note. The Middle East is in disequilibrium because it has experienced not just the Gulf War, but more profound change—the impact of the end of the cold war. While the region remains a shatterbelt, it is now tilting geostrategically towards the Maritime Realm.

Still highly divided and immature, the regional system can gain in cohesiveness as forces promoting spatial interaction acquire greater salience. A coordinated effort between the United States and the European Community, as well as widespread fears throughout the Middle East, can create the conditions for long-term equilibrium. The reference is to equilibrium not order, for equilibrium is dynamic and order is

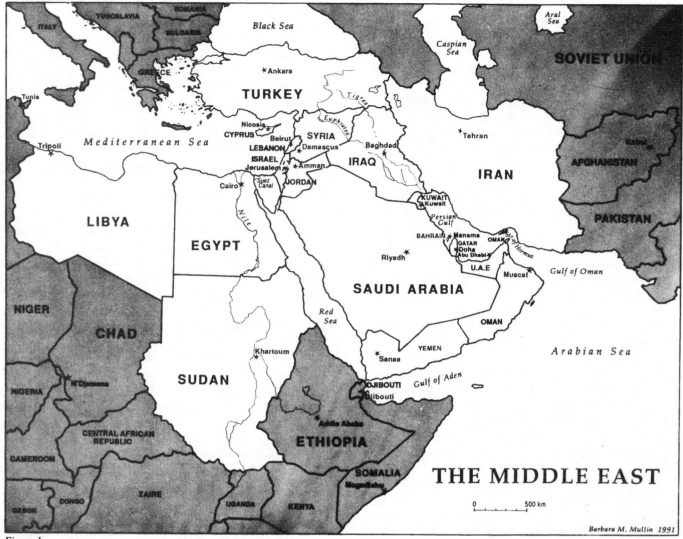

Figure 1.

static. The Gulf War has accelerated the process of geopolitical change within the region, but it is not the basic cause of such change. The war occurred because the Middle East had already begun to experience a fundamental geopolitical restructuring.

It is the end of the cold war that is responsible for the changing geopolitical status of the Middle East. Indeed, if it were not for the end of the cold war, we would not have had a "hot war." The diminished role of the USSR created a power vacuum. Disappearance of the restraining hand of the former Soviet Union gave the Iraqi dictator the opportunity to seize Kuwait and to seek domination over his neighbors to the south. Soviet disengagement also enabled the United States to move quickly on both the political and military fronts to respond to the Iraqi aggression. The Soviet Union continues to play a role in Middle East diplomacy, but lacks the military, economic, and ideological capacity to go beyond its efforts to mediate the Gulf War, which while they failed were not wholly without influence.

Geopolitical Evolution of the Middle East Region

When we speak of the Middle East, we refer to the area that

extends from Turkey through Iran, and from Libya through the Sudan (Figure 1). This is a relatively new geographical region. There was no Middle East in its current outlines in either the folk or the scholarly regional consciousness of the 1920s and early 1930s. While the term was introduced by Alfred Mahan in 1902 and used by Sir Mark Sykes in British parliamentary speeches in 1916, it did not gain currency until the Second World War (Lewis 1966). Then, in 1941, the British created the Middle East Supply Centre, to embrace a single-purpose logistics and transportation region.

For most of its five millennia of recorded history, this juncture of three continents was not a distinct region. The Nile and Mesopotamia were worlds unto themselves, meeting each other in the Levant in times of war and peace. Overlapping and injecting themselves into these areas were powers from the adjoining highlands and deserts of Western and Central Asia, as well as from the interpenetrating Mediterranean and Arabian seas.

The Forging of a Region

Several modern-era forces led to the creation of the Middle East as we now know it. They include Great Britain's strategy for empire building, the discovery of oil, and the evolution of transportation and communications technology. They overlay the political-cultural frameworks of Arabism, Islam, and the Ottoman Empire.

By 1800, the British had become the dominant commercial force in the Persian/Arab Gulf, for the purpose of supporting and enhancing their interests in India. Then in 1839 Aden was acquired to reinforce the route to South Asia.

At the western end of the region, Britain began to build a power bastion in Egypt. British and Turkish troops forced Napoleon out of Egypt in 1801, thwarting French plans to use the country as a military base against India. British influence grew rapidly, culminating in the building of the Suez Canal (1869) and the military occupation of Egypt (1882).

In effect, Great Britain had brought together the Persian/Arab Gulf, the southern Arabian peninsula, and the Nile in a military, economic, and political framework. In addition, it extended its sphere of influence into southern Persia, countering the Russian pressures upon Persia, which Britain perceived as a threat to India.

The 1909 discovery of petroleum in southern Persia (Khuzistan), followed by oil finds in northern Iraq in 1927, triggered another region-creating force. This is expressed in today's system of oil-producing and transporting states that embraces every part of the Middle East. To transport Iraqi petroleum, the Iraq Petroleum Company built a pipeline in the mid-1930s from Kirkuk to the port of Haifa and a new refinery at Haifa Bay. This was the forerunner of a vast network of region-linking lines that has subsequently emerged.

A third early example of region formation was the railroad. The Germans acquired a concession for the Anatolian-Baghdad railway as early as 1902. The plan was stymied by British resistance. However, British engineers began to build an Iraqi system during the First World War. By the 1930s, Baghdad was connected to the Mediterranean via Mosul and the Taurus section of Turkey. At the western end, Egypt was connected by rail to Turkey via Palestine, Lebanon, and Syria when the Haifa-Beirut line was completed in 1943.

Britain's building of the port of Haifa (1929-33), and its deep dredging in 1930 of the Shatt al Arab, on which Basra is located, was further evidence of a colonial military and economic strategy to knit the region together. All of these efforts, when added to the power of the ideal, if not the reality of the pan-Arabism that was stirred by the Arab Revolt against the Ottomans (1916), were the underpinnings of the modern Middle East.

When the Middle East assumed definable regional form after the Second World War, what kind of region was it? It was clearly what we geographers describe as a shatterbelt. A shatterbelt is a region that is highly fragmented politically as a result of internal divisiveness and of the pressures of external great powers in pursuit of their own geostrategic interests (Cohen 1973).

Superpower competition made the Middle East a shatterbelt. The previous imperialistic domination by Britain and France, each of which had carved out its discrete sphere, provided a stabilizing influence. In contrast, the competition between the United States and the USSR for regional influence destabilized the scene. The external shatterbelt element was provided by the region's unique global location. The Middle East lies between the world's two geostrategic realms—the trade-dependent Maritime and the Eurasian Continental. This location and the new-found petroleum wealth made it a natural battleground between the two superpowers. Each lined up allies and satellites in the struggle for regional dominance. The fact that the superpowers dealt with highly unstable states whose governing regimes were frequently overthrown and switched alliances added to the general regional turmoil.

While superpower competition was fierce, it was constrained by the fear of war and its nuclear consequences. The United States and Soviet Union acted in concert to bring an end to hostilities in two major instances: in the 1956 Sinai War and in the October 1973 war between Israel and Egypt. In both cases, the superpowers resisted being drawn into situations which might escalate into nuclear war.

The cold war fanned existing hatreds among the region's different national, religious, and ethnic groups through the supply of arms, equipment, and resources by the Soviet Union and the United States, and by other western and eastern bloc nations seeking political and economic gain. This supported the ambitions and rivalries of Middle Eastern states bent upon achieving regional or local hegemony. The instability stemming from conflict and shifts in alliances kept the region in constant geopolitical upheaval.

Now, four decades after its emergence as a shatterbelt, the Middle East is on the verge of a major change in geostrategic affiliation. It is tilting towards the Maritime Realm. The collapse of communism in Eastern Europe and the former USSR and the erosion of Soviet central authority have eliminated the new Commonwealth of Independent States as a major military and economic player within the region, at least for the foreseeable future, and relegated it to the important but not decisive role of mediator. Heartlandic Eurasia will remain sensitive to events along its 1,400-mile land and sea borders with Turkey and Iran, but its era of broad and enveloping regional penetration through naval and air bases in the Red and Arabian seas and the Eastern Mediterranean seems over.

In the geopolitical reordering now underway, it is Maritime Europe that is likely to become America's equal. While the United States still has a decisive military voice in the Western alliance, economic factors are changing the world power balance. Europe has an even greater strategic, economic, and historic stake in the Middle East than does the United States. It is dependent on the region for oil, is a major importer of Middle Eastern products and capital, and exports goods, capital, and other financial services to the countries of the region. The economic vitality and political structure of the European Community should lead to a more vigorous and independent pursuit of its Middle East interests in the post-Gulf War era. Evidence of this in the months following the war's end include Britain's initiative in formulating the Kurdish "safe haven" concept and Germany's lead in developing a Western presence in Iran to aid the refugees there.

If the United States and Maritime Europe can agree on certain vital issues like the Arab-Israeli conflict, the future of Iraq, and the acceptance of Iran as a full participant in shaping the region's future, the Middle East will probably shift to the Maritime orbit. If, however, there is serious disagreement between these two major power centers, then the region will remain a shatterbelt, particularly since this will provide Russia and/or Muslim Central Asian states with a new window of opportunity for intervention. Whatever the case, because the Maritime geostrategic realm is no longer dominated by a "Pax Americana," it is not possible for the United States to take unilateral action in reshaping the map of the Middle East.

With the end of the Gulf War, the military arsenals that were built up within the region, in part by exploiting the United States/Soviet rivalry, will seriously impede efforts to resolve the other conflicts that engulf the region. If the arms race is renewed, rather than dampened, destabilization will persist. However, if the outside powers draw up and fully comply with post-war agreements to restrict arms sales to all of the region's states, the capacities of hostile parties to use war rather than negotiations to address their differences will be considerably reduced.

Middle Eastern Regional Characteristics
Regional Immaturity
The internal divisions that contribute so greatly to Middle Eastern regional immaturity cannot be quickly or easily erased. While social and religious differences are the roots of much of the region's fragmentation, part of the problem is economic inequality. The uneven distribution of natural resources—land, fresh water, soils, and minerals; imbalances in population density, manpower pools, and educational/technological levels; and inequalities in health and living standards—are sources of continuing tensions within and among the nations of the Middle East.

Until now, the wealthy states of the Middle East have done little to allay poverty in other lands. Most of their gifts and loans have been devoted to supporting militaristic adventures (Iraq's invasion of Iran) and terrorism (the PLO's activities against Israel), or to humanistic relief efforts (maintenance of refugee camps). Very little has gone to genuine economic development projects. A positive result of the Gulf War is that the "have not" nations of the anti-Iraqi Arab coalition, Egypt and Syria, will be the immediate beneficiaries of some wealth-sharing, unless they squander it on arms purchases.

Parts-to-Whole Relationships
In spite of all the past divisions, and those likely to persist into the future, we can begin to anticipate the Middle East as a geopolitically unified, if not yet cohesive, unit. The parts that constitute the region cannot disentangle themselves, they cannot turn their backs upon one another. Thus, there cannot be a "Persian/Arab Gulf strategy," there can only be a "Middle Eastern strategy"—for the United States and the United Nations, for Iraq and Egypt, for Turkey and Iran, for Jordan and Israel.

Even if it suited some countries to try to opt out of the region, they could not. Turkey cannot arbitrarily declare itself European, nor can Israel and Egypt shift from being Middle Eastern countries to being Eastern Mediterranean by describing themselves as such in their tourist brochures. As long as Anatolia is home to half the Kurdish people, Turkey cannot detach itself from events in Iran and Iraq. As long as it has a northeast Mediterranean Sea coast and especially the Bay of Iskenderun, it cannot escape its geographical role as a land bridge for pipelines from Mesopotamia and ultimately Iran, or as a base for the U.S. warplanes that were used against Iraq. As long as it remains the source of the Tigris and Euphrates rivers, its water policies will be a major focus for interaction with Syria and Iraq.

Egypt was the lead Arab nation among those in the Middle East confronting Iraq. It will be in the forefront of the Arab states involved in enforcing a Gulf peace, because of its size, military power, traditions of regional leadership, and stake in regional stability.

Israel's present and future is inextricably entwined with, not simply the Palestinian Arabs, but also with the war and peace strategies of Egypt, Jordan, Lebanon, Syria, Iran, and Iraq. The Scuds that landed in Greater Tel Aviv, Haifa, and the south were launched from western Iraq, not from the sea to Israel's west.

Forces of Regional Interaction
There is no doubt that centrifugal forces have torn the region apart, especially with the advent of modern nationalism. The important question is: can centripetal forces develop enough momentum to override the divisive character of the region.

Of the various geopolitical forces that have the near-term capability for giving greater unity to what is becoming a holistic region are regional migrations, capital flows, and water and petroleum transitways. These are modern expressions of historical movement—unifying ties that are more tangible than the regional overlay of Arabism and Islam, which, because of internal tensions, cannot guarantee the region's integration.

Of course, pan-Arabism and Islam strongly influence the nature of Middle Eastern politics. However, they do not define the region geopolitically. During the Gulf War, rallies in the Arab states of northwest Africa may have expressed popular solidarity with Iraq, but the fate of the Maghreb is strategically and economically tied to Western Europe. In fact, France's greatest fears were that the anger of Algerian masses against French participation in the Gulf War would strengthen Muslim fundamentalism and increase political unrest to the point that there would be new large-scale Algerian migration to France. Immigration, not the rupture of Franco-Algerian relations, remains the concern, as is also the case for Spain. Similarly, Muslims in Pakistan may have vented their anger at the American military effort in the Persian/Arab Gulf, but Pakistan's geopolitical destiny is with South Asia.

Migration and Capital Flows
Two elements flow relatively freely throughout the region,

people and capital, even in times of war. Because of its locale as a global transitway, the Middle East has been a focus for migration throughout the ages. Today's important Middle Eastern migrations are essentially intra-regional, with the exception of Turks, who have sought temporary work in Western Europe, and Jews from Europe, Asian-North African lands, and the former Soviet Union, who have found haven in Israel.

While there is some daily transborder movement in the region, macro-intraregional migration is the more significant integrator. Egyptians have migrated to Iraq, Kuwait, Saudi Arabia, and Libya in the millions. It is estimated that Egyptian workers left 12 billion dollars behind as they fled Kuwait. Prior to the invasion of Kuwait, there were two million Egyptian workers in Iraq. More than half have left, and the exodus is likely to continue as the shattered Iraqi economy has less need for foreign labor. On the other hand, the number of Egyptians in Saudi Arabia has increased to over one million; over 700,000 Egyptians were issued visas from September 1990 to March 1991 to replace Yemenis and Palestinians. Hundreds of thousands of Palestinians had successfully integrated themselves into the Saudi Arabian and Gulf states' economies, parallel or subsequent to being absorbed into the mainstream of their first countries of refuge—Jordan, Lebanon, Syria, and Egypt. The number of Yemeni working in Saudi Arabia and the Gulf countries may have approximated four million in recent years, and one million Sudanese work outside their country's borders.

The Gulf crisis wreaked havoc with all migrants, forcing many to flee. When the crisis is over, many of the Middle Eastern foreign workers will doubtlessly return. However, Palestinians now hold far fewer jobs in Kuwait than before, as they have been permanently replaced by Asians or Egyptians, or social change within Kuwait brings more Kuwaitis into the job market. The same holds true for Palestinians in Saudi Arabia. The chronic labor shortages of the oil-rich states will require labor in the rebuilding of their economies, and widespread unemployment and underemployment in the sending countries will be a compelling "push" factor. But the welcome mat has been pulled from under the Yemenis and the Palestinians, at least until the outlines of a peace are clear.

Raw statistics of migrant numbers mask the significance of these individuals to the economies of their home countries. For Egypt, cash remittances either sent home to relatives or brought back as savings have amounted to four billion dollars per annum or up to eight percent of gross national product—Egypt's single largest source of revenue. Cash remittances have also been vital to the economies of Yemen and Jordan—in the case of Yemen, 40 percent, and Jordan, 34 percent of gross national product.

Water

A major element in both the economy and the politics of the region is water. There are several measures that could rationalize the supply of water for the whole region, but only if there is genuine peace.

For example, Turkey's freshwater withdrawals are under 10 percent of potentially available water. In order to increase its usable supply, Turkey has recently completed the Ataturk Dam on the Euphrates River. This is part of a 20-billion-dollar network of 22 dams along the Euphrates and Tigris rivers, which will provide irrigation and hydroelectric power to Turkey's dry southeast. These dams also represent a potential for exporting stored water by pipeline to the arid, dusty plains of northern Syria and Iraq, when political conditions warrant, and even to Jordan and Saudi Arabia.

Libya now withdraws far more water than is available to it from groundwater and rainwater—which means that it relies upon expensive distillation processes. Under peaceful political conditions, Egypt could pipe to Libya some of the several billion cubic meters of water that now escape into the Mediterranean, or, alternately export it to Gaza and Israel's Negev. The amount of the diversion would depend upon how much water is needed to maintain a strong flow in the Nile distributary to prevent siltation from occurring where the delta meets the Mediterranean.

Better use of irrigation waters in Egypt, such as through the drip technology that Israel has developed, could free Nile water for export to other countries. So would technologies for reducing evaporation loss from waters stored in Lake Nasser.

Israel is highly dependent on the headwaters of the Jordan River, but makes very limited use of the Yarmuk River, which forms the border between Syria and Jordan, and Israel and Jordan, before emptying into the Jordan River. At present, there are few reservoirs to capture and store most of the Yarmuk's winter floodwaters. A cost-effective measure would be to divert the Yarmuk to the Sea of Galilee in Israel, from which the waters could then be fed southward to Jordan's Ghor Canal, benefiting both countries. Jordan would remain the major user of the Yarmuk, and would have far greater quantities available to it than it now has (Ben-Shahar 1989). Another mutually beneficial water project could involve Lebanon and Israel in using the waters of the Lower Litani River that now flow out to the sea largely unused. Finally, as a result of a persistent drought in Israel, the shipment of fresh water in tankers from Turkey to Israel is being discussed by leaders of both countries.

These shared water projects are not possible in the present climate of tension and conflict. Only if the nations of the region place the mutual benefits of Middle Eastern economic integration above the potential gains of war can economic development receive the priority that it deserves.

Oil Pipelines

When we think of the Middle East, we think oil. The network of oil pipelines, existing and proposed, operational and non-operational, forges a critical set of regional links, as indicated on the map (Figure 2) (McCaslin 1989). These ties reflect such geographical variables as distance, terrain and optimal exit points, as well as international prices and political conditions. The direction taken by pipeline routes may also, in the future, reflect environmental hazard concerns, although to date these have not been significant factors.

The earliest transregional pipeline was, as previously

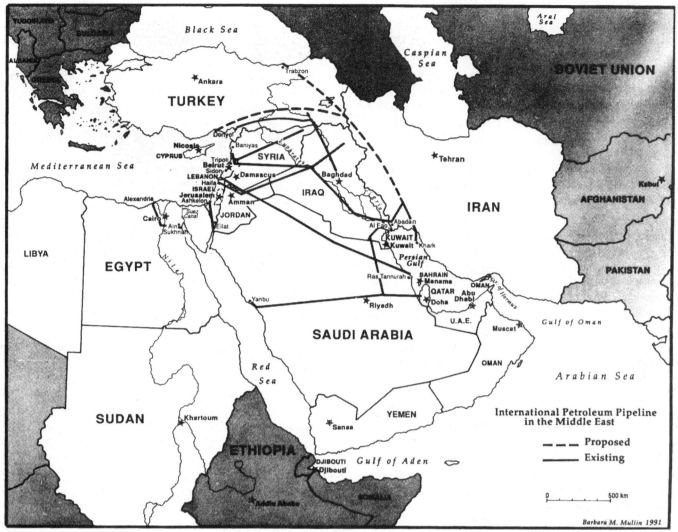

Figure 2.

noted, the Iraq Petroleum Company's (IPC) 12-inch line from Kirkuk to Haifa. It spawned a major refinery and petrochemical industry in Haifa Bay. The second major line, the 30/31" TAP-line, was completed in 1947. Originating in northeastern Saudi Arabia, at Ras Tannurah on the Gulf, the line was to have terminated in Haifa. However, the 1948-49 Arab-Israeli conflict caused TAP-line to be diverted to Sidon in Lebanon. Its capacity of 25 million tons has never exceeded 17 million, and it is now unused. In place of its prewar line to Haifa that was closed by that conflict, the IPC built two lines to Tripoli in Lebanon in 1952 (12 inch and 16 inch), to which was later added a 30-inch line to Banias in Syria.

Other key routes that have been developed in recent years are the Trans-Saudi 48-inch pipeline from Buqayq in northeastern Saudi Arabia to Yanbu on the Red Sea, built in 1981, and the line from Rumailah in southern Iraq, which was linked to the Yanbu carrier. The flow of oil from Iraq has been shut off by the Saudi government as a result of the current crisis.

When the Suez Canal was closed to Israel-bound shipping (from 1949 to 1975), a pipeline to carry Iranian oil was built from the Gulf of Aqaba port of Eilat to Israel's Mediterranean port of Ashdod. It is no longer operational. Also, when

Iraq's Persian/Arab Gulf tanker traffic was interdicted by its war with Iran and tensions with Syria, Iraq developed a pipeline from Kirkuk in its northern fields to Cizre in Turkey and then west via the piedmont region of southern Turkey to the Bay of Iskenderun on the Mediterranean. It is now closed to Iraq. A second line was planned northward from Kirkuk to link up with a Turkish pipeline from Turkey's Batman oil fields to Iskenderun. Finally, when the Suez Canal was blocked during the 1967-75 period as a result of the Arab-Israeli conflict, Egypt built a bypass line that took oil from tankers off-loading at the Gulf of Suez and carried it to Alexandria on the Mediterranean via Helwan.

Suggested routes include a line from the oil fields of southwestern Iran to Urmiyeh in northern Iran and west across the border to join up with the Turkish network, or north for a distance of 300 miles across Anatolia to Trabzon on the Black Sea. Other proposed projects, dependent upon resolution of the Arab-Israeli conflict, would reopen TAP-line, building an extension from Mafraq in Jordan to Haifa, and reopen the Eilat-Ashkelon line, extending it northward to Haifa. The latter line would take oil shipped by tankers from Yanbu.

What is unique to the Middle East is that its historic

political centers have persisted into contemporary times, resulting in a greater number of evenly balanced power centers than in other parts of the world. There are six such power centers—Egypt, Iraq, Iran, Israel, Syria, and Turkey. Even if Saddam had succeeded in keeping Kuwait, his mastery over the Arab world, let alone the Middle East, was hardly assured. Egypt is a powerful competitor to any Arab state that seeks to dominate the Arab world. Its peace agreement with Israel enabled it to gain support from the U.S. and freed it to devote political and military energies to other regional affairs, particularly in Arabia and the Levant. Now, of course, the aftermath of the Gulf war has increased Saudi dependence on Egypt. Moreover, Iraq's defeat has benefited Syria. Having backed the wrong side among the Arabs in the Iraq-Iran war, Syria has regained its regional stature by lining up on the allied side. The defeat of Iraq has strengthened Syria's influence in Lebanon and will enable it to regain a prominent role in Jordanian and Palestinian affairs. Even a defeated Iraq will remain a regional power.

Whether the Baath party maintains control (with or without Hussein), or whether some kind of new, federated state emerges from the ashes of war, Iraq will remain an important regional presence. It has oil, water, agricultural resources, and a population large enough to provide all of the needed human resources without the burdens of overpopulation.

In the six-power competition in the region, there will continue to be shifting alliances among Egypt, Syria, and Iraq, and between these and subordinate states. For the present, however, an Egyptian-Syrian pact could have the desirable effect of facilitating an agreement between Syria and Israel over the Golan, and between Israel, the Palestinian Arabs, and Jordan over independence for the West Bank and Gaza. The prospects for a West Bank confederation with Jordan are strong as an aftermath of the emergence of a West Bank entity.

The process of striking a balance among these six regional powers is complex. There are profound difficulties in finding a security arrangement that will stabilize the regional balance of power. While the United States and other outside agents can help in the process, particularly by pressing for elimination of nuclear, chemical, and biological weapons and reduction of armaments, they cannot guarantee against continued turbulence. The challenge will be to manage regional turbulence, since it cannot be eliminated.

A litmus test of genuine commitment to post-war reduction of arms transfers to the Middle East is whether the United States itself will refrain from a new round of highly sophisticated conventional arms transfers to its Arab allies and Israel in the name of making up for war-time material loss and wear. The region has accounted for 60 percent of all world arms exports and a continuation of this trend can only be destabilizing. Given the U.S. administration's intentions to promote such sales, it is up to the U.S. Congress to exercise the necessary wisdom and courage to block such a move.

Important to the arms flow, and therefore the balancing process, is America's stand on OPEC oil-pricing. There is strong sentiment within the American administration to support an OPEC policy designed to raise prices on the grounds that this will protect United States domestic oil production and provide the Gulf Arab states with necessary funds for their rehabilitation efforts. Historically, the Arab OPEC states have used their oil profits for massive military arms purchases. There is no evidence that they will not continue to do so. American opposition to OPEC price-rigging would, on the other hand, contribute realistically to reducing the regional arms race. Lower oil prices would also give major relief to Third World oil-importing nations that are now the chief victims of high petroleum costs.

Planning for a post-war balance must take into account the ephemeral nature of alliances between and among the regional powers. Since the 1950s every one of the states concerned has abrogated formal or tacit alliances with one or more other regional power or powers. Former enemies have joined hands to balance off other regional states, e.g., Egypt and Israel, Egypt and Iraq, Syria and Egypt.

When the two opposing superpowers were the base supports of the regional seesaw, the shifts among the six regional powers were rapid and unpredictable, but some degree of equilibrium was maintained because there were two stabilizers, not one dominating superpower (Figure 3). For the moment, there is only one balancer, the United States, and the region is in high geopolitical flux.

In the post-Gulf War world, there are likely to be two major external stabilizers again—this time the United States and the European Community. Since they are allied, rather than opposing forces, the shifts of the seesaw should be less frequent and less violent. The former USSR will have only a secondary role in this balancing process. Moreover, the future influence of the outside powers is likely to be more indirect, with front-line, peace-keeping efforts being assigned to a combination of inter-Arab and United Nations forces. It would be a mistake, in the immediate post-war period, for the United States to retain some sort of military presence to help assure Gulf security, unless that presence is shared with its allied European powers. Just as there was need for a Western-Arab coalition for war, so will there be a need for the West to act in unison in the peace.

As Europe once again becomes a major player in the Middle East, it is likely to pursue its own agenda far more aggressively, filling the vacuum left by the partial withdrawal of the former Soviet Union. A greater European presence will probably be felt in both Iran and a rehabilitated Iraq, facilitating a "cold peace" between the two former enemies. Hostility towards the United States is likely to move both Iraqi and Iranian regimes in the direction of Europe; also, the oil industries of the two countries are strongly oriented to Europe. Egypt and Syria's pairing should become stronger; Turkey can be expected to seek a position that bridges American and European interests. Israel, while tied to the American camp, will have to be more responsive to European economic and, therefore, political pressures.

Relations among the six regional states are affected, not only by their ties to outside powers, but also by their relationships with smaller neighboring countries and peoples. Over-

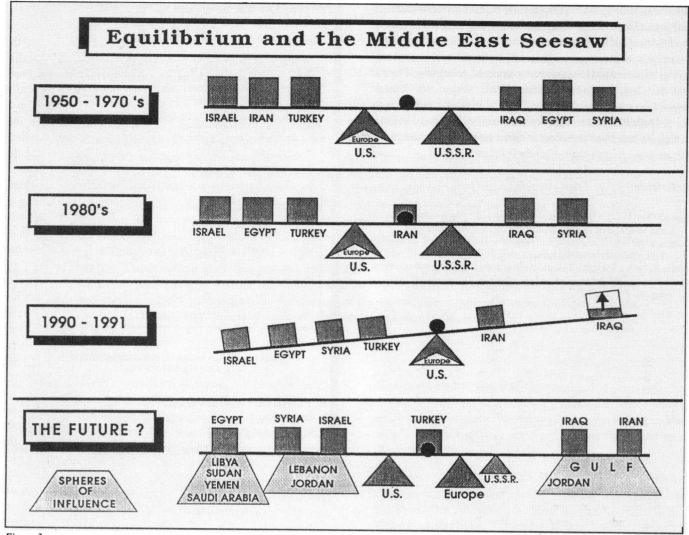

Figure 3.

lapping spheres of influence lead to tensions and potential conflict. The map of local spheres shows that all of the regional powers, save Turkey, have military and economic interests in their neighbors, which add to the complexity of balancing the system, but also contribute certain safety valves to the process. Mutual interests or vulnerabilities have led to tacit understandings between rivals, such as Israel and Syria in South Lebanon, and Israel and Iraq and Jordan at the northern end of the Gulf of Aqaba.

Jordan, Lebanon, and the Persian/Arab Gulf remain the most volatile areas of influence overlap, and the role of external powers, essentially those of the Maritime Realm, including Japan, can help to minimize the friction of overlap. Although geopolitical speculation is risky, I believe that the post-War Middle East will follow geopolitical developmental principles of specialization, integration, and hierarchy.

Conclusion

Historical events still play a key role in shaping the new Middle Eastern order. But the past is only a guide to the future. New resources, new technologies, and new modes of societal organization present challenges for which history cannot provide answers.

Would-be pharaohs and Nebuchadnezzars are unlikely to succeed in dominating the entire Middle East because it is geostrategically part of a larger realm which sets limits on the ambitions of regional overlords. The region also has a geopolitical structure whose patterns and features are shaped by a multiplicity of evenly balanced power cores that offer little comparative advantage to those who would seek gain through war. Around these nodes a network is being woven that can enhance the interests of individual states while furthering regional needs. The shock of this war has brought home the acute need to deal with water, food, and energy resources, to safeguard fragile environments, and to further the development of regional transitways in an atmosphere of peace, if not harmony.

The resources that were mobilized to wage the Gulf War were immense. If only a portion of the materials, money, diplomatic energy, and human costs would be dedicated to the war's aftermath, the peace can be won—in the Gulf, in the West Bank and Gaza, in Lebanon, in Kurdistan, and in the Sudan. A single outside power, the United States, cannot impose its designs on the region. However, the combination of

American, European, Japanese, and former Soviet assistance and guarantees, reinforcing the finely balanced strengths of the Middle East's indigenous regional powers, can help forge a new geopolitical order. This will be based upon a regional power balance that is set within a changed geostrategic world and tied together by a sharing of oil, water, and human resources across open borders. This is a historic opportunity for the Middle East to achieve the stability that has eluded it during its brief but turbulent modern history as a shatterbelt.

Let us hope that the opportunity is seized.

References

Ben-Shahar, H., G. Fishelson, and S. Hirsch. 1989. *Economic cooperation and Middle East peace.* London: Weidenfeld and Nicholson.

Cohen, S. B. 1973. *Geography and politics in a world divided.* 2d ed. New York: Oxford University Press.

Lewis, B. 1966. *The Middle East and the West.* New York: Harper Torchbooks.

McCaslin, J. 1989. *International oil encyclopedia.* Tulsa: Pennwell Publishing Co.

Welcome to the Borderlands

"Los Dos Laredos" and other cities along the U.S.–Mexican border are dynamic consumer markets, and maquiladoras *(U.S.–owned assembly plants) are responsible for much of the growth. Border families in Mexico and the U.S. are larger, younger, and more brand-loyal than other U.S. families. A proposed* free trade agreement *could bring even more rapid change to the borderlands.*

Blayne Cutler

Blayne Cutler is a contributing editor of American Demographics.

It's an ordinary Saturday for Mexican factory worker Miguel Sanchez. First he rounds up his mother, sister, brother, and pregnant wife. To save grocery money, they walk over the local bridge rather than take the bus. They buy sundries— toilet paper, bread, beer, soda, and cigarettes—then divide up the bags for a much slower walk home in the heat and haze.

Sanchez hardly notices that the local bridge crosses the Rio Grande, or that his neighborhood grocery store is in the United States. For him the U.S.–Mexico border is just a division between neighborhoods.

"Most people think the Rio Grande is an international border," says Steve Harmon, an administrator at Laredo State University, whose office sits less than five miles from Mexico. "To us, it's just a stream of water we both drink from."

In Washington, D.C. and Mexico City, the border defines where one set of laws ends and another begins. But the recent growth of bi-national metropolitan areas shows how far the laws of politics can diverge from the laws of economics. Millions of Americans and Mexicans live in border cities; they buy from the same stores, watch the same television shows, and work for the same companies.

LOS DOS LAREDOS

"We're more like Minneapolis and St. Paul than the U.S. and Mexico says Laredo city manager Peter Vargas. Laredo and Nuevo Laredo ("New" Laredo) are typical twins that straddle the Texas/Mexico border. Families there were not forced to choose sides until Texas became part of the U.S., in 1845. Nearly everyone in Laredo, Texas, is related to someone in Nuevo Laredo, Mexico, says Dianne Freeman, director of the local Chamber of Commerce.

Today, 350,000 people live in "Los Dos Laredos," with about 218,000 in Mexico and 132,000 in the U.S. Almost all (96 percent) of the U.S. residents are of Hispanic origin.

"You don't even have such a completely Hispanic population in Spain," says Michael Landeck, director of the Institute for International Trade at Laredo State University.

In Laredo's old town square along the river, a bronze statue of war hero Ignacio Zaragoza faces longingly south. Billboards advertising "Mall del Norte" (Mall of the North) face the same direction. "It is because our market is northern Mexico," says Peter Vargas. "The economy here is geared to serve the Mexican consumer." For decades, shoppers from Nuevo Laredo to as far south as the industrial city of Monterrey have exchanged their pesos for dollars and their wages for U.S. goods. The metropolitan zone of Monterrey, home to almost 3 million Mexicans, is about 150 miles from Laredo.

Years of cross-border shopping frenzies have taken their toll on the shops of Laredo. On the north end of San Bernardo Avenue, one of many self-styled import-export clearinghouses claims to "Buy, Sell, Trade Anything." Border Electronics and Tools displays an assortment of TVs, lawn equipment, car stereos, and CB radios. Down on Flores Avenue, giant spools of uncut fabric keep at least five shop owners living and dying by the dollar-peso exchange rate. When the peso was originally devalued in 1982, 800 stores in Laredo went bust.

But scenes like these—five-minute passport studios, pawnshops, and cheap motels along a quirky old strip—are no longer the main event in Laredo. The city's new retail centers are clusters of chain stores like Wal-Mart, Sams, and HEB of California. The stores and malls hug four-lane highways that could be anywhere in the United States. Laredo's population on the U.S. side grew 32 percent between 1980 and 1990.

The fuel behind border growth can be described in one word: *maquiladoras.* For 25 years, the Mexican government has allowed maquiladoras, or foreign-owned assembly plants, to operate along the border without paying tariffs, as long as all of the assembled products are shipped back across the border. U.S. duties are paid only on the "value added" in Mexico. The cheaper peso and the rising cost of Asian labor

North American Demographics

The seven major areas of population density that lie on both sides of the U.S.-Mexico border are home to 7.8 million people.

(population of selected metropolitan areas in the United States and municipios in Mexico, 1980, 1990, and percent change 1980–90)

U.S. metropolitan area / Mexican municipio	1990	1980	percent change 1980–90
San Diego, CA	2,461,050	1,876,870	31.1%
Tijuana	742,608	480,000	54.7
TOTAL	3,203,658	2,356,870	35.9
Imperial County, CA (nonmetro)	109,940	92,790	18.5%
Mexicali	602,390	532,000	13.2
TOTAL	712,330	624,790	14.0
Tucson, AZ	654,980	534,470	22.5%
Nogales	107,119	71,000	50.9
TOTAL	762,099	605,470	25.9
Las Cruces, NM/El Paso, TX	737,830	580,590	27.1%
Juarez	727,679	591,000	23.1
TOTAL	1,465,509	1,171,590	25.1
Laredo, TX	132,190	100,290	31.8%
Nuevo Laredo	217,912	203,286	7.2
TOTAL	350,102	303,576	15.3
McAllen-Edinburg-Mission, TX	416,660	286,460	45.5%
Reynosa	281,618	220,000	28.0
TOTAL	698,278	506,460	37.9
Brownsville-Harlingen, TX	281,210	211,780	32.8%
Matamoros	303,392	249,000	21.8
TOTAL	584,602	460,780	26.9
TOTAL-ALL AREAS	7,776,578	6,029,536	29.0%

Source: U.S. numbers from Woods & Poole Economics, Washington, D.C.; Mexican numbers from Instituto Nacional de Estadística, Geografía, e Informática, preliminary results from the 1990 Mexican census

attracted companies like Ford, General Motors, and Packard Electric. In Mexico, they found a country that combines Third World labor costs with overnight delivery to many U.S. markets.

MAQUILADORAS

The number of maquiladoras (or maquilas) in Nuevo Laredo grew from 10 in 1984 to 93 in mid-1990. "There was no industrial base in Laredo. It was always a trade town," says Frank Leach, executive director of the Laredo Development Foundation. Now the maquiladoras operate on the Mexican side of the border, and many of their managers live just over the bridge. Leach estimates that the Mexican plants now employ 24 percent of Laredo's work force and add $190 million to the local economy. Similar accounting, on a much larger scale, applies to the biggest centers of maquiladora

activity: Las Cruces-El Paso/Ciudad Juarez, Brownsville-Harlingen/Matamoros, and San Diego/Tijuana.

Mexico's 1,800 maquiladoras directly employ almost 500,000 Mexicans. Leach estimates that maquiladoras also indirectly support over 1 million American jobs.

Not everyone is happy with this bonanza, however. Labor leaders in the U.S. argue that maquila jobs steal higher-paying work from Buffalo and Detroit. Environmental groups say that Mexico's more lenient toxic waste laws will leave border towns with serious long-term public health

The fuel behind border growth can be described in one word: "maquiladoras."

problems. Recent tests on both sides of the border at Nogales indicate groundwater contaminated with high levels of cadmium, chromium, arsenic, and other pollutants, according to *Tucson Weekly.*

Mexican unions add that relative to the U.S., maquiladoras offer extremely low pay. The average wage rate in 1988 at Mexican maquilas was about $1 an hour, including benefits. The average annual turnover can exceed 100 percent, estimates 20-year maquila veteran manager Mark Pease, whose 3,600 employees turn out about 37 million pairs of slippers for R. G. Barry Company of Columbus, Ohio. Pease closed his plant in Juarez after turnover there topped 330 percent.

"Every time they build a plant over there, it's really closing one over here," says Joe B. Finley, Jr., a Laredo-area rancher since 1947. "We're exporting jobs."

Some experts say that even if some American jobs cross the border, the balance is positive for American workers. "Some of the low-level jobs we don't want are being done for us, freeing us for the high-skilled work," says Landeck of Laredo State University. "You are always gaining if you take away the simple jobs. Let's concentrate on robotics."

Estimating the size and scope of border markets is a complicated business. It's hard to find someone who'll tell you the value of Mexican deposits in border banks, for example, but it is clear that border banks are quite profitable. International Bank of Commerce (IBC), the largest bank headquartered on the border, has accumulated $1.4 billion in assets after 25 years of business. Some of that money is Mexican deposits and investments. The bank services both sides of the border, says president Dennis Nixon. "We look at ourselves as on the north and south sides of the city," he says.

Dallas and New York City banks see the border as an arena for international finance. Nixon's bank succeeds by emphasizing hometown service. "You can't do business by remote control on the border," says Nixon. "Executives from Dallas are not comfortable in Laredo. They find it peculiar when people hand you half their money in pesos and half in dollars." Nixon's bank welcomes that mix.

Business leaders are reluctant to talk about it, but corruption and smuggling also add huge sums to the border economy. Some immigrants to the U.S. hire "coyotes," or "guides," who demand hundreds of dollars to take them across the line. More than 1 million illegal immigrants were stopped along the U.S.–Mexico border in fiscal year 1990, up nearly 200,000 from fiscal 1989.

Also during fiscal year 1990, drugs valued at $126 million were seized along the Laredo sector of the Texas border alone, says Joe Garza, a chief patrol agent for the U.S. Border Patrol. What was the value of the drugs that got away? Says Garza, "That's like asking a traffic cop how many speeders he didn't get."

The legal and economic differences between the U.S. and Mexico have a powerful effect on the demand for all kinds of products and services. San Diegoans cross into Tijuana for cheap dental work, prescription medication, and Saturday night entertainment. Maquila workers in Ciudad Juarez come to El Paso for language training, household appliances, and especially to give birth to their children, who gain immediate U.S. citizenship if born on U.S. soil.

"There's a kind of commerce that happens on the border that doesn't happen in other places," says Leach. "It requires skills that are not needed anywhere else."

BORDER FAMILIES

Despite their promise, consumer markets along the border have received little attention from businesses and researchers. "Comparatively little demographic investigation of the border has been done," says Ed Fernandez, a researcher at the Census Bureau. "I tried to find out what makes the border different." Fernandez identifies 25 counties in California, New Mexico, Arizona, and Texas as the borderland. Some 5.2 million people live in these U.S. counties, up 30 percent from just over 4 million in 1980. About 35 percent of the border's population is Hispanic.

On average, residents of the borderlands are younger, poorer, and more likely to come from a large family than are other Americans. According to Market Statistics' 1989 *Survey of Buying Power Data Service,* the median age of women in the McAllen-Edinburg-Mission, Texas, metropolitan area was 28.5 in 1988, compared with 33.5 for women nationwide. The median age of men in the Laredo MSA [Metropolitan Statistical Area] was 25, compared with 31 nationwide. Almost half (48 percent) of Laredo residents were aged 24 or younger in 1988, compared with 37 percent across the country.

The downscale demographics of border markets hide their potential. Except for San Diego, all U.S. border cities rank low in disposable income. The median household effective buying income in Laredo was $13,900 in 1988, according to the 1989 *Survey of Buying Power.* In Dallas, the median household effective buying income was $27,000. And yet, Laredo had one of the highest per household retail spending of all MSAs in Texas ($34,000). Much of the money spent in Laredo comes from Nuevo Laredo.

In Arizona, about one-quarter of border residents are Hispanic, says Fernandez. Only 17 percent of California's border population is Hispanic, compared with a statewide average of 19 percent. Half of New Mexicans who live by the border are Hispanic, compared with 37 percent for the state. Texas is the most Hispanic border of all: three-quarters of the 1.2 million people who live in the border region stretching from El Paso to Brownsville are Hispanic. This area is home

The Hot Line

About 4.8 million people live in densely populated areas along the north side of the 2,000-mile U.S.-Mexico border. Another 3 million live on the Mexican side.

(U.S. border counties and metropolitan areas, and Mexican municipios and cities, 1990)

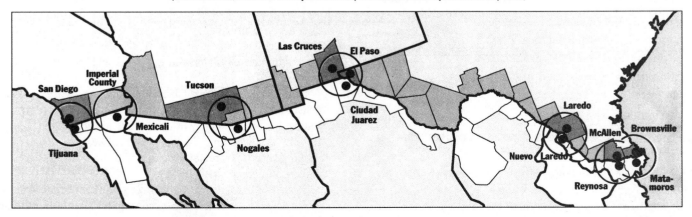

Source: Woods & Poole Economics, Washington, D.C., INEGI, Mexico, D.F.

NOME TAKES AIM AT SOVIET CUSTOMERS

The U.S. borders on three countries. Our nuclear weapons still target one of them. Increasingly, so do our businesses.

Businesses in Nome, Alaska—252 miles from Provideniya, U.S.S.R.—are getting ready for glasnost. They accept rubles from Soviet visitors, even though the money may as well have been printed at Parker Brothers.

The Russian ruble is officially worthless outside the Soviet Union, but you can use it to buy VCRs, vodka, and volleyballs in Nome.

The Alaskans want to help the Soviets switch to a market economy. They hope to stake a claim now and cash in when the vast market opens up. It's a gamble, but many Nome businesses are happy to have the chance.

"We want to get consumer goods from our businesses in western Alaska to the Soviet Far East," says Jim Stimpfle of the Nome Chamber of Commerce. Eventually, he hopes, the currency will be used to buy valuable Soviet raw materials.

For the moment, businesses like Rasmussen's Music Mart in Nome are sitting on 10,000 worthless rubles. Owner Leo Rasmussen sells the rubles to tourists for about $1 each, but he doesn't have 10,000 customers.

Rasmussen takes rubles to foster good relations. Sometimes he gets disheartened, like the time a customer bought 4,000 rubles' worth of stereo equipment. "It's nice to help goodwill, but not when your goodwill goes out the door and you don't have anything to sell to paying customers," Rasmussen says. "That guy probably sold those items for three to five times what he paid."

Economists estimate the ruble could become convertible within two years. The exchange rate set by the Soviet Union for tourists is six rubles to one dollar, and many Nome businesses take four to one. Rasmussen offers a three-to-one rate. Bering Air—which runs a charter service between Nome and Provideniya—offers a one-to-one rate, and donates its rubles to Soviet charities.

Nome and Provideniya are both remote outposts in the Land of the Midnight Sun. Provideniya, a city of 5,000 people, relies on military bases and a deep-water port for its economy. Gold mines drive the Nome economy, says city manager Polly Prehal. Nome has 8,293 people in a census area of 23,871 square miles, many of them native Alaskans, according to state demographer Greg Williams. Per capita income in the Nome area was $14,346 in 1988. That's $5,000 less than the rest of the state, despite a much higher cost of living. And there are no highways connecting Nome to the rest of Alaska.

Stimpfle hopes the ruble effort will be matched by goodwill gestures from Soviet policymakers. He would like to see Soviet citizens arriving with 72-hour visas and permission to bring rubles in and out. "The economy here needs an enormous spark," he says. "An open border and tourism development could do it."

—Dan Fost

---

to 8 percent of all Texans, but 30 percent of the state's Hispanics.

The Hispanic share of the border population is growing, just as it is in the rest of the U.S. In five of the seven metropolitan areas Fernandez designates as "areas of border influence" (Tucson, San Diego, Las Cruces, El Paso, and Laredo), Hispanic growth exceeded non-Hispanic growth by about two to one between 1980 and 1985. But the number of non-Hispanics grew faster than Hispanics in McAllen-Edinburg-Mission and Brownsville-Harlingen, mostly because of

The downscale demographics of border markets hide their potential.

rapid industrial development and the in-migration of Anglo workers.

Border families are larger than the average American family. Average household size stood at 2.6 across the U.S. in 1988, but it was 3.7 in the Laredo MSA, according to the *Survey of Buying Power.* The share of households with six members or more was 3 percent in Dallas in 1988 and less than 4 percent nationwide, but it was 9 percent in El Paso, 13 percent in Brownsville-Harlingen, and 15 percent in Laredo.

The effect large families have on local spending patterns reveals itself on the shelves of any border grocery store. While the rest of the country markets microwave dinners for one, HEB in Laredo plays up the "family size" of detergent, ice cream, and 50-pound sacks of rice.

"Families along the border are much more united here than they are anywhere else," says Landeck. "People graduate from college, and they take a pay cut to come back home. The family nucleus, including parents and grandparents, is strong."

Family ties and strong brand loyalty are two important consumer characteristics along the border, according to Landeck's recent survey of consumers in Monterrey, Nuevo Laredo, Laredo, and Corpus Christi. Mexican consumers "think there is a greater risk in not using a brand than do American consumers," he says.

Even if the wages of maquiladora workers seem low by U.S. standards, workers are occasionally able to spend like middle-class Americans, says Frank Leach. Of the 30,000 maquila workers in Nuevo Laredo, he estimates that 20,000 have held their jobs for two or three years. "As soon as they make money, their spending patterns change," he says. "First, it goes to the kids. Then they buy consumer goods."

Customers are so plentiful on the border that most businesses have hardly needed to advertise themselves. Says Landeck, "Most of the business done on the border is done in cash. There is no market research, no identification of target markets, no segmentation, and no understanding of the consumer. When you open up your door and people immediately walk in to buy 15 TV sets, why do you need to know them?"

FREE TRADE AGREEMENT

This crude but profitable business environment may change soon. U.S. and Mexican negotiators are working on a free trade agreement similar to the one that now exists between the U.S. and Canada.* If they succeed, border towns like Laredo may lose their competitive edge.

Mexico is already our third-largest international trading partner. U.S. firms sold $24.7 billion in goods to Mexico in 1989, up 20 percent from 1988, and imported $27.1 billion in Mexican goods, up 16 percent. The reason for the growth is simple: Mexico has petroleum and cheap labor, and we have consumer goods that an increasing share of Mexicans can afford.

*See "North of the Border," American Demographics, April 1990, page 45.

Tariff restrictions now prevent many U.S. businesses from marketing to Mexican consumers. They are also hampered by Mexico's poor roads, primitive phone system, bad water, and inefficient banks. Mexican president Carlos Salinas de Gortari is now trying to sell state-owned enterprises such as airlines and insurance companies to private firms. If free trade comes, these enterprises will probably be purchased and rapidly modernized, linking businesses here to markets down there.

Not everyone is enthusiastic about free trade with Mexico. A grassroots "industrial retention" movement has already begun in some midwestern cities, and many Mexicans are wary of increasing the American influence on their economy. The legislative debate over a free trade pact will be fierce in both countries.

Free trade won't be a bed of roses for Laredo, either. "It will be beautiful in macroeconomic terms," says Landeck. "But now you come to the details: who will be the winners and who will be the losers? If the free trade agreement means that the maquiladora laws will become generalized throughout Mexico, what will keep the maquiladora on the border?"

Without careful planning, Laredo could lose its status as a gateway and turn into a truckstop. But the locals say they will continue to thrive whether the maquiladoras move to the interior or not. They live by two rules: stick together, and pull economic opportunity out of political adversity wherever possible.

"We're brackish. We're not the United States and we're not Mexico. We're different," says IBC's Dennis Nixon. "We think we gather the best of both cultures."

RECLAIMING CITIES FOR PEOPLE

Cities can be made more humane—at least physically—through improvements in urban design. Among them are ways to welcome people back to the city's heart and make nature a part of the urban scene.

MARCIA D. LOWE

Marcia D. Lowe is author of Worldwatch Paper 105, Shaping Cities: The Environmental and Human Dimensions.

The suburban shopping mall inspires little ambivalence: you either despise it or adore it. To some architecture critics, it is a sterile monstrosity; to environmentalists, it is a blight on the landscape and testament to our car-dependence. But to many others, the mall is a favorite place. One group of Americans—teenagers—spends more time there than anywhere else but school and home. Chroniclers of popular culture have even extolled the mall as a reincarnation of the traditional small-town Main Street.

Like it or not, this sprawling, climate-controlled structure is a symbol of our times. Once a distinctly American phenomenon, the suburban shopping mall has proliferated throughout the industrial world and is now invading developing countries. It's no wonder. The shopping mall offers a reprieve from the harsh landscape of the modern city. People are attracted to its array of bright, new stores, its trees and flowers, fountains, benches, and sunlight pouring in through skylights. There are precious few other places in our cities, towns, and suburbs where one can stroll around freely and enjoy these amenities—all without the threat of getting hit by a car or victimized by crime.

Why are so many urban areas hostile to humans? They were planned and designed that way—not planned to be hostile, but to have features that, when all is said and done, turn out to be inhospitable. After decades of accommodating automobiles first and foremost, city landscapes are scarred by dangerous traffic, acres of concrete and asphalt, and towering, garish signs aimed only at viewers speeding by in cars. Once-lively downtowns have lost residents to the suburbs. The provision of parks, squares, and other public spaces has been neglected until even the few remaining ones are often uninviting and unsafe. Surely Michelangelo had something else in mind when he exulted, "I love cities above all."

With better urban planning and design, city dwellers could enjoy the qualities they find appealing not only in a shopping mall, but everywhere around them—not just where they shop, but where they live, work, play, and go to school. Streets could welcome pedestrians safely, and people could reach shops and entertainment conveniently without having to get in the car. In animated town centers, sunlight would stream down not through skylights but through a canopy of trees. Best of all, this environment would not be artificially created but would have the richness of diversity, informality, and spontaneity that city-lovers like Michelangelo have revered.

Of course, making cities more humane requires far more than just a redesign of urban spaces. Deep social alienation and the decline of central cities result from the formidable forces of racial and class discrimination and income disparities. All metropolitan dwellers deserve a more welcoming physical environment. Wiser planning of urban space also can help spark the vital economic development that would address more fundamental social problems, which in turn can help stem the flight of people (and tax base) to the suburbs.

Neither Expressway Nor Parking Lot

Architecture critic Lewis Mumford was a young man when early this century he fixed attention on the expanding role of the automobile. Dismayed by the approaches car-infatuated urban planners were already taking, the budding humanist admonished, "Forget the damned motor car and build cities for lovers and friends."

Today it is obvious that few heeded Mumford's advice. City streets are so ruled by motor traffic that they form impenetrable barriers. In the United States, it's not unusual for a single intersection to have two identical "convenience" stores of the same chain at opposite corners. The high-speed, heavy traffic divides the intersection into entirely distinct markets. In many places, people cannot even get to destinations that are within easy walking distance without the protection of an automobile.

Even in countries where few can afford cars, speeding motor traffic shreds the city. In India, a country with one automobile for every 408 people (compared with at least one for every two people in the United States), a foreign visitor recently complained of having to hire a taxi just to cross a street.

In cities all over the world, automobile traffic needs to be restrained. The London Planning Advisory Council concluded in a recent study that traffic restraint is "the only way of improving the environment of central and inner London." Many European cities have redesigned roads to "calm" traffic. Typically, this entails posting reduced speed limits and introducing strategically placed trees, bushes, flower beds, or play areas along or in the roadway—gentle inducements that make drivers proceed slowly and yield the right-of-way to pedestrians, cyclists, and children at play. Traffic calming is most common in Germany and the Netherlands, and is gaining ground on residential streets and main roads throughout northern Europe, Australia, and Japan.

Many of the world's large cities have reserved their thriving centers exclusively for people on foot—from the bazaars of Northern Africa to the marketplaces of Asia and Latin America to the robust pedestrian cores of European capitals. These auto-free city centers work best where there is a variety of shops, cafes, galleries, and other attractions for people to spend their leisure time. Shade-giving plants and comfortable chairs and tables invite people to linger and enjoy the space. Typically, emergency vehicles and cars belonging to local residents are the only motor vehicles allowed to enter the area. Delivery trucks are admitted only during certain hours.

As the United States' mixed experience with downtown pedestrian malls shows, not all city centers make successful auto-free havens. While pedestrian zones are flourishing in a few cities (including Boston; Boulder, Colorado; and Ithaca, New York), several others have failed because they either were poorly designed or were implemented in already-declining downtowns. Many pedestrian malls have been reopened to cars. One lesson learned is that in smaller cities where public transport to downtown is a less-than-adequate alternative, it may be more appropriate to restrain car traffic than to ban it altogether. A comfortable pedestrian-dominated zone can be created by slowing the motor traffic, expanding the space for walkers, and installing traffic lights that allow pedestrians plenty of time to cross safely.

If it is a mistake to turn much of the city into a de facto expressway, it is equally foolish to make the remainder a giant parking lot. The amount of space devoted to automobile parking is one of the most important determinants of a city's dependence on cars. Extravagant parking facilities also have unintended negative effects. Massive expanses of parking create a hostile setting and deter people from walking by creating long distances between buildings. And the lure of a convenient parking space leads many people to choose driving, even where the finest public transport is available.

Zoning and building codes often enforce an overabundance of parking. In the United States, minimum parking require-

Table 1. Car Parking Spaces and Residents in Central Area, Selected Cities

	Parking Spaces in Central Area	Residents in Central Area	Parking Spaces per Resident
Singapore	25,327	157,300	0.16
New York	144,926	506,100	0.29
Amsterdam	20,400	69,400	0.29
Tokyo	112,998	337,644	0.33
Paris	187,000	548,620	0.34
Munich	45,750	77,172	0.59
London	138,843	179,000	0.78
Copenhagen	31,400	38,571	0.81
Sydney	28,151	4,400	6.40
Los Angeles	80,074	9,516	8.41
Detroit	52,400	4,046	12.95
Houston	64,194	2,145	29.93

Source: Peter W.G. Newman and Jeffrey R. Kenworthy, *Cities and Automobile Dependence: An International Sourcebook* (Brookfield, Vermont: Gower Press, 1989).

ments are particularly lavish—often forcing new commercial developments to devote more space to auto parking than to the building's actual floor area. Some cities in developing countries also have gone to self-defeating extremes. In Delhi, the zoning code demands that commercial and civic buildings provide abundant parking at ground level. In Lima, buildings in up to 10 percent of the city center had been torn down for parking lots by the early 1980s.

Several cities world-renowned for their urban charm and attractiveness are notable for their modest parking allotments. This may be more than coincidence. For example, Paris and Amsterdam each provide more habitat for people than for automobiles—roughly one parking space for every three people living in the central area (see Table 1). Conversely, in some of the world's least inviting metropolises, an inordinate share of central city space is given over to parking lots. Detroit has almost 13 parking spaces for every central-area resident, and Houston has nearly 30.

Several major European cities are winning valuable city space back from parked cars. Paris plans to *remove* some 200,000 parking spaces. Geneva prohibits car parking at workplaces in the city's center (motivating commuters to use the excellent public transport system), and Copenhagen bans all on-street parking in the downtown core. These cities have found that if car parking has to be provided, it is best placed underground or behind buildings. This saves space, helps keep the cities lively and compact, and makes buildings more accessible to people on foot.

Soul of the City

Anyone who savors the vitality of urban life cannot resist a great city's informal gathering spots. Vienna has its coffee houses; London, the corner pubs; New York, its Central Park. From the sidewalk cafes of Paris to the tea houses of Shanghai to the plazas of Buenos Aires, local spaces designed expressly for people to get together and enjoy themselves serve a similar function. "Public space," writes Peter Newman, associate professor of environmental science at Murdoch University in Australia, "defines the character and soul of a city."

By that measure, many urban areas are dull company, indeed. Particularly in North America, local zoning codes often preclude diversity in land uses (say, mingling homes with offices, shops, restaurants, and theaters). Yet, it is precisely this kind of diversity that makes urban neighborhoods not only convenient but convivial places to live. Although the strict separation of land uses is perhaps most pronounced in the United States, other countries, including some in the developing world, have followed suit. Asian cities like Bombay and Bangkok were more dynamic and had greater internal variety before they adopted Western-style zoning.

Ray Oldenburg, a professor of sociology at the University of West Florida in Pensacola, observes that the homogeneity wrought by compartmentalized land-use planning robs neighborhoods of public spaces that are "essential to community." He notes, "Beginning with a resolve 'to promote the health, safety, morals, and general welfare of the inhabitants of _____,' zoning ordinances do as much to promote loneliness, alienation, and the atomization of society."

In addition to a greater variety of land uses, people-friendly cities spring from more careful design of city streets and non-commercial public spaces. Studies of street life in cities worldwide confirm that certain common elements make up a welcoming urban scene. Among them are abundant trees and bushes and car-free spaces for people to walk or sit together.

In their attempts to attract people to a given

area, planners and designers often repeat familiar mistakes. Author and urban social critic William H. Whyte has noted how cities can avoid these pitfalls. For example, to accommodate a greater number of people on a street, it is more effective to expand the pedestrian space, not the room given large vehicles carrying one person each. Relegating street vendors to a single area makes less sense than allowing them to space themselves in small, frequent clusters. A park or courtyard is much safer if it is not walled off but rather made visible from the street. Spikes placed on ledges to ward off "undesirables" provide less public security than do comfortable surfaces that entice people to idle.

In her 1961 classic, *The Death and Life of Great American Cities*, Jane Jacobs advanced the notion that by providing "eyes on the street," people-filled areas become less vulnerable to crime. She reasoned, "The safety of the street works best, most casually, and with least frequent taint of hostility or suspicion precisely where people are using and enjoying the city streets voluntarily." Compact developments work best to welcome pedestrians and foster conviviality, wrote Jacobs. Also important are short city blocks and a diversity of building types that offer shops, services, galleries, offices, and housing for people of different ages.

Downtown Living

In the 1950s and 1960s, massive urban highways were built in many industrial world cities, displacing entire neighborhoods and hastening the demise of old downtowns in the process. Since the 1970s, local governments in the United States and several West European cities have been trying to lure people back to declining city centers. William Whyte warns against one approach to urban revival that several large U.S. cities are now taking: building giant, fortress-like commercial complexes. These edifices clash with the city's downtown and repel people with their solid, blank walls at street level. Whyte has observed that visitors to a new convention center in Houston drive from the freeway to the center's parking garage, walk through enclosed skyways to various offices, shops, hotels, and restaurants, and then drive off "without ever having to set foot in Houston at all."

Despite the popularity of this "citadel" approach with developers, the surest way to liven up a downtown is to have people make themselves at home there—literally. Residents, especially if they choose to live at the heart of the city, can make it come alive.

Luckily, the convenience and liveliness of a genuinely urban setting is enough to attract residents. In their 1984 book, *Beyond the Neighborhood Unit*, authors Tridib Banerjee and William C. Baer describe a study conducted in Los Angeles among upper-, middle-, and low-income families that shows that people want more diversity in their residential neighborhoods than zoning codes allow. All groups showed a strong desire to have markets, drugstores, libraries, and post offices in their residential areas. When asked to rank desirable characteristics in a neighborhood, the respondents rated sociability first and friendliness second. Convenience was placed above property safety, and quiet was ranked last.

Often there is ample space to make way for additional apartments and town homes even in well-established, older downtowns. Abandoned and underused manufacturing structures in older industrial cities can be recycled into imaginative combinations of residences and commercial space. Several cities in England, including Norwich, Ipswich, and Cambridge, are in the midst of converting great amounts of vacant space in commercial buildings for residential use under the slogan "living over the shop."

Cities also can create land banks of tax-delinquent property in a process that can turn vacant property in declining inner-city neighborhoods into decent, affordable housing. In land banking, local governments break the typical cycle in which the city forecloses on derelict property and sells it—often to speculators who let the land become tax-delinquent again. Properties held in a land bank can be resold specifically for socially productive uses such as housing.

Applied successfully in several North American cities—including St. Louis, Missouri; Saskatoon, Saskatchewan; and Edmonton, Alberta—land banking not only has increased these cities' supply of affordable residential land but also has made existing urban space more compact. Camden, New Jersey, a declining industrial city with more than 3,500 vacant properties, turns dilapidated homes over to nonprofit groups that rehabilitate them. Block by block, the city's neighborhoods are being revitalized. In Louisville, Kentucky, the Land Bank Authority cancels back taxes on abandoned lots and sells them for $1 each to nonprofit developers who build low-income housing. Land banking serves a dual purpose: It helps the city eliminate blight and generate taxes on abandoned land, and it boosts the supply of available, affordable homes.

Urban Nature

It is generally agreed that natural green space is pleasing to the eye and soothing for the spirit. A city's trees, bushes, vines, and flowers also can provide habitat for birds and other wildlife, ease air pollution, and even bring urban temperatures down. Far from a luxury, nature is vital to a humane city. Anne Whiston Spirn, chair of the Department of Landscape Architecture and Regional Planning at the University of Pennsylvania, challenges the view that pieces of nature should be added to the urban landscape as mere adornments. "The city is part of nature," she stresses. "Nature in the city must be cultivated, like a garden. It must not be ignored or subdued."

A growing number of cities are conserving their remaining wild spaces and creating nature reserves. Great Britain's Nature Conservancy Council, a central government agency, now supports reserve projects in more than 60 urban areas. A share of the funding is specifically aimed at planting greenery in inner cities and public housing projects, which typically lack such connections with nature.

Many cities are linking stretches of verdant space along rivers, canals, or old rail lines into continuous paths for cycling, horseback riding, jogging, and walking. For urbanites, these "greenways" bring fresh air and nature closer to home. If designed carefully, they also can create corridors for protecting wildlife. In the United States, greenways in Washington, D.C., Seattle, and elsewhere have become major routes for bicycle commuters. An estimated 500 new greenway projects (led largely by citizens' groups) are currently in the works in this country. Many Dutch, Danish, and German cities and towns are connected by extensive path networks, and footpaths are common in much of Europe. The city of Leicester, England, is planning to convert an abandoned rail line into the Great Central Way, a car-free route that will bisect the entire city from north to south.

In developing countries, green space often provides opportunities for economic development and even survival. Urban agriculture, particularly in the developing world, produces essential food and fuel on a commercial scale. City farms produce eucalyptus in Addis Ababa, livestock in Hong Kong, vegetables in Seoul, and firewood in Lae (Papua New Guinea). Urban agriculture can provide important insurance against interruptions in supplies of food or fuel and can give the poor more direct control over meeting their basic needs. Many Asian cities produce a considerable share of their own food through urban agriculture. China is outstanding both in growing food within urban boundaries and in recycling the city's sewage and solid wastes into fertilizer for cropland. Several Chinese cities produce at least 85 percent of their vegetable supply in greenbelts that also recycle composted garbage and human waste.

Communities in many industrial world cities use food-growing plots to restore greenery to barren neighborhoods and to supplement the diets and incomes of inner-city residents. Copenhagen's citizens cultivate "allotment" gardens at the city's edge, and schoolchildren in The Hague learn to grow vegetables in communal tracts. Urban gardens are common in many cities in Eastern Europe and the Soviet Union. New York has 700 to 800 community gardens. More than 2 million gardeners in U.S. cities, including many elderly, young, and disabled workers, tend neighborhood plots, rooftop gardens, and solar greenhouses. In a particularly successful gardening project initiated by a regional food giveaway program in Peoria, Illinois, people who once received free food are now reciprocating by cultivating vegetables for the same program.

Where city environments are too polluted for growing food safely, a better means of reestablishing links to the countryside is to devote space to farmers' markets. Open-air markets are a primary source of fruits and vegetables in cities throughout Asia, Africa, and Latin America. They enable farmers to sell directly to consumers, expanding their clientele and capturing profit otherwise lost to opportunistic brokers. Low-income city dwellers, unable to pay prices charged by commercial grocery outlets, can benefit from the fresh food available at such markets.

Farmers' markets are making a comeback

in the United States. In 1976, New York City's Council on the Environment—a citizens' organization that works out of the mayor's office—initiated a system of "Greenmarkets." Currently operating at 10 sites year-round and 20 in the summertime, the Greenmarkets aim to preserve farmland and help struggling farmers in the counties north of the city while making fresh fruits and vegetables available in city neighborhoods. The markets offer many New Yorkers their only chance to get local produce without journeying to the suburbs.

If Walls Could Talk

In a more people-friendly city, municipal officials, planners, and other decision-makers would set new priorities: the city is not to be a throughway or storage area for motor vehicles, nor is it to be a place from which people escape to find pleasant surroundings. It should be a place to make one's home.

City streets would become habitats for humans and plant life; land-use regulations would encourage and celebrate diversity, not reinforce homogeneity; public spaces would recapture the ground they've lost; and nature would become an integral part of the urban landscape.

An important ingredient in this vision of a humane city is a planning and design process that involves the public. Experience has made painfully clear the repercussions of imposing urban land-use decisions without community influence or consent. In the worst extreme, insensitive planning can obliterate whole neighborhoods, as happens in slum clearance projects thinly veiled as urban renewal. But even in less conspicuous cases, planners can easily err in trying to "clean up" what they view as problems, thus sterilizing a community's inherent charm or failing to respect its history and character. A more participatory planning process helps avert such losses and is more likely to address the concerns of the elderly, handicapped, children, and other groups with special needs.

Neighborhood organizations can be tapped to provide an effective liaison between citizens and the city administration and planning council. In several U.S. cities, including Atlanta, Cincinnati, and Washington, D.C., neighborhood organizations play a legal role in zoning and other land-use decisions. Baltimore gives funding and technical assistance to its neighborhood groups to facilitate their participation in the urban planning process.

Some urban dwellers have acted without waiting for such a formal arrangement. In Dakar, the capital of Senegal, a group of young citizens recently set out to make their city a more welcoming place. After clearing sand from the streets, pruning trees, draining pools of stagnant water, and removing litter, the volunteers turned to beautifying their public spaces. They repainted sidewalks and put up homemade decorations in communal places. As the centerpiece of their effort, the youths created a profusion of "Talking Walls"—walls and sidewalks emblazoned with spirited murals depicting national heroes, cartoon characters, and scenes of everyday life.

In most of the world's urban areas, it would be rare to find such a spontaneous display of the will to make a more livable city. But if each city were to create its own version of Dakar's Talking Walls, these would no doubt have much more to say than the uniform walls of a suburban shopping mall.

The aral sea basin

A critical environmental zone

V. M. Kotlyakov

V. M. KOTLYAKOV is director of the Institute of Geography at the Soviet Academy of Sciences in Moscow.

Among today's severely damaged ecological zones, the Aral Sea region is one of the most notorious.[1] The Aral, a large, desert-bound sea in south central Asia, was brought to life by the abundant Amu Darya and Syr Darya rivers, which drain into the Aral Sea basin from tributaries deep in Soviet central Asia and Kazakhstan (see Figure 1). But the sea, once comparable in size to one of the larger Great Lakes in North America, has been shrinking at an alarming rate over the last 30 years.

The process taking place in the Aral region may be called anthropogenic desertification because the sea is, in effect, drying up as a consequence of the development of irrigated agriculture in the basin. Its river deltas and other natural habitats, the local climate, and regional hydrology have all been similarly affected. Moreover, these negative ecological changes have been ac-

companied by grave socioeconomic costs: deteriorating human health, increasing unemployment as resource-based economic activity declines, and decreasing production of cotton, rice, and other agricultural crops. The fate of the Aral basin bears watching by other regions in the world prone to desertification.

In the 1950s, the Aral Sea covered an area of 66,000 square kilometers, containing 1,064 cubic kilometers of water with a mean salinity of 1.0 or 1.1 percent. At the time, the mean depth of the Aral was 16 meters, and 90 centimeters of water evaporated from its surface annually. The sea's water balance was maintained by an annual inflow of 56 cubic kilometers of water from the Amu Darya and Syr Darya rivers, supplemented by 5 cubic kilometers of atmospheric precipitation.

But early in the 1960s, rapid development of irrigated agriculture began, and water withdrawals from the Amu Darya, Syr Darya, and their tributaries greatly increased. By 1965, 4.5 million hectares in Soviet central Asia were irrigated, consuming 50 to 55 cubic kilometers of water annually. In the next

quarter century, another 2.6 million hectares were irrigated, using an additional 50 cubic kilometers of water annually.

Since the early 1960s, the area of irrigated lands has increased by 50 percent in the Uzbek and Tadzhik republics, by 70 percent in the Kazakh Republic, and by 140 percent in Turkmenia. Today, a total of 7 million hectares in the Aral region are irrigated, with annual withdrawals of 60 cubic kilometers of water from the Amu Darya and 45 cubic kilometers from the Syr Darya.

With the rapid expansion of irrigated lands, the total inflow of water to the Aral decreased sharply, and the sea's level dropped (see Table 1). Between 1961 and 1970, the mean decline was 0.21 meters per year. The annual decline was 0.58 meters during the 1970s and as much as 1.09 meters during the 1980s. By 1990, the sea level had fallen more than 14 meters from the 1960 level, the area of the sea had shrunk by more than 40 percent, and its volume had decreased by more than 60 percent.

The mean inflow of river water to the

From *Environment*, Vol. 33, No. 1, January/February 1991, pp. 4-9, 36-38. Reprinted with permission of the Helen Dwight Reid Educational Foundation. Published by Heldref Publications, 1319 Eighteenth St., NW, Washington, DC 20036-1802. © 1991

Aral Sea between 1971 and 1980 was only 16.7 cubic kilometers per year, and early in the 1980s, it practically ceased altogether. In 1987 and 1988, there was a relative abundance of water: the Syr Darya delivered 1.2 and 6.2 cubic kilometers of water to the Aral, and the Amu Darya delivered 8 and 16 cubic kilometers. In 1989, the inflow from the Syr Darya dropped to 4.2 cubic kilometers, and that of the Amu Darya fell to only 1.1 cubic kilometers.

By the end of 1989, the Aral Sea had receded into two separate parts. The level of the southern "Greater Sea" dropped to 38.6 meters, and that of the northern "Lesser Sea" dropped to 39.5 meters. Now the area of the Greater Sea is about 33,500 square kilometers with a volume of 310 cubic kilometers and mean salinity of 3.0 percent. The Lesser Sea covers approximately 3,000 square kilometers with a volume of 20 cubic kilometers. The salinity of its waters varies in different parts from 1.8 to 3.5 percent because of the freshening effect of water from the Syr Darya.

The trend is clear. Without drastic measures, the Aral Sea is destined to become a small brine lake of between 4,000 and 5,000 square kilometers.

Ecological and Health Impacts

As a consequence of the physical changes, the Aral Sea now has almost completely lost its productive fisheries. The Aral once contained more than 20 species of fish, including 12 game species. Now all but a few have died out as shallow spawning grounds have dried up and food reserves have disappeared. If the present trend in salinization continues, by the year 2000, the salinity of the sea water will reach 3.8 to 4.2 percent. This high concentration will exclude the possibility of colonization by new species of fish and the organisms they feed on that might have acclimatized to the saline conditions.

The exposed sea bottom of the Aral is becoming a salty (solonchak) badlands, a source of constant wind-blown sand. Dust storms that are so powerful that they can be observed from space have become common in spring. In addition to fine particles of soil, the dust contains salts—primarily sulfates and chlorides, which are poisonous to plants. The dust clouds, which stretch for 150 to 300 kilometers, are particularly harmful to desert oases and pastures during spring flowering.

During the last 25 years, the duration of dust storms in the coastal areas of the sea has increased 2 to 3 times.[2] Based on preliminary estimates, the Aral region is the source of from 15,000 to 75,000 tonnes of atmospheric dust annually. Each year, each hectare in the lower Aral basin receives 340 kilograms of dry salt and an additional 180 kilograms of salt in precipitation for a total of 520 kilograms of salt deposited per hectare. The mineral content of precipitation in the basin was 3 times higher in the 1980s than it was in the 1960s and 1970s.

Before its level dropped, the Aral Sea influenced the climate of the land within 100 to 200 kilometers of the sea. But as the sea and its river deltas have dried up, that influence has diminished, and the continentality of the basin's climate has increased—that is, the climate has become more like that of a continental interior, with decreased precipitation and greater swings in temperature. Early in the 1980s, the difference between the mean monthly temperatures of January and July increased by 1.5° to 2° C. Frosts are now more likely to occur later in spring and earlier in fall. The frost-free period in the Amu Darya delta, a cotton-growing region, has shortened to as few as 170 days, far fewer than the minimum of 200 frost-free days required for growing cotton.

As spring floods on the Amu Darya and Syr Darya stopped and the deltas dried up, the natural landscapes of the region were destroyed. During the past 20 years, the 550,000 hectares of reeds on the floodplains of the Amu Darya shrunk to 20,000 hectares, and the productivity of pastures and haying grounds dropped by roughly 20 percent. More than 50 lakes in the Amu Darya delta—about 10 percent of the region's lakes—have dried up, as have the floodplain forests known as *tugai*. As a result, the region's biodiversity has been greatly impoverished. Only 168 of 319 species of bird continue to nest in the delta, and only 30 of 70 mammalian species remain.

A considerable part of the mineralized and polluted drain water from the surrounding expanse of irrigated fields returns to the Amu Darya and Syr Darya. This run-off has critically lowered the water quality in the river deltas. The problem has been exacerbated by the salinization of croplands because water is used to wash away the salts. The mean consumption of water between 1978 and 1982 in Uzbekistan was about 17,500 cubic meters per hectare. In Karakalpak and Khorezm, it reached 36,000 cubic meters per hectare.

In the lower reaches of the Amu Darya, the mineral content of the river water from run-off of drain water and fertilizers is already about 1.5 grams per liter. In some years, the mineral content of the water in the lower stretches

TABLE 1
HYDROLOGIC PARAMETERS OF THE ARAL SEA FROM 1960 TO 1989

Year	Sea level (meters)	Sea area (thousand square kilometers)	Sea volume (cubic kilometers)	Mineral content (grams per liter)	Total river run-off into sea (cubic kilometers)
1960	53.3	67.9	1,090	10.0	40
1965	52.5	63.9	1,030	10.5	31
1970	51.6	60.4	970	11.1	33
1975	49.4	57.2	840	13.7	11
1980	46.2	52.4	670	16.5	0
1985	42.0	44.4	470	23.5	0
1989	39.0	37.0	340	28.0	5

SOURCE: D. B. Oreshkin, "Aral'skaya Katastrofa," (The Aral Catastrophe), *Nauka o Zemle*, no. 2 (1990):41.

3. THE REGION

of the Syr Darya reaches 2.5 grams per liter. One liter of water from these rivers contains up to 6 grams of phosphates, 3 milligrams of ammonia, 2 milligrams of nitrites, and 6 milligrams of nitrates. The chlorinated hydrocarbon pesticides applied to agricultural fields in the basin can be detected in the water of both rivers.

Leaked and surplus irrigation water often ends up not back in the rivers but in local depressions and irrigation reservoirs at the periphery of irrigated zones, where the water evaporates. Such reservoirs have become particularly numerous along the lower reaches of the Amu Darya, where more than 40 large water basins annually lose from 6 to 7 cubic kilometers of water to evaporation. The largest reservoir of Sarykamysh receives 3 to 4 cubic kilometers of irrigation wastewater annually, increasing its total volume to 30 cubic kilometers.

Because of these reservoirs, bogs and boggy patches not typically found in desert habitats are forming. The bogs displace valuable pastures, which are adapted to scant precipitation. The bog water itself is wasted through evaporation, thereby increasing the salinization of the land.

The anthropogenic desertification of the Aral and its region, initially an ecological problem, in recent times has turned into a human health problem.

Much illness has resulted from the use of highly toxic chemical pesticides, which accumulate in drainage waters that flow into the rivers and contaminate local drinking-water supplies. The deterioration of the drinking water, furthered by inadequate purification plants, piping, and sewage systems, has greatly impaired the health and sanitary conditions of the people in the Aral region. In the last 15 years, the incidence of typhoid fever has increased almost 30 times and that of hepatitis, 7 times.[3] There are also considerable increases in the incidences of kidney disease, gallstones, and chronic gastritis. Many babies are born weak and ailing, and child mortality is more than 50 per

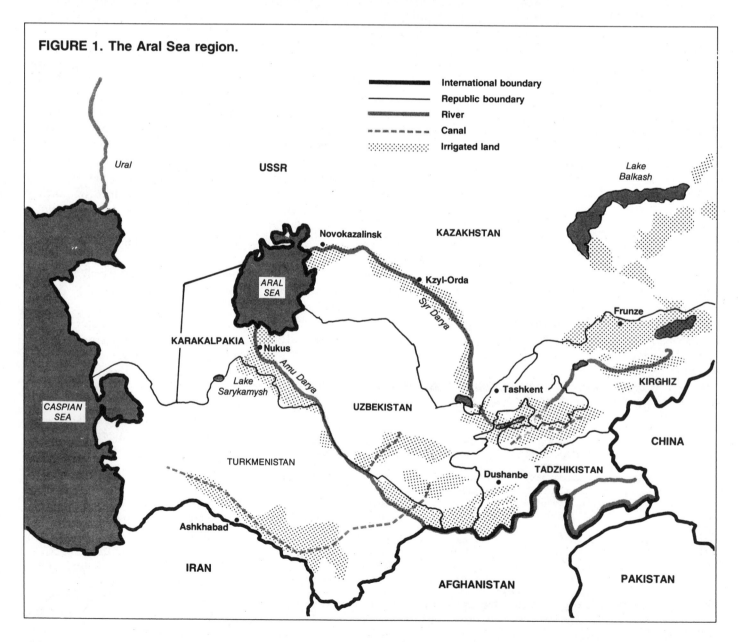

FIGURE 1. The Aral Sea region.

1,000 births. A recent appeal by scientists to improve the health of the region's children appears in the box on this page.

The Path of Development

The desertification of the Aral region is a direct result of the socioeconomic development of Soviet central Asia and a considerable part of Kazakhstan. For example, the current direction of agricultural development in the region is toward water-intensive production. The agriculture is characterized by a predominance of cotton monoculture, an absence of crop rotation, and overly large plantations of water-intensive rice. Marginal lands are being developed, producing low harvests but abundant salt-laden runoff. As a result, despite the increase in irrigated areas, the gross yield of cotton in the central Asian republics and in Kazakhstan (excluding the Tadzhik Republic) has not increased. For example, Uzbekistan produced 3.9 million tonnes of raw cotton in 1970 and 5.3 million tonnes in 1975, but only 5.1 million tonnes in 1985.[4]

Despite the rapid salinization of soils because of excessive water application with insufficient drainage, attempts to economize water are inefficient. The accepted quotas of water consumption per irrigated hectare are excessive because specific soil and plant types and the actual consumption and mineralization of irrigation water are not fully taken into account. Often, the actual volume of water supplied to fields is even greater than the quotas; in different central Asian republics, the excess varies from 20 to 100 percent. Yet as mineralization of the river water used for irrigation increases, irrigation quotas are raised even further, resulting in still greater water withdrawal. An increase in water mineralization by 0.1 grams per liter is offset by applying an additional 1,000 cubic meters of water per hectare.

If the anthropogenic desertification of the Aral region is not stopped, all environmental factors will be affected. The Aral Sea will be replaced by a 60,000-square-kilometer salt-and-sand

WOMEN SCIENTISTS' APPEAL FOR IMMEDIATE ACTION TO SAVE CHILDREN IN THE ARAL ECOLOGICAL CRISIS REGION[1]

We, the women participants in the first international symposium, "The Aral Crisis: Origins and Solution," held in Nukus, Karakalpak ASSR [Autonomous Soviet Socialist Republic], on 2–5 October 1990—specialists in the areas of ecology, medicine, geography, sociology, and demography—have concluded from analysis of data gathered from field observations that the Aral Sea region is one of ecological catastrophe that is especially hazardous to children.

The mortality rate of children in the Karakalpak ASSR is among the highest in the world and is growing each year: from 47.3 percent in 1978 to 59.8 percent in 1989. The maternal death rate in Karakalpakia has tripled in the last 5 years. More than 80 percent of women suffer from anemia, and every third [pregnant] woman gives birth prematurely. The dispensary system has revealed that in 1989–1990 almost 70 percent of the children in Karakalpakia fall into health categories II and III—i.e., [they are] practically sick. The number of children suffering from nervous and psychological disorders has tripled in the last 2 years.

The primary reasons for so sharp a decline in public health as to threaten the survival of the population are degradation of the environment, qualitative and quantitative depletion of the supply of potable water, microbial contamination of water, pesticide concentration, and protein and vitamin deficiency.

The situation is aggravated by the lack of necessary medical services and ecological education. We are extremely concerned by the slow pace, bordering on criminal, of [remedial] action. The knowledge already accumulated is sufficient to justify urgent action. We demand immediate action to save the population of the Aral region and other ecological disaster areas.

We call upon the government and the peoples of the USSR and the republics of central Asia and Kazakhstan, at all administrative levels of the country and the region; we call upon the United Nations, its specialized institutions—UNICEF [United Nations International Children's Emergency Fund], WHO [World Health Organization], and UNEP [United Nations Environment Programme] —and all organizations concerned with problems of public health and survival; and we call upon all of the world's women to render emergency assistance to save the lives of children in the Aral region and to declare this region an ecological disaster zone. During 1990–1991,

- provide the population with clean drinking water and food as well as necessary medical assistance;
- hasten the development and implementation of a plan of action for solving the Aral problem;
- immediately shut down all hazardous production;
- place under tight control and decrease use of all pollutants, toxic chemicals, radioactive materials, and ozone-depleting substances;
- provide *glasnost* [complete reporting] of ecology-related information;
- spread energy-, water-, and resource-conserving technology; and
- prohibit use of child labor in cotton fields.

We send our appeal from Nukus, the burning center of ecological disaster, but we know that similar problems are emerging in an ever increasing number of areas encompassing the whole world.

Working women in all fields and at all levels—teachers, physicians, engineers, writers, and artists—we must step up our efforts to maintain normal environmental conditions to ensure the health of our children.

Here in Nukus, we are creating a committee of women scientists and other specialists, "Mothers, Save Your Children." The primary purpose of the committee is to gather and disseminate knowledge needed for emergency action to restore the environment in the Aral region and the other most degraded areas in the world.

We will not allow the killing of our children!

Please direct inquiries about membership in the committee or suggestions for enhancing the effectiveness of our work to Nina Maksimovna Novikova, Coordinator, "Mothers, Save Your Children" Committee, USSR Academy of Sciences Commission on Ecological Problems, Sadovaya-Chernogradskaya 13/3, Moscow 103064, USSR.

1. Before being edited, this appeal was translated by Katya Partan.

RESOLUTIONS OF THE INTERNATIONAL SYMPOSIUM "THE ARAL CRISIS: ORIGINS AND SOLUTION"[1]

The international symposium "The Aral Crisis: Origins and Solution" was held on 2–5 October 1990 in Nukus, Karakalpak Autonomous Soviet Socialist Republic. More than 200 leading specialists and scientists participated. Among these were 27 foreign scientists from the United States, England, Canada, Australia, Germany, and Spain, representing international research institutions specializing in hydrologic, health/sanitation, and biomedical problems in arid zones. Also in attendance were about 100 scientists from institutes of the USSR Academy of Sciences and the academies of sciences of Uzbekistan, Turkmenia, Tadzhikistan, and Kirgizia and from Moscow University and other leading institutions of higher learning in the USSR, and more than 50 representatives from the conservation committees of the central Asian republics and Kazakhstan, the USSR State Hydrometeorological Center, social and economic organizations, and foreign companies.

The symposium included presentations of more than 40 papers and reports, as well as a Nukus–Muynak–Aral field trip. The presentations focused on the deteriorating ecology of the region, the resultant declining health of its population, decreasing efficiency in economic management, and growing social tension. The medical/hygienic and socioeconomic aspects of the Aral problem were discussed in connection with environmental pollution, improper use of pesticides and fertilizers, insufficient medical facilities, and the lack of a modern water-supply system. Attention was also focused on imprudent water use, insufficient water resources, and conservation and restoration of the Aral Sea.

The scientists of the central Asian republics and Kazakhstan reached a unanimous conclusion as to the reasons for and ways to resolve the Aral crisis:

1. The scientists consider it necessary to request that the supreme soviets of the central Asian republics and Kazakhstan, as well as the Supreme Soviet of the USSR, declare the lower reaches of the rivers in the Aral basin to be an ecological disaster area and confer a corresponding status on the region.
2. The fundamental causes of the Aral crisis are erroneous choices of strategies for developing the region's productive forces; extensive development of water resources and agriculture; low standards for planning, construction, and operation of irrigation systems; and indiscriminate use of chemicals in agriculture.
3. Ecological restoration is impossible without stabilizing the level of the Aral Sea. To accomplish this, it will be necessary to perfect a complete water-management system.
4. A combination of urgent measures is necessary to restore the ecological conditions in the Aral Sea basin. The most urgent of these must be completed during the period 1993 to 1995, as delaying their implementation may have unforeseen and irreversible consequences. These measures are economic/organizational in nature and are aimed at restructuring industry and agriculture. Immediate measures must be taken to improve the health of the population and the quality and quantity of potable water and to impose strict controls on the use of pesticides and fertilizers.
5. To solve the problem, the following measures must be implemented in stages:
  • Strictly limit intake of water by the republics, with consideration for the need to introduce water-conserving technology in all areas of the economy.
  • Strictly prohibit expansion of the land area under irrigation and new siting of water-retaining industry, thus freeing river flow for preserving the Aral Sea.
  • Immediately revise the structure of cultivated land and crop composition to limit rice culture, remove low-yielding irrigated land from cultivation, and end cotton monoculture, thus freeing the land for development of orchards, viticulture [vineyards], and seed alfalfa.
  • By the end of 1992, complete the removal of outflow from the region's [water] collection/drainage networks to the sea and reassess the continued use of lowland, valley, and container reservoirs, which lose 5 cubic kilometers of water annually.
6. Concomitantly, it is necessary to accelerate a comprehensive reconstruction of land-reclamation systems based on progressive water conservation and irrigation technology and coupled use of ground and surface water, which in many cases will provide waste-free use of irrigation waters.

7. Use of newly available water for unorganized flooding of high deltas and creation of polders are inadvisable. Feed crops should be grown only on land that has long been used for pasture or been under irrigation.
8. Construction of collectors to divert outflow to the Aral and Caspian seas from collection-drainage networks in the middle basins of the Syr Darya and Amu Darya is impermissible.
9. Current projects for redistributing flow between basins are scientifically unfounded. Fundamental research is needed to provide water to the Aral basin by alternate means.
10. Inasmuch as incomplete study of many important problems has made decisionmaking impossible, it is necessary to strengthen basic research and to create a specially funded comprehensive program to solve the Aral problem.
11. The leadership of the union's sovereign republics of central Asia and Kazakhstan is requested to create an interrepublic committee to coordinate joint activities, including scientific investigations, to solve the Aral Sea problem.
12. It is recommended that the republics of the union create scientific research subdivisions in their respective academies of sciences for ecological and water problems.
13. A program should be developed and implemented to increase ecological awareness in the population at large and to train personnel skilled in the area of ecology.
14. The conference considers it expedient to enlist the cooperation of international and foreign scientific organizations as well as individual foreign scientists in researching the Aral problem.
15. The participants in the international symposium support requesting the Union of Red Cross and Red Crescent Societies of the USSR to assist the population of the Aral region in improving [medical] service, the health of mothers and children, and supplies of good quality drinking water, and [they] appeal to governments, social organizations, and foundations to assist in these humanitarian activities.

---

1. Before being edited, these resolutions were translated by Katya Partan.

desert generating dust that will harm the desert oases. Salt that formerly went to the Aral, which was the main salt repository for the whole region, will be deposited in the oases, which will shrink under the burden of salt and dust and, eventually, disappear.

The Search for Remedies

The ecological problems of the Aral Sea and its basin have worried scientists since the end of the 19th century. The Institute of Geography of the Soviet Academy of Sciences has participated in research on the Aral since 1976. At that time, many specialists thought that the ecological and socioeconomic consequences of the development of large-scale irrigation in the Aral Sea basin—primarily, the drop in the sea's level—would be insignificant and ignorable. However, geographers at the institute thought that the socioeconomic consequences could be serious and must be taken into account. Recent events have shown that position was correct.

It is now realized that the problems of the Aral basin affect the health and life of a vast territory. Their solution requires new approaches and decision-making based on a profound and integrated analysis of the numerous direct and feedback relationships between the sea and the land; the fate of one cannot be divided from the other.

The Aral region is in a state of ecological crisis in which the natural processes and ecological links have been disturbed so profoundly and the degradation has become so serious that the human population can no longer live and work as it once did. An ecological catastrophe is fast approaching wherein the degradation would be irreversible and life and work in the region impossible. The urgency of the situation must be realized. Only a short time remains to provide the population with clean drinking water, to ban the application of highly toxic pesticides, and to clean up the mineralized and polluted drainage and river waters.

Also, the current practice of adjusting water use to meet the demands of agricultural production development must change. On the contrary, the problems of water use should be the deciding factor in agricultural production plans. The role of cotton in the country's economy also should be reconsidered to increase the production of foodstuffs and to introduce more salt-resistant and less water-intensive crops.

The water resources of Soviet central Asia and Kazakhstan must be inventoried. Special attention should be given to increasing the efficiency of water-economizing systems, decreasing irrigation quotas by 15 to 20 percent, optimizing drainage systems to reduce water consumption by several thousand cubic meters per hectare, and improving regulation of run-off. The water freed in this way should be used primarily for resolving ecological imbalances to reinstate the normal conditions of life and health for local populations. If necessary, the water supply for agriculture could be increased by introducing water-saving technologies such as drip irrigation, by planting protective forest belts, by decreasing evapotranspiration, and by desalinizing water resources, primarily drainage waters. A combination of measures for increasing the productivity of irrigated and nonirrigated lands is needed to increase agricultural production without expanding irrigated areas or water consumption.

Saving the Aral region and resolving its tangle of ecological, hydrological, and socioeconomic problems will require a new attitude and an entirely new approach to economic activity throughout Soviet central Asia. Agriculture and other branches of the economy should be allowed to return to their traditional forms, which have been destroyed over the past decades, and attitudes toward water use should be changed. A number of new economic laws already have been adopted to change economic and social relationships in the USSR in principle; changes in practice will naturally take time.

Global Connections

The problems of the Aral Sea region are not unique on Earth. They are part of the universal problem of desertification that is occurring in many regions of the world, especially in Africa.[5] Hence, the search for a solution to the complex problems of the Aral region is of general and global significance.

Indeed, international organizations already have begun to work cooperatively on the problems of the Aral Sea basin. In cooperation with Soviet specialists, the United Nations Environment Programme has set up a research project and formed a small working group on the problems of the Aral Sea. The group is expected to provide specific recommendations for action in two years. The executive committee of the International Geographical Union also has formed a research group that is working on the problem of critical ecological zones. The investigations of this group will embrace several critical zones, but one of the most important will be the Aral Sea region.

This past fall, an international symposium on the state of the Aral Sea basin was held in Nukus, the capital of Karakalpak. (The symposium resolutions may be found in the box on the previous page.) The symposium, organized by the Aral Research and Coordination Center with the participation and assistance of the International Geographical Union Study Group on Critical Environmental Zones and Global Change, was an important landmark in cooperation on the Aral. The symposium launched a serious effort to gather the data needed to conceptualize scientifically grounded socioeconomic development in Soviet central Asia. In the end, such thinking can be the only way to resolve the problems of the Aral Sea.

NOTES

1. P. P. Micklin, "Dessication of the Aral Sea: A Water Management Disaster in the Soviet Union," *Science* 241 (1988):1170-76.
2. T. I. Molosnova, O. I. Subbotina, and S. G. Chanysheva, *Klimaticheskiye Posledstviya Khozyaistvennoi Deyatel Nosti v Zone Aral'skogo Morya* (Climatic Consequences of Economic Activity in the Zone of the Aral Sea) (Leningrad: Gidrometeoizdat, 1987).
3. D. B. Oreshkin, "Aral'skaya Katastrofa," (The Aral Catastrofe), *Nauka o Zemle*, no. 2 (1990):41.
4. See also, F. I. Khakimov, *Pochvenno-Meliorativnye Usloviya Opustynivaniya Del't* (Soil-Meliorative Conditions of Desertification of Deltas) (Pushchino, 1989).
5. H. E. Dregne, "Aridity and Land Degradation," *Environment*, October 1985, 16.

Lash of the Dragon

*After millennia of struggles to subdue it,
China's mighty Yellow River remains untamed*

Daniel Hillel

In June 1938, the Japanese army was advancing rapidly westward across northern China. Kaifeng, the capital of ancient China, had been overrun and the railroad linking Beijing with Hankou, the temporary capital in the south, was threatened. Desperate to stop the invaders and buy time, Chiang Kai-shek turned to the Yellow River (Huang He), ordering the destruction of the river's dikes to drown the enemy. On its course across the North China Plain, the river flows on a ridge of its own making between levees, so that it resembles a giant aqueduct. The river's bed, perched precariously thirty feet or more above the surrounding country, had always posed the threat of natural disaster, but now it was to be used as a weapon of mass destruction.

At a particularly vulnerable point on the river, forty-five miles west of Kaifeng, Gen. Shang Chen supervised the mining of the dikes with dynamite and then waited for the summer rains. As the invading army closed in, the general ordered that the dynamite be detonated. At first, the breach was only several hundred feet long, but the force of the rushing water continued to tear away the embankment, widening the gap to a quarter mile. As the water spilled out across the flat plain, the broad channel that had conveyed it to the northeast was reduced to a stream. The unleashed river was now cutting a new course, seeking a more direct route to the sea. Following the almost imperceptible slope of the plain, the flood spread out and moved toward the southeast along several paths. The waters overran a tributary of the Yangtze River (Chang Jiang) and finally entered the sea more than 400 miles south of the Yellow River's former outlet.

From a military point of view the maneuver was a success. Part of the invading army was destroyed and its heavy equipment mired in thick mud. The flooded area remained an effective barrier throughout the war, shifting the Japanese offensive to the south. For the farmers living on the floodplain, however, the breaching of the dikes greatly magnified the suffering already caused by the war. The floodwaters advanced at five miles an hour across a broad front, inundating some 21,000 square miles. Only treetops and the eaves of temples and houses were visible above the muddy water. With no high ground and with all escape routes cut off by the rising water, nearly a million Chinese perished. The crops in this fertile region were also destroyed, and the land was smothered beneath a thick blanket of silt and sand. The change in the river's course disrupted irrigation over a vast region, so that countless others suffered from famine during the following years. The river continued to run rampant throughout the war and was not returned to its old course until 1947.

The river's dikes had been intentionally breached on several other occasions, for example, in 1642, toward the end of the Ming dynasty (1368–1644), when Gen. Gao Mingheng used the tactic near Kaifeng in an attempt to suppress a peasant uprising. Natural floods were common, however, occurring whenever torrential summer rains raised the river above its embankments. Historical records show that the river has breached its dikes more than 1,500 times during the last 2,500 years, often with devastating results. The river has also repeatedly carved a new course—about once every century. The scars of no fewer than fifteen ancient riverbeds are still discernible in aerial photographs. The Yellow River's destructive power is so great that it has long been called "China's sorrow." Although the Chinese have kept the river at bay for the last few decades, the battle goes on, and victory seems as elusive as ever.

Nevertheless, the Chinese regard the Yellow River with reverence. Just as Egypt's civilization is the "gift of the Nile," so too Chinese culture arose on the fertile land created by the Yellow River. Over time, the river has deposited thick, rich soils where there once were vast lakes and marshland, forming the North China Plain—the largest delta in the world.

The region's temperate climate is well suited to agriculture. In the north, however, the rainfall is sparse and notoriously variable from one year to the next. Since ancient times, farmers have overcome this deficiency by turning to the Yellow River for irrigation. Major irrigation canals to divert water from the river were first constructed about 200 B.C. Evidence of the ancient hydraulic works is etched on the face of the land, and some of the ancient canals, perhaps 2,000 years old, are still in use. The flat terrain and the deep soil, coupled with the availability of river water for irrigation, enabled the Chinese to transform the land into a farming region second to none in extent and productivity.

The ability to manage the river and prevent floods had long been a criterion of good government. The emperor was believed to carry a heavenly mandate and responsibility to control the river, and failure to do so would deprive him of legitimacy. The task was not easy, however. In addition to the network of canals, the Chinese have had to build dikes on top of the river's natural levees to prevent the river from cresting over its banks and changing its course entirely. At times, as many as a million workers were conscripted to shore up these defenses. When breaches did occur, the workers piled bundles of kaoliang stalks (a type of sorghum) or rocks into the gap to stem the hemorrhage.

The river, which at times has the consistency of thick lentil soup, owes its unruly behavior and name to the heavy burden of silt it carries to the sea. The Yellow River is the muddiest of the world's major rivers; its churning waters may contain as much as 34 percent silt by weight. The two billion tons of sediment it carries each year are four times the Mississippi's load.

From *Natural History*, August, 1991, pp. 28-36. © 1991 by the American Museum of Natural History. Reprinted with permission.

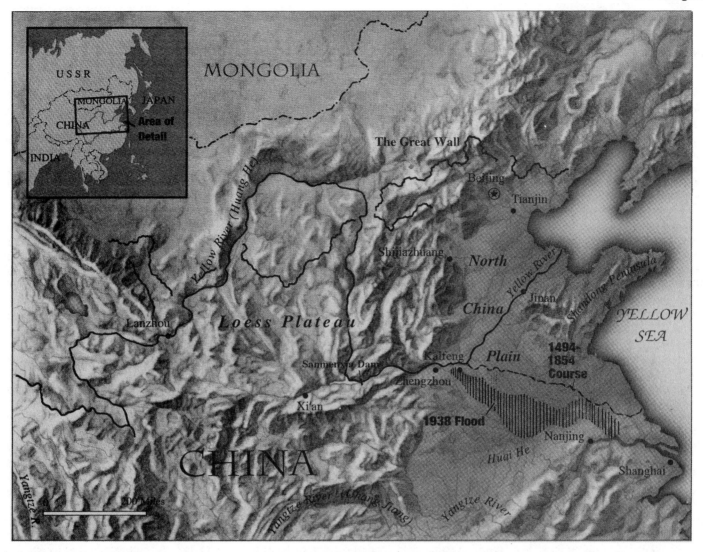

The map shows three courses of the Yellow River: the current one; the path of the 1938 flood; and the route prior to the river's shift to the north in 1854—one of the many abandoned channels that cross the North China Plain.

When the river finally reaches its estuary, it spills its sediment into the Yellow Sea, which is itself tinted by the pale golden silt. While the Mississippi extends its delta some six miles into the Gulf of Mexico every century, the Yellow River's delta can grow by the same amount in just a few years.

After coursing through the mountains, the river emerges onto the flat North China Plain, some 400 miles short of the sea. At this point, the river becomes sluggish and loses some of its capacity to carry particles in suspension. About a quarter of the river's silt settles to the riverbed, filling in the channel. As is usual with heavily silt-laden rivers, the Yellow River builds a bed above the surrounding plain and flows between the broad levees that form naturally when high waters deposit silt on the banks. Left to run free, the river would regularly abandon the raised channel and

establish a new course, until it, too, rose above the plain. For more than half a million years, the river meandered across northern China in this fashion, distributing its load evenly across the plain.

Shoring up the river's natural levees with more than 700 miles of dikes has required enormous investments of time and labor and has proved again and again to be only a temporary expediency. As silt is deposited along the bottom of the channel, the river's capacity to transport water is diminished. Each year the riverbed rises by as much as three inches. When the summer's torrential rains fall on the river's catchment, the tributaries surge, and as they converge into the main channel, the river swells. At this point it may burst through the confining levees built so laboriously by hand, once again inundating huge tracts of land and causing human suffering on a colossal scale.

The yellowish silt that makes this river so ungovernable comes from northern China's highlands. The river originates on the border of Tibet, 3,000 miles from the sea, and on its turbulent and tortuous course east and north, it winds through the loess plateau, named for its distinctive yellowish, wind-blown soil. For more than a million years, strong winds have swept up dust from the vast deserts of Central Asia and laid it down in north-central China. Similar loess deposits are found in other parts of the world, such as the Great Plains of the United States, but nowhere do they reach the thickness that they do in China. Over the millennia, layers of loess accumulated, forming a blanket of soil several hundred feet deep—the deepest soil in the world.

Loess is remarkably homogeneous and devoid of stones. Although it contains very little organic matter, its high mineral content makes loess very fertile when properly cultivated. But because it is loosely packed, loess is extremely vulnerable to erosion. When a handful of dry loess is crumbled, it reverts to a fluffy powder easily carried by swirling air currents. When pelted by raindrops or lapped by running water, the soil appears to dissolve, forming a muddy suspension. Although the mean annual precipitation on the plateau is less than eighteen inches, most of it falls in July and August in violent downpours that scour the exposed surface and wash away the delicate soil, resulting in a desolate "badlands" terrain of barren hills, deep gullies, and steep cliffs.

Although the tiny grains of loess are not cemented by humus, the jagged particles pack together and interlock, so that vertical walls cut into the soil hold up without slumping—as long as the material remains dry and undisturbed. The soil supports sheer cliffs several hundred feet high, into which people have, since ancient times, dug caves for dwelling and grain storage. When saturated with water, however, the loess becomes unstable and may suddenly collapse. Where undercut by streams, huge chunks of the cliffs crash down and may temporarily block or divert an entire river. Where the Yellow River's tributaries have washed away much of the loess, only great bluffs remain.

According to legend, the Chinese learned how to cultivate crops from Houji, who was born a son of heaven and became an official in charge of agriculture on the banks of the Qishui River in the southern part of the loess plateau. Records from the Han period, some 2,000 years ago, reveal that the area was once wetter and covered with vegetation, so that agriculture and livestock flourished. Before the dawn of civilization, the vegetation growing on the plateau protected it against rapid erosion. Prolonged exploitation, however, has completely destroyed the natural ecosystem that once protected the soil. In the 1700s, the growth of the region's population led to the widespread cutting of forests and use of the plow in farming, greatly accelerating the rate of erosion. (Half of China may have once been forested, compared with today's 8 percent.)

Nevertheless, cultivation of the loess plateau continues. The fertile soil and the mild climate make it one of China's major centers of grain production, despite the unreliable and occasionally destructive nature of the rainfall. Inadequate rainfall has led the farmers of this region to further expand the area under cultivation in an effort to compensate for the low yield of their fields. This expansion further denudes the land and accelerates erosion.

In addition to farming the arable land, the people of the region graze their domestic animals, mainly sheep, on the steepest slopes. Because the ownership of these rangelands is often unclear, herders tend to increase the number of animals grazing, with little regard for the pastures' capacity to recover from overgrazing.

Added to these problems is the shortage of fuel for cooking and heating. Because they cut down most of the trees on the hills long ago, and because they are often too poor to buy fuels at the market, the region's farmers have resorted to burning the inedible portion of their crops, which in earlier times they had returned to the land to replenish the soil's fertility.

In recent years, the Chinese have intensified their efforts to moderate and harness the Yellow River at its source. The government has undertaken massive engineering projects along the upper reaches of the river, including flood-control dams that are also used for irrigation and the generation of hydroelectric power. The first such attempt ended in disaster. Gradual siltation of reservoirs is a major problem worldwide, but the reservoir created by the Sanmenxia Dam (completed in 1960 in the last gorge before the river enters the plain) was almost completely filled with silt in four years. Extensive modifications of the dam, including converting eight tunnels housing turbines for power generation into channels for silt-laden water to drain through, have temporarily restored a third of the reservoir's capacity to hold floodwaters. Another large dam is planned downstream from Sanmenxia, and fourteen more dams upstream in hopes of controlling the largest flood, which might be expected to occur once in a thousand years. Rapid siltation, however, is likely to limit the effectiveness of these efforts. Drains at the base of dams create a swift current that scours away some of the accumulating silt, but most of the load is deposited upstream of the dams, where the river enters the reservoir, so that siltation will quickly reduce their capacity. Large engineering projects, therefore, may not be the answer to the river's problems.

With government encouragement, the farmers themselves have been trying to reduce runoff and erosion from the uplands, a Sisyphean effort that involves cutting terraces into the steep slopes. Most of the work is done by hand. Each field plot is leveled, and the edge of each is marked by a low earthen embankment, about half a foot above the plot. This height is considered sufficient to impound the water during heavy rains, promoting infiltration and minimizing runoff. From time to time, however, torrential rains scour the embankments of these terraces. Occasionally, cracks form in the vertical walls between the step-terraces, causing portions of hillsides to collapse. So far these terracing efforts have resulted in only a slight reduction of the erosion, and the task is a never-ending, arduous endeavor.

Although the Yellow River has been controlled within the ever-rising dikes for several decades, the hazard of inundation increases as the channel's capacity to carry water diminishes. Despite the immense effort devoted to their reinforcement, the dikes along the lower course of the river are now even more vulnerable to destruction by surging waters. Dike breaches took place in 1954 and 1958, and as the channel continues to rise, a catastrophic flood may occur at any time. Historical records show that peak flow rates exceeding 106,000 cubic feet per second have occurred in previous centuries—twice the peak flood discharge in 1982, which was the highest in recent decades. Because such peak flow rates are likely to recur sooner or later, the inhabitants of the plain are living on borrowed time.

As long as the loess plateau remains largely denuded of its natural vegetation and heavily cultivated, the accelerated runoff and erosion will continue to exacerbate the problem. Reestablishing a dense and stable cover of grasses, shrubs, and trees would protect the slopes of erodible soil from the torrential rains. But the farmers of the region cannot be expected to give up their cropland and pastures unless they are offered an alternative livelihood. With China's population of more than 1.1 billion growing at 1.4 percent per year, the problem is growing more intractable with each passing year.

Even in the absence of human disturbance, the loess plateau would continue to erode naturally. The silt would accumulate in the lowlands at a slower rate, but the process is inexorable. The Chinese are indeed facing an unrelenting force of nature that must be accommodated, since it cannot be completely tamed. The ancient, capricious dragon still lurks in the Yellow River, which continues to wear away the plateau and then tumbles onto the plain on its way seaward, as contemptuous as ever of human efforts to control it.

Low Water in the American High Plains

David E. Kromm

David E. Kromm, professor of geography at Kansas State University, has authored and coauthored numerous articles and one book on water management issues.

Depletion of the Ogallala Aquifer once threatened to return the region to a Dust Bowl, but much is being done to conserve groundwater today.

Water defines many regions. It determines their character and sustains their well-being. Ironically, this is especially true of areas with an inherent water scarcity. What would Southern California and central Arizona be like without water imported from distant sources? The green fields and golf courses would give way to dry plains and hills, populated more by grazing animals than by people.

A huge dry area in the middle of the United States depends no less on water to transform a dusty outback into a productive garden. This is the High Plains overlying the Ogallala Aquifer.

There exists widespread concern that America's largest underground water reserve is drying up. The vast Ogallala Aquifer that underlies 1,374,000 square miles of the High Plains from west Texas northward into South Dakota has been partially depleted as more than 150,000 wells pump water for irrigation, municipal supply, and industry. In some areas the wells no longer yield enough water to make irrigation possible. In others there remains sufficient water, but it lies 300 or more feet below the surface. The cost of lifting water from such depths makes it uneconomical for many uses.

Touching base with some key words in the groundwater vocabulary helps tell the Ogallala story more precisely. One is *aquifer*. An aquifer is a zone of water-saturated sands and gravels beneath the earth's surface. It is not an underground lake or river. It is the porous rock structure that contains the water that we tap with wells. Some aquifers release water to wells readily, whereas others hold the water tightly. This affects the rate at which water can be withdrawn and is called the *specific yield* of a well. *Depth to water* describes the vertical distance from ground level to the aquifer. Together with the volume and quality of water, the depth and specific yield define the economic limit of water withdrawal. The fresh water in an aquifer is called *groundwater*, in contrast to surface waters such as rivers and lakes. Over half the nation's population depends on groundwater for its drinking water. Although there are several aquifers, the Ogallala ranks as the main formation in the High Plains aquifer system. Most people in the region refer to the entire system as the Ogallala.

The unconsolidated sand and gravel that form the Ogallala aquifer were laid down by fluvial deposition from the Rocky Mountains about 10 million years ago. More recent, near-surface deposits

This article appeared in *The World & I*, February 1992, pp. 312-319. Reprinted with permission from *The World & I*, a publication of The Washington Times Corporation. © 1992.

3. THE REGION

of the late Tertiary and Quaternary ages compose the High Plains aquifer.

The volume of available water varies from place to place. Nebraska has about two-thirds (65 percent) of the High Plains groundwater, followed by Texas (12 percent), and Kansas (10 percent). The total water in storage is enough to fill Lake Huron or to cover the state of Colorado to a depth of 45 feet. Physical characteristics of the aquifer result in one farmer in a county having little or no access to water while another has substantial reserves. Ogallala water moves gradually from west to east, following the general slope of the land surface. This movement does not affect use or depletion, as it takes nearly a century and a half to flow a mere 10 miles.

Changing views

When Stephen Long explored the region in the early nineteenth century, he discovered a sea of grass that contrasted with the trees that dominated landscapes to the east. Few lakes or rivers can be found. These signs of aridity led Long to name the area "The Great American Desert." Settlers on their way to rainy Oregon viewed the plains as worthless territory that had to be crossed in order to reach a land more promising. Even those few who attempted to break the prairie sod found it tough to plow. The native Americans and buffalo herds were initially displaced by cattle ranching, not by cultivation.

A whole new perception of the plains burst forth in the middle of the nineteenth century as railroads penetrated the region and touted its virtues. The promoters encouraged the development of farming communities that would depend on the rails for goods and the delivery of grain to eastern markets. New steel plows cracked the soil and many boost-

The Ogallala Aquifer

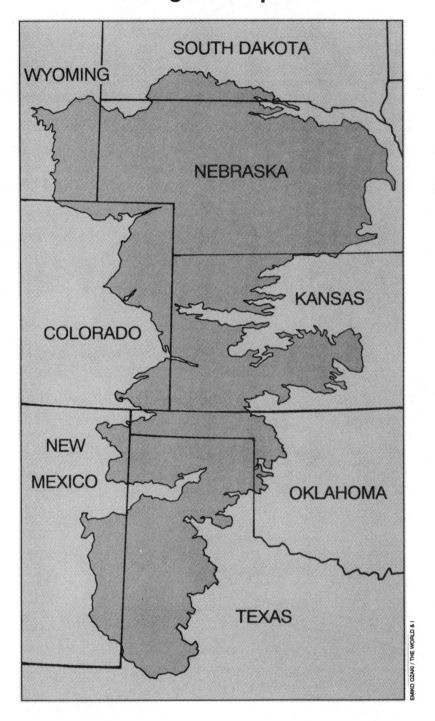

EMIKO OZAKI / THE WORLD & I

ers came to believe that "the rain follows the plow." Once the ground was broken and crops planted, the aridity would end. Hopeful farmers took up homesteads from the government or bought lands directly from the railroad companies.

There ensued frequent dry years and just enough wet ones

to allow many of the farmers to hang on. The rosy view of the High Plains as a garden was soon tarnished but never wholly abandoned. This changed with the 1930s. Everything went wrong. Depression weakened markets and reduced capital availability and drought destroyed the crops. A new name was coined for the

■ **The promotional railroad poster, 1887: the advantages of settling the High Plains.**

High Plains—the Dust Bowl. Outmigration swelled and the popular image was that of skies darkened by blowing dirt, desolate farmsteads surrounded by drifting soil, and impoverished families attempting to leave with whatever they could carry in their dilapidated vehicles. John Steinbeck chronicled the depth of misery in *The Grapes of Wrath*.

Emergence of a new Corn Belt

Even as the last settlers were taking up land in the 1890s, irrigation was beginning. It occupied small patches dependent on surface sources such as the Arkansas River or shallow wells. The Dust Bowl brought renewed interest in soil and water conservation and encouraged irrigation to create a stable moisture supply. Farmers suffered greatly from climatic variability in the High Plains. In a given year it might be a land of bountiful rain or a sun-drenched and parched desert.

Irrigation expanded in the 1930s and even more so in the dry years of the 1950s, as new technologies made possible large-scale tapping of the Ogallala aquifer. Pump engines came on the market that could lift large quantities of water from sources deep underground. New watering methods, such as center-pivot sprinklers, allowed irrigation of sandy and hilly areas. The green circles seen from aircraft flying more than 30,000 feet above the surface became the most prominent feature in the High Plains. The perception of the 1960s and 1970s was of a new Corn Belt flourishing in the land of the underground rain. Thanks to the Ogallala, stability had at last come to the High Plains.

Or had it? For the most part the Ogallala is a fossil water aquifer. It was formed millions of years ago and has been minimally recharged since. Only in a few areas such as the Nebraska, Sandhills and the sand-sage prairie of southwestern Kansas

was new water being added. Elsewhere, whatever was pumped out was forever gone. Wells sunk into zones with limited saturation thickness dried up early. Many were in west Texas, where large-scale irrigation began. By the late 1970s the whole High Plains region seemed threatened again, no less than it had been in the dirty thirties. Between 1978 and 1987, the irrigated area harvested in the High Plains sharply fell from nearly 13 million acres to 10.4 million acres.

Depletion ranks as such a serious problem that quality and streamflow issues are easily overlooked. In its natural state, the High Plains aquifer is generally of high quality. It is suitable for a wide array of uses without filtration. Where soils are sandy, however, precipitation can reach the water table, bringing with it nitrates from fertilizers and other surface contaminants. Abandoned water wells directly link pollutants and the aquifer. Where irrigation development is intense, decline in stream flows often follows. In Kansas, more than 700 miles of once permanently flowing rivers are now dry. The blue band on maps representing the Arkansas River in western Kansas hides the reality of a dry streambed.

An integrated agribusiness economy

Fears of impending disaster were real as the regional economy had expanded and grown relatively prosperous on the basis of irrigated agriculture. Except for the Platte River valley in Nebraska and Colorado, and minor surface sources in other places, groundwater sustained irrigation. In the southern reaches of the High Plains, irrigated cotton supports gins, oil mills, and a denim factory. Throughout most of the region farmers cultivate forage crops to

■ Center-pivot irrigation has permitted the short-term cultivation of sandy, sloping land. Such areas are suitable only for grazing in the long term.

feed cattle. Much of the cattle industry centers on huge feeder lots that collectively consume many tons of grain daily. The cattle in turn supply large meat processing plants, some of which are among the largest in the world.

Through the multiplier effect, irrigated agriculture affects all aspects of the regional economy. Where irrigation prevails, most rural towns seem vital, alive, and healthy. Where irrigation is absent, much less economic activity occurs, far fewer people are employed in commerce, and a decreasing volume and array of goods and services are available. Boarded-up storefronts provide visual evidence of the decline. One measure of civic progress in these communities is the tearing down of no longer needed buildings. Another is keeping the weeds under control in empty lots where structures once stood.

Observing this depopulation and decay, Deborah and Frank Popper of Rutgers University have recommended that the entire area be turned over to native grasses and animals. They

call for a buffalo commons. The Poppers see a natural process of people moving away, with the population becoming so small and scattered that basic educational, medical, and commercial services could no longer be provided. To be fair, it should be noted that extensive cultivation and cattle ranching would continue to occupy the land, leaving little space for buffalo. Nonetheless, the reduction in the number of towns and farmsteads would approximate the appearance of the High Plains of more than a century ago. This general contraction of people and services occurs where dryland farming prevails. It could become the norm for irrigated areas, where scarcity results from high pumping costs as water is lifted from ever-greater depths, reduction in specific yield, or lack of water as wells are physically depleted.

The High Plains has exemplified what Wes Jackson, director of the Land Institute in Kansas, calls "the failure of success." The success of irrigation was sowing the seeds of its own destruction through depletion. Relying on a single water source

also creates vulnerability. Take away the aquifer and the High Plains is without sufficient precipitation or surface water to sustain intensive agriculture or most other activities.

Responding to water scarcity

Few residents in the High Plains are standing idly by, merely watching a drama of decline being played out as irrigation becomes less and less a part of the regional economy. Water conservation has become more prevalent and is seen as the answer to sustaining irrigation and the integrated agribusiness economy it supports. Municipalities, factories, and other water users have lessened water use, but the major concern remains irrigation. More than 90 percent of the water consumed in the High Plains goes to irrigation. If irrigators are able to significantly reduce water consumption, for most there should be adequate and accessible water for many decades to come.

What are irrigators doing to conserve water? In 1988–90, with support from the Ford Foundation, the author and fellow geographer Stephen E. White looked into this question. A questionnaire was mailed to 1,750 irrigators in Texas, Oklahoma, Kansas, and Nebraska, asking them what water-saving practices they had adopted. Over 40 percent (709) returned a completed survey. Those who responded cited over 21 conservation practices now in wide use throughout the High Plains. Some of these techniques are: periodically checking pumping plant efficiency, planting drought-tolerant crops, replacing open ditches with underground pipes, and recovering runoff from fields. On the center-pivot sprinkler system, low-pressure heads are being installed on drop tubes to reduce overwatering and wind drift.

The days of large amounts of water draining off the fields into ditches or leaching downward beyond the root zone are largely gone.

Several of the widely used practices are sophisticated, information-based techniques that can be highly effective. Scheduling irrigation based on moisture need serves as an example. The goal is to ensure that just enough water is in the root zone of a plant to accomplish the management objectives of the farmer. Soil water levels and plant stress are monitored so that the farmer can determine how much water to apply and when. The days of large amounts of water draining off the fields into ditches or leaching downward beyond the root zone are largely gone.

Institutions play a major role in how the Ogallala and other aquifers are used and who controls the decisions affecting water use. Water law provides an example. Two general forms of water rights exist in the United States, and both function in the High Plains. In Texas, *riparian water rights* prevail, wherein the owner of land overlying an aquifer has the right to use the groundwater beneath. In the remaining states in the region differing forms of the *appropriation doctrine* exist. Under appropriation, the first party to establish the right to use a specified amount of groundwater (or surface water such as a river) has priority of use in times of scarcity. Junior rights may be reduced or even curtailed when they endanger senior rights. The doctrine may be summarized as "first in time, first in right." It has served the western states well, in that insufficient water is available to provide enough for all landowners, as is necessary to apply the riparian doctrine. Most

states have developed a priority in water use, with domestic and livestock consumption usually ranking first.

These water rights are administered by various levels of organizations and agencies. Water largely falls under state control, but substate or local districts are often permitted by law. These authorities are formed usually to ensure that existing rights, often for irrigation, are protected.

The authority of the local district varies from largely educational, establishing programs to advocate water conservation, to requiring a land-use plan or metering of water flow for all irrigators with water rights. Local districts frequently have enforcement authority. As a rule, a main goal of both state and local groups is to protect existing water rights, most of which are held by irrigators. Increasingly, water markets

■ **Different application strategies are helping conserve irrigation water.**

Top: **This standard center-pivot sprays water up into the wind.**

Right: **These low-energy precision application sprinklers have low-pressure heads attached to drop tubes, thereby minimizing wind drift.**

COURTESY OF DAVID KROMM

COURTESY OF DAVID KROMM

are being introduced that allow rights to be sold to other users. A farmer, for example, may sell a water right to a municipality or to a developer converting land from agricultural to residential use.

Much remains to be done, but the institutions and technologies of the High Plains show that the region is responding effectively. Most of the states now have water or resource management districts that are responsible for ensuring wise use of the Ogallala. A variety of policies have emerged. In various areas, wells are spaced so as not to interfere with each other, flow meters are required to measure water use, and drilling of new irrigation wells is prohibited. Although planned and orderly depletion constitutes the primary goal, one groundwater management district in western Kansas is considering a zero-depletion policy.

Innovation in irrigation technology has been striking. Many of the water-saving practices and devices used by farmers in the region and elsewhere were developed or first applied in the High Plains. Most are manufactured in the area. Land-grant universities throughout the region lead the way in developing techniques to conserve water. Nonprofit groups

■ This field of soybeans is irrigated using a "surge flow system" that wets the soil with intermittent impulses to reduce infiltration. Water-saving practices such as this are now being adopted by many irrigators tapping the waters of the Ogallala Aquifer.

such as the Nebraska Groundwater Foundation support public education efforts. The High Plains ranks as a global center of technology and institutions for improving efficiency in irrigation.

What next?

Although numerous writers continue to cite the High Plains as a region where significant ground-water depletion occurs and where irrigation is doomed by its own excesses, current practices and conditions no longer support these views. Because of institutional and technical innovation, educational campaigns, and the land and water stewardship of most farmers, sustainable irrigation appears likely in the High Plains. The Great American Desert will not reappear as the Buffalo Commons.

The Key To Understanding The Former Soviet Union

Cultural geography is more powerful than boundaries

Joel Garreau

Joel Garreau, a Washington Post *staff writer, is the author of "The Nine Nations of North America," which helped shape the "Beyond Boundaries" project, His latest book is "Edge City: Life on the New Frontier."*

In Alaska, people feel so remote from the rest of America that they refer to it as "The Lower Forty-Eight" or, more telling, just "Outside." There's a place just like that in the former Soviet Union. On the Pacific Coast, especially the Kamchatka Peninsula, people feel they live in a land of adventure of mythic proportions so separate from the rest of their countrymen that they refer to a trek to the population centers back West as "visiting the Continent."

In Appalachia and the Ozarks, hillbillies have been the butt of jokes for generations.

There's a place just like that in the former Soviet Union. If a Russian wanted to suggest that somebody was stupid, he would ask "Where are you from, the Urals?"

New Orleans is famous as a land of wit and anecdote and rogues and crooks—a joyful warm-water city that is the mouth of export for the inland Breadbasket.

There's a place just like that in the former Soviet Union. It's called Odessa.

In fact, all across the former Soviet Union, these parallels to America abound. And they often explain the origin of events in the news more clearly than do the political maps of places like Belarus or Ukraine to which Americans are just beginning to adjust. So say four Soviet and five American geographers who are collaborating to produce a new explanation of the underpinnings of the former Soviet Union. Appropriately for a time when old frontiers have evaporated, it is called "Beyond Borders" and is scheduled to be published in Russian and English, the latter by Macmillan next year.

Their approach is based on the idea that all countries have underlying patterns of loyalties, futures, indus-tries, histories, climates, resources and politics. These realities can be mapped as cultural and economic regions. These functional regions, in turn, frequently are far more significant than the arbitrary boundaries and surveyors' mistakes that usually make up politically defined borders. And these constituent realities of the former Soviet Union have far more parallels to North America than most Americans realize.

In other words, in a New World Order in which nations and boundaries are changing at a dizzying pace, what these geographers are trying to do is map the human values that endure—those that have taken centuries to produce and are not likely to change any time soon. For these factors, they believe, are the sources of current events. Americans, for example, know that Houston and the American Southwest are vastly different places from Boston and New England, and that elections and fortunes can and do pivot on that fact. Similarly, the Muslim portion of the former Soviet Union near Iran is profoundly different from the European gateway of St. Petersburg, so near Scandinavia. And that kind of functional reality will only become more important as empires dissolve.

The "Beyond Borders" geographers hasten to add that they are not trying to map the way they think the world should work. For example, by their reckoning, Ukraine is tugged at by four separate worlds, and it is not clear which is winning, or how this turbulent political situation will play out. Not a comforting thought about a place with nuclear weapons that wants to control the Black Sea fleet.

What they are claiming, however, is that this is the way the former Soviet Union really does work—at some profound level—and they believe Americans will readily recognize the analogies to their own continent. They add that those who ignore these distinctions and divisions stand to lose power, money, influence—and elections.

The American authors of "Beyond Borders" are Thomas Baerwald of the National Science Foundation, Kathleen Braden of Seattle Pacific University, Stanley

3. THE REGION

Brunn of the University of Kentucky, John Florin of the University of North Carolina at Chapel Hill and Wilbur Zelinsky of Penn State. The Moscow State University contingent includes Raymond Krishchyunas, Aleksei Naumov, Aleksei Novikov and Sergei Rogachev. Collectively they divide the United States and the former Soviet Union into 11 comparable regions:

■ **The Core.** In the Soviet Union, this is the region around Moscow that corresponds to the Russian state in the 15th century. It was what Russia was all about before expansion extended its hegemony across 11 time zones. The analogous core in America is that portion of the Eastern seaboard that includes Philadelphia, where the Republic was founded, and New York City. The big difference is that the Soviet system could and did command that Moscow would be the undisputed economic and political center. Rival capitals were not allowed to flourish in a way comparable to, say, Los Angeles—or Washington.

■ **The North.** This is Russia's New England, and St. Petersburg (a/k/a Leningrad) is Boston. Once the capital of the czars, it remains the home of a self-anointed intellectual elite, an historic center of learning and literature, and a bastion of high-brow culture highly influenced by the fashions and ideas of Europe. On both continents, this region is a cradle of democracy. Where Boston had its Tea Party, St. Petersburg produced revolutions twice (1905 and 1917). In this harsh physical environment, such traits as self-reliance, frugality, independence and cultural distinctiveness are prized, as well as a tradition of political liberalism.

■ **The Heartland.** The industrialized area of Russia east of Moscow from Gorky (now Nizhniy Novgorod) to the Ural Mountains is comparable to America's chain of Midwest industrial cities from Chicago to the Appalachians. Both are centered on muscular waterways—the Volga River and the Great Lakes. Both are the center of their respective auto industries. Both were the staging area for the settlement of the frontier. Both are the bedrock of their country's mercantile culture. But most critically, both hold a central position in the national mythology: If brawny, brawling, but cheerful and outgoing places like Chicago are a symbol of what it means to be all-American, then Samara and Kazan are roughly equivalent in the former Soviet Union.

■ **The South.** Like Dixie a land of romance, tradition and gentle, beguiling climate, it too is stained by massive historical social injustice. In the former Soviet Union, this was serfdom, which was basically slave labor. This legacy, which destroyed pride in work for both the master and the slave, killed initiative, making the area conservative and subdued. Many inhabitants still don't care about any issues beyond their garden plot. Cheap demagoguery is a tradition, but new currents of engagement are stirring. This place includes much of Ukraine. The Ukranian capital of Kiev, once in decline, is now dynamic, suggesting parallels to Atlanta. At the same time, this is a place that has supported an aristocratic, educated elite. The Russian portion of The South is second only to the New England-like North in terms of literature, being home to Turgenev and Tolstoy. It also was marked by a chivalrous code of behavior, and to this day its inhabitants are disproportionately represented in the officer corps. After the serfs were emancipated in 1861, unable to adjust to a free labor market, this area declined economically. At the time of the 1917 Revolution, this area mostly supported the royalist forces of the old order against the Communists of the North. In fact, it still harbors tensions against the North, seen as the historic exploiter.

■ **The Breadbasket.** Like our Breadbasket, this land of prairies/steppes has rich black soil, as much as seven feet deep. The Soviet Breadbasket has a wheat climate more like Canada's than the corn climate of Iowa. Before communism, this part of Russia, Ukraine and Kazakhstan fed Europe. And it may again. Interestingly, former Soviet President Mikhail Gorbachev is from here, as were a striking number of Communist Party leaders. This is thought to be because a Breadbasket almost by definition is a land of self-made men. After all, the Cossacks—Russia's cowboys—were from here.

■ **The Crossroads.** As in southern Illinois, southern Indiana and western Kentucky, Belarus is neither really East nor West, North nor South. It is a land of split allegiances, poor land, and failed attempts at creating important cities. Not for nothing is its capital, Minsk, a Borscht Belt joke. Belarus is a faceless place—which maybe explains why its capital is an ideal capital for the new Commonwealth. It accepted communism far more than its neighbors, the Baltics to the north and western Ukraine. Those places, by contrast, are psychologically similar to Poland and other parts of Eastern Europe and are oriented west, away from Russia.

■ **The Old Mountains.** The Urals, like the Appalachians, are fairly easily passable—they are not much of a geographic barrier. Nonetheless, they trapped the least adventurous settlers on their way to the frontier, and today it is an exhausted and socially depressed land caught in industrial serfdom. This place was so remote that the Soviets moved much of their military industry here after the German invasion in World War II. But undercapitalization has left the industrial base obsolete.

■ **The Tropical South.** Like south Florida and the Caribbean Rim, this is a land of smugglers, drug runners, questionable dealings—and nice weather. A region of swamps only reclaimed in the 20th century, it was the center of shadow capitalism under communism, very good at profiting hugely from loopholes in the system. This place includes Crimea, and extends along the shore of the Black Sea to Georgia. Its major recreational center of Sochi is, well, Miami.

■ **Mexistan.** Like the American Southwest and Northern Mexico, this is a transition zone between two cultures—the European Russians to the north, and the Asian Muslims to the South. (The Central Asian zone beyond Mexistan—all those places like Turkmenistan, Uzbekistan, Tajikistan and Kirgizstan—is really part of the same Muslim core as Iran and Afghanistan.) Alma Ata, the Kazakh capital, is roughly the equivalent of Houston. The region is bilingual in Russian and various Turkic dialects. It is semidesert—agriculture cannot exist without irrigation. In fact, this place is so much like Southwest Texas that the local expression for "How do you do" translates literally as "How is your livestock?" There is significant industrialization because the Russians thought the locals here were culturally more open to training than those farther south. This is also the main proving ground for nuclear weapons, as well as the site of a lot of missiles with nuclear warheads presumably aimed at China.

■ **The Land Ocean.** Like the "flyovers" of America's Intermountain West, this vast place is full of precious mineral wealth—oil, uranium, coal, copper—and just about no people. Wyoming is Siberia minus the prison camps.

■ **Pacific Gateway.** Like the residents of the northern Pacific Rim of America in Oregon, Washington and Alaska, the residents of the Russian Pacific Gateway think of themselves as islanders—historically isolated from the bulk of their country. As in Alaska, which it faces across the Bering Straits, wages here can be high, as can prices. The region is also seen as a place of adventure, and intense relationships to the land.

The "Beyond Borders" geographers do not feel that all parts of the United States and the former Soviet Union are analogous.

For example, they isolate California as a world entirely unto itself.

But you knew that.

Many of the apparent aberrations they identify nonetheless resolve themselves if you include in your mental map Canada and Mexico.

For example, a Baltic nation like Lithuania with an overwhelmingly Catholic population, a history of being absorbed into an alien empire only at the point of a gun, a long-standing sense of nationhood, and a never-conquered sentiment for separatism, has a perfectly legitimate North American doppleganger—Quebec.

(Indeed, Quebec separatists are closely watching how swiftly and seemingly easily the European Community and the United States recognize the former Soviet republics as sovereign nations.)

And if Mexistan is a buffer between the First and Third Worlds, like the region that includes the American Southwest and Northern Mexico, the more purely Muslim region near Iran is analogous to central or even southern Mexico.

The practical significance of these observations is simple—it helps to explain or even predict the news. Without in any way suggesting that civil conflict is looming in the former Soviet lands, geographer Zelinsky nevertheless turned to the American Civil War as a moment that defined underlying realities of this country in extraordinarily clear ways.

"Look at a monstrosity like Maryland," he says. "That's a collection of historical accidents. It's at least three or four different places. Well, as long as there is a United States, Maryland will probably remain intact. But empires can divide. We had a comparable situation in 1860, 1861 in this country." He points out that during the American Civil War, "Virginia, one of the major states in the Confederacy just could not hold together. If we'd had a map like the one for this project back then, we could have anticipated that a West Virginia was aborning. We could also have realized that the mountaineers of Eastern Tennessee, or the Free State in Northern Alabama, would not go along with Jeff Davis. That's Appalachia.

"Even today, there is the likelihood that Quebec will opt for secession. What will happen to the large Anglo minority that inhabits the southwest corner? It's like Lithuania, with a very large minority population that is Russian or Polish.

"The point is that when we're talking about the former Soviet Union or even Russia, we're not talking about some homogeneous block of territory. To understand this really important commonwealth [is] . . . to understand they have powerful internal tensions. The word 'Ukraine' means 'The Borderland.' There are similar divisions in Kazakhstan, Ukraine and Georgia.

"There was a civil war within Missouri during our Civil War. It was every bit as bloody as Kansas. That was a matter of a state being drawn in total ignorance of cultural realities.

"If there had been a cultural geographer around in the 1850s, that would have been no surprise."

Spatial Interaction and Mapping

Geography is the study not only of places in their own right, but of the ways in which places interact. Places are connected by highways, airline routes, telecommunication systems, and even thoughts. Geographers study these forms of spatial interaction because they are an important part of their work.

In "Transportation and Urban Growth: The Shaping of the American Metropolis," Peter Muller considers transportation systems, analyzing their impact on the growth of American cities. The next article deals with spatial interaction in the form of Hispanic migration in the United States. The next two articles provide excellent examples of significant changes in spatial interaction and accessibility. The Rhine-Main-Danube Canal and the "Chunnel" will work to bring the countries and regions of Europe closer together.

How could geographers even begin to describe in words the detailed spatial pattern of, for instance, the major agricultural regions in the world? Even photographs could not do the job adequately, because they literally capture too much of the detail of a place. There is no better way to present many of the topics analyzed in geography than with maps. Maps and geography go hand in hand. Although maps are used in other disciplines, their association with geography is the most highly developed.

A map is a graphic that presents a generalized and scaled-down view of particular occurrences or themes in an area. If a picture is worth a thousand words, then a map is worth a thousand (or more!) pictures. There is simply no better way to "view" a portion of Earth's surface or an associated pattern than with a map. How else could

we see the entirety of South America, for example, in one glance?

The first two articles on mapping deal respectively with the history of cartography and the way maps can both broaden one's perspective and possibly reinforce ethnocentrism. The last two articles deal with current developments in the growth of geographic information systems (GIS) and their applications to land-use planning and management.

Looking Ahead: Challenge Questions

How would you describe the spatial form of the place in which you live? If you live in a town or city, why was that particular location chosen?

How does your hometown interact with its surrounding region? With other places in the state? With other states? With other places in the world?

How are places "brought closer together" when transportation systems are improved?

What problems occur when transportation systems are overloaded?

How will public transportation be different in the future? Will there be more private autos in the next 25 years, or fewer?

Are you a good map reader? Why are maps useful in studying a place? How will the new technologies in satellite imagery improve mapping?

What changes in geography will be brought about by GIS innovations?

Unit 4

Transportation and Urban Growth

The shaping of the American metropolis

Peter O. Muller

In his monumental new work on the historical geography of transportation, James Vance states that geographic mobility is crucial to the successful functioning of any population cluster, and that "shifts in the availability of mobility provide, in all likelihood, the most powerful single process at work in transforming and evolving the human half of geography." Any adult urbanite who has watched the American metropolis turn inside-out over the past quarter-century can readily appreciate the significance of that maxim. In truth, the nation's largest single urban concentration today is not represented by the seven-plus million who agglomerate in New York City but rather by the 14 million who have settled in Gotham's vast, curvilinear outer city—a 50-mile-wide suburban band that stretches across Long Island, southwestern Connecticut, the Hudson Valley as far north as West Point, and most of New Jersey north of a line drawn from Trenton to Asbury Park. This latest episode of intrametropolitan deconcentration was fueled by the modern automobile and the interstate expressway. It is, however, merely the most recent of a series of evolutionary stages dating back to colonial times, wherein breakthroughs in transport technology unleashed forces that produced significant restructuring of the urban spatial form.

The emerging form and structure of the American metropolis has been traced within a framework of

From FOCUS, Summer 1986, pp. 8-17. Reprinted by permission of The American Geographical Society.

Horse-drawn trolleys in downtown Boston, circa 1885. **(Library of the Boston Athenaeum)**

four transportation-related eras. Each successive growth stage is dominated by a particular movement technology and transport-network expansion process that shaped a distinctive pattern of intraurban spatial organization. The stages are the Walking/Horsecar Era (pre-1800–1890), the Electric Streetcar Era (1890–1920), the Recreational Automobile Era (1920–1945), and the Freeway Era (1945–present). As with all generalized models of this kind, there is a risk of oversimplification because the building processes of several simultaneously developing cities do not always fall into neat time-space compartments. Chicago's growth over the past 150 years,

for example, reveals numerous irregularities, suggesting that the overall metropolitan growth pattern is more complex than a simple, continuous outward thrust. Yet even after developmental ebb and flow, leapfrogging, backfilling, and other departures from the idealized scheme are considered, there still remains an acceptable correspondence between the model and reality.

Before 1850 the American city was a highly compact settlement in which the dominant means of getting about was on foot, requiring people and activities to

tightly agglomerate in close proximity to one another. This usually meant less than a 30-minute walk from the center of town to any given urban point—an accessibility radius later extended to 45 minutes when the pressures of industrial growth intensified after 1830. Within this pedestrian city, recognizable activity concentrations materialized as well as the beginnings of income-based residential congregations. The latter was particularly characteristic of the wealthy, who not only walled themselves off in their large homes near the city center but also took to the privacy of horse-drawn carriages for moving about town. Those of means also sought to escape the city's noise

Electric streetcar lines radiated outward from central cities, giving rise to star-shaped metropolises. Boston, circa 1915. (Library of the Boston Athenaeum)

and frequent epidemics resulting from the lack of sanitary conditions. Horse-and-carriage transportation enabled the wealthy to reside in the nearby countryside for the disease-prone summer months. The arrival of the railroad in the 1830s provided the opportunity for year-round daily commuting, and by 1840 hundreds of affluent businessmen in Boston, New York, and Philadelphia were making round trips from exclusive new trackside suburbs every week-day.

As industrialization and its teeming concentrations of working-class housing increasingly engulfed the mid-nineteenth century city, the deteriorating physical and social environment reinforced the desires of middle-income residents to suburbanize as well. They were unable, however, to afford the cost and time of commuting by steam train, and with the walking city now stretched to its morphological limit, their aspirations intensified the pressures to improve intraurban transport technology. Early attempts involving stagecoach-like omnibuses, cable-car systems, and steam railroads proved impractical, but by 1852 the

first meaningful transit break-through was finally introduced in Manhattan in the form of the horse-drawn trolley. Light street rails were easy to install, overcame the problems of muddy, unpaved roadways,

---

Before 1850 the American city was a highly compact settlement in which the dominant means of getting about was on foot, requiring people and activities to tightly agglomerate in close proximity to one another.

---

and enabled horsecars to be hauled along them at speeds slightly (about five mph) faster than those of pedestrians. This modest improvement in mobility permitted the opening of a narrow belt of land at the city's edge for new home construction. Middle-income urbanites flocked to

these "horsecar suburbs," which multiplied rapidly after the Civil War. Radial routes were the first to spawn such peripheral development, but the relentless demand for housing necessitated the building of crosstown horsecar lines, thereby filling in the interstices and preserving the generally circular shape of the city.

The less affluent majority of the urban population, however, was confined to the old pedestrian city and its bleak, high-density industrial appendages. With the massive immigration of unskilled laborers, (mostly of European origin after 1870) huge blue-collar communities sprang up around the factories. Because these newcomers to the city settled in the order in which they arrived—thereby denying them the small luxury of living in the immediate company of their fellow ethnics—social stress and conflict were repeatedly generated. With the immigrant tide continuing to pour into the nearly bursting industrial city throughout the late nineteenth century, pressures redoubled to further improve intraurban transit and open up more of the adjacent countryside. By the late 1880s that urgently needed mobility revolution was at last in the making, and when it came it swiftly transformed the compact city and its suburban periphery into the modern metropolis.

The key to this urban transport revolution was the invention by Frank Sprague of the electric traction motor, an often overlooked innovation that surely ranks among the most important in American history. The first electrified trolley line opened in Richmond in 1888, was adopted by two dozen other big cities within a year, and by the early 1890s swept across the nation to become the dominant mode of intraurban transit. The rapidity of this innovation's diffusion was enhanced by the immediate recognition of its ability to resolve the urban transportation problem of the day: motors could be attached to existing horsecars, converting them into self-propelled vehicles powered by easily constructed overhead wires. The tripling of average speeds (to over 15 mph) that resulted from this invention brought a large band of open land beyond the city's perimeter into trolley-commuting range.

The most dramatic geographic change of the Electric Streetcar Era was the swift residential development of those urban fringes, which transformed the emerging metropolis into a decidedly star-shaped spatial entity. This pattern was produced by radial streetcar corridors extending several miles beyond the compact city's limits. With so much new space available for homebuilding within walking distance of the trolley lines, there was no need to extend trackage laterally, and so the interstices remained undeveloped. The typical streetcar suburb of the turn of this century was a continuous axial corridor whose backbone was the road carrying the trolley line (usually lined with stores and other local commercial facilities), from which gridded residential streets fanned out for several blocks on both sides of the tracks. In general, the quality of housing and prosperity of streetcar subdivisions increased with distance from the edge of the central city. These suburban corridors were populated by the emerging, highly mobile middle class, which was already stratifying itself according to a plethora of mi-

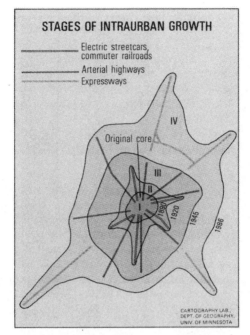

STAGES OF INTRAURBAN GROWTH

——— Electric streetcars, commuter railroads
——— Arterial highways
------ Expressways

IV

Original core

III

II

I 1890
1920
1945
1986

CARTOGRAPHY LAB.
DEPT. OF GEOGRAPHY.
UNIV. OF MINNESOTA

nor income and status differences. With frequent upward (and local geographic) mobility the norm, community formation became an elusive goal, a process further retarded by the grid-settlement morphology and the reliance on the distant downtown for employment and most shopping.

The ready availability and low fare of the electric trolley now provided every resident with access to the intracity circulatory system, thereby introducing truly "mass" transit to urban America.

Within the city, too, the streetcar sparked a spatial transformation. The ready availability and low fare of the electric trolley now provided every resident with access to the intracity circulatory system, thereby introducing truly "mass" transit to urban America in the final years of the nineteenth century. For nonresidential activities this new ease of movement among the city's various parts quickly triggered the emergence of specialized land-use dis-

tricts for commerce, manufacturing, and transportation, as well as the continued growth of the multipurpose central business district (CBD) that had formed after mid-century. But the greatest impact of the streetcar was on the central city's social geography, because it made possible the congregation of ethnic groups in their own neighborhoods. No longer were these moderate-income masses forced to reside in the heterogeneous jumble of rowhouses and tenements that ringed the factories. The trolley brought them the opportunity to "live with their own kind," allowing the sorting of discrete groups into their own inner-city social territories within convenient and inexpensive traveling distance of the workplace.

By World War I, the electric trolleys had transformed the tracked city into a full-fledged metropolis whose streetcar suburbs, in the larger cases, spread out more than 20 miles from the metropolitan center. It was at this point in time that intrametropolitan transportation achieved its greatest level of efficiency—that the bustling industrial city really "worked." How much closer the American metropolis might have approached optimal workability for all its residents, however, will never be known because the next urban transport revolution was already beginning to assert itself through the increasingly popular automobile. Americans took to cars as wholeheartedly as anything in the nation's long cultural history. Although Lewis Mumford and other scholars vilified the car as the destroyer of the city, more balanced assessments of the role of the automobile recognize its overwhelming acceptance for what it was—the long-awaited attainment of private mass transportation that offered users the freedom to travel whenever and wherever they chose. As cars came to the metropolis in ever greater numbers throughout the interwar decades, their major influence was twofold: to accelerate the deconcentration of population through the development of interstices bypassed during the streetcar era, and to push the suburban frontier farther into the countryside, again producing a compact, regular-shaped urban entity.

Afternoon commuters converge at the tunnel leading out of central Boston, 1948.   (Boston Public Library)

While it certainly produced a dramatic impact on the urban fabric by the eve of World War II, the introduction of the automobile into the American metropolis during the 1920s and 1930s came at a leisurely pace. The earliest flurry of auto adoptions had been in rural areas, where farmers badly needed better access to local service centers. In the cities, cars were initially used for weekend outings—hence the term "*Recreational* Auto Era"—and some of the earliest paved roadways were landscaped parkways along scenic water routes, such as New York's pioneering Bronx River Parkway and Chicago's Lake Shore Drive. But it was into the suburbs, where growth rates were now for the first time overtaking those of the central cities, that cars made a decisive penetration throughout the prosperous 1920s. In fact, the rapid expansion of automobile suburbia by 1930 so adversely affected the metropolitan public transportation system that, through significant diversions of streetcar and commuter-rail passengers, the large cities began to feel the negative effects of the car years before the auto's actual arrival in the urban center. By facilitat-

---

Americans took to cars as wholeheartedly as anything in the nation's long cultural history.

---

ing the opening of unbuilt areas lying between suburban rail axes, the automobile effectively lured residential developers away from densely populated traction-line corridors into the suddenly accessible interstices. Thus, the suburban homebuilding industry no longer found it necessary to subsidize pri-

vately-owned streetcar companies to provide low-fare access to trolley-line housing tracts. Without this financial underpinning, the modern urban transit crisis quickly began to surface.

The new recreational motorways also helped to intensify the decentralization of the population. Most were radial highways that penetrated deeply into the suburban ring and provided weekend motorists with easy access to this urban countryside. There they obviously were impressed by what they saw, and they soon responded in massive numbers to the sales pitches of suburban subdivision developers. The residential development of automobile suburbia followed a simple formula that was devised in the prewar years and greatly magnified in scale after 1945. The leading motivation was developer profit from the quick turnover of land, which was acquired in large parcels, subdivided, and auctioned off. Understandably,

Central City-Focused Rail Transit

The widely dispersed distribution of people and activities in today's metropolis makes rail transit that focuses in the central business district (CBD) an obsolete solution to the urban transportation problem. To be successful, any rail line must link places where travel origins and destinations are highly clustered. Even more important is the need to connect places where people really want to go, which in the metropolitan America of the late twentieth century means suburban shopping centers, freeway-oriented office complexes, and the airport. Yet a brief look at the rail systems that have been built in the last 20 years shows that transit planners cannot—or will not—recognize those travel demands, and insist on designing CBD-oriented systems as if we all still lived in the 1920s.

One of the newest urban transit systems is Metrorail in Miami and surrounding Dade County, Florida.

Major Activity Center

Metrorail

Freeway

Main Highway

0 1 2 3 4 5 6
Miles

CARTOGRAPHY LAB., DEPT. OF GEOGRAPHY, UNIV. OF MINNESOTA

It has been a resounding failure since its opening in 1984. The northern leg of this line connects downtown Miami to a number of low- and moderate-income black and Hispanic neighborhoods, yet it carries only about the same number of passengers that used to ride on parallel bus lines. The reason is that the high-skill, service economy of Miami's CBD is about as mismatched as it could possibly be to the modest employment skills and training levels possessed by residents of that Metrorail corridor. To the south, the prospects seemed far brighter because of the possibility of connecting the system to Coral Gables and Dadeland, two leading suburban activity centers. However, both central Coral Gables and the nearby International Airport complex were bypassed in favor of a cheaply available, abandoned railroad corridor alongside U.S. 1. Station locations were poorly planned, particularly at the University of Miami and at Dadeland—where terminal location necessitates a dangerous walk across a six-lane highway from the region's largest shopping mall. Not surprisingly, ridership levels have been shockingly below projections, averaging only about 21,000 trips per day in early 1986. While Dade County's worried officials will soon be called upon to decide the future of the system, the federal government is using the Miami experience as an excuse to withdraw from financially supporting all construction of new urban heavy-rail systems. Unfortunately, we will not be able to discover if a well-planned, high-speed rail system that is congruent with the travel demands of today's polycentric metropolis is capable of solving traffic congestion problems. Hopefully, transportation policymakers across the nation will heed the lessons of Miami's textbook example of how not to plan a hub-and-spoke public transportation network in an urban era dominated by the multi-centered city.

developers much preferred open areas at the metropolitan fringe, where large packages of cheap land could readily be assembled. Silently approving and underwriting this uncontrolled spread of residential suburbia were public policies at all levels of government: financing road construction, obligating lending institutions to invest in new homebuilding, insuring individual mortgages, and providing low-interest loans to FHA and VA clients.

Because automobility removed most of the pre-existing movement constraints, suburban social geography now became dominated by locally homogeneous income-group clusters that isolated themselves from dissimilar neighbors. Gone was the highly localized stratification of streetcar suburbia. In its place arose a far more dispersed, increasingly fragmented residential mosaic to which builders were only too eager to cater, helping shape a kaleidoscopic settlement pattern by shrewdly constructing the most expensive houses that could be sold in each locality. The continued partitioning of suburban society was further legitimized by the widespread adoption of zoning (legalized in 1916), which gave municipalities control over lot and building standards that, in turn, assured dwelling prices that would only attract newcomers whose incomes at least equaled those of the existing local population. Among the middle class, particularly, these exclusionary economic practices were enthusiastically supported, because such devices extended to them the ability of upper-income groups to maintain their social distance from people of lower socioeconomic status.

Nonresidential activities were also suburbanizing at an increasing rate during the Recreational Auto Era. Indeed, many large-scale manufacturers had decentralized during the streetcar era, choosing locations in suburban freight-rail corridors. These corridors rapidly spawned surrounding working-class towns that became important satellites of the central city in the emerging metropolitan constellation. During the interwar period, industrial employers accelerated their intraurban deconcentration, as more efficient horizontal fabrication methods replaced

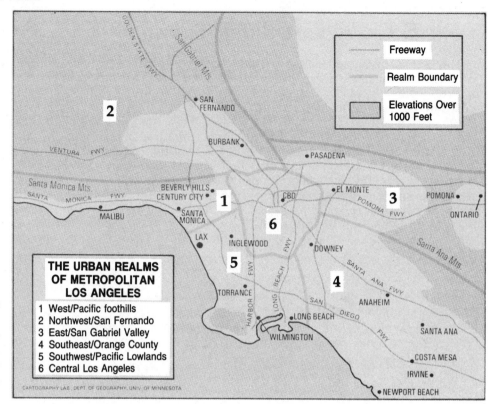

THE URBAN REALMS OF METROPOLITAN LOS ANGELES
1 West/Pacific foothills
2 Northwest/San Fernando
3 East/San Gabriel Valley
4 Southeast/Orange County
5 Southwest/Pacific Lowlands
6 Central Los Angeles

CARTOGRAPHY LAB, DEPT. OF GEOGRAPHY, UNIV. OF MINNESOTA

older techniques requiring multi-storied plants—thereby generating greater space needs that were too expensive to satisfy in the high-density central city. Newly suburbanizing manufacturers, however, continued their affiliation with intercity freight-rail corridors, because motor trucks were not yet able to operate with their present-day efficiencies and because the highway network of the outer ring remained inadequate until the 1950s.

Retail activities were featured in dozens of planned automobile suburbs that sprang up after World War I—most notably in Kansas City's Country Club District, where the nation's first complete shopping center was opened in 1922.

The other major nonresidential activity of interwar suburbia was re-

tailing. Clusters of automobile-oriented stores had first appeared in the urban fringes before World War I. By the early 1920s the roadside commercial strip had become a common sight in many southern California suburbs. Retail activities were also featured in dozens of planned automobile suburbs that sprang up after World War I—most notably in Kansas City's Country Club District, where the nation's first complete shopping center was opened in 1922. But these diversified retail centers spread slowly before the suburban highway improvements of the 1950s.

Unlike the two preceding eras, the postwar Freeway Era was not sparked by a revolution in urban transportation. Rather, it represented the coming of age of the now pervasive automobile culture, which coincided with the emergence of the U.S. from 15 years of economic depression and war. Suddenly the automobile was no longer a luxury or a recreational diversion: overnight it had become a necessity for commuting, shopping, and socializing, essential to the successful realization of personal opportunities for a rapidly expanding majority of the metropolitan population. People snapped up cars as fast as the reviving peacetime automobile industry could roll them off the assembly lines, and a prodigious highway-building effort was launched, spearheaded by high-speed, limited-access expressways. Given impetus by the 1956 Interstate Highway Act, these new freeways would soon reshape every corner of urban America, as the more distant suburbs they engendered represented nothing less than the turning inside-out of the historic metropolitan city.

The snowballing effect of these changes is expressed geographically in the sprawling metropolis of the postwar era. Most striking is the enormous band of growth that was added between 1945 and the 1980s, with freeway sectors pushing the metropolitan frontier deeply into the urban-rural fringe. By the late 1960s, the maturing expressway system began to underwrite a new suburban co-equality with the central city, because it was eliminating the metropolitanwide centrality advantage of the CBD. Now *any* location on the freeway network could easily be reached by motor vehicle, and intraurban accessibility had become a ubiquitous spatial good. Ironically, large cities had encouraged the construction of radial expressways in the 1950s and 1960s because they appeared to enable the downtown to remain accessible to the swiftly dispersing suburban population. However, as one economic activity after another discovered its new locational flexibility within the freeway metropolis, nonresidential deconcentration sharply accelerated in the 1970s and 1980s. Moreover, as expressways expanded the radius of commuting to encompass the entire dispersed metropolis, residential location constraints relaxed as well. No longer were most urbanites required to live within a short distance of their job: the workplace had now become a locus of opportunity offering access to the best possible residence that an individual could afford anywhere in the urbanized area. Thus, the overall pattern of locally uniform, income-based clusters that had emerged in prewar automobile suburbia was greatly magnified in the Freeway Era, and such new social variables as age and life-

style produced an ever more balkanized population mosaic.

The revolutionary changes in movement and accessibility introduced during the four decades of the Freeway Era have resulted in nothing less than the complete geographic restructuring of the metropolis. The single-center urban structure of the past has been transformed into a polycentric metropolitan form in which several outlying activity concentrations rival the CBD. These new "suburban downtowns," consisting of vast orchestrations of retailing, office-based business, and light industry, have become common features near the highway interchanges that now encircle every large central city. As these emerging metropolitan-level cores achieve economic and geographic parity with each other, as well as with the CBD of the nearby central city, they provide the totality of urban goods and services to their surrounding populations. Thus each metropolitan sector becomes a self-sufficient functional entity, or *realm*. The application of this model to the Los Angeles region reveals six broad

---

The new freeways would soon reshape every corner of urban America, as the more distant suburbs they engendered represented nothing less than the turning inside-out of the historic metropolitan city.

---

realms. Competition among several new suburban downtowns for dominance in the five outer realms is still occurring. In wealthy Orange County, for example, this rivalry is especially fierce, but Costa Mesa's burgeoning South Coast Metro is winning out as of early 1986.

The legacy of more than two centuries of intraurban transportation innovations, and the development patterns they helped stamp on the landscape of metropolitan America, is suburbanization—the growth of the edges of the urbanized area at a rate faster than in the already-developed interior. Since the geographic

extent of the built-up urban areas has, throughout history, exhibited a remarkably constant radius of about 45 minutes of travel from the center, each breakthrough in higher-speed transport technology extended that radius into a new outer zone of suburban residential opportunity. In the nineteenth century, commuter railroads, horse-drawn trolleys, and electric streetcars each created their own suburbs—and thereby also created the large industrial city, which could not have been formed without incorporating these new suburbs into the pre-existing compact urban center. But the suburbs that materialized in the early twentieth century began to assert their independence from the central cities, which were ever more perceived as undesirable. As the automobile greatly reinforced the dispersal trend of the metropolitan population, the distinction between central city and suburban ring grew as well. And as freeways eventually eliminated the friction effects of intrametropolitan distance for most urban functions, nonresidential activities deconcentrated to such an extent that by 1980 the emerging outer suburban city had become co-equal with the central city that spawned it.

As the transition to an information-dominated, postindustrial economy is completed, today's intraurban movement problems may be mitigated by the increasing substitution of communication for the physical movement of people. Thus, the city of the future is likely to be the "wired metropolis." Such a development would portend further deconcentration because activity centers would potentially be able to locate at any site offering access to global computer and satellite networks.

Further Reading

Jackson, Kenneth T. 1985. *Crabgrass Frontier: The Suburbanization of the United States.* New York: Oxford University Press.

Muller, Peter O. 1981. *Contemporary Suburban America.* Englewood Cliffs, N.J.: Prentice-Hall.

Schaeffer, K. H. and Sclar, Elliot. 1975. *Access for All: Transportation and Urban Growth.* Baltimore: Penguin Books.

HISPANIC MIGRATION
AND POPULATION REDISTRIBUTION
IN THE UNITED STATES

The US Hispanic population has grown rapidly over the last two decades and remains geographically concentrated in nine states. Redistribution away from core states through internal migration has been largely offset by heavy immigration to traditional areas of Hispanic concentration. Geographical patterns of Hispanic migration show broad similarities to overall patterns of population redistribution in the United States. New York and California serve as key spatial redistributors or pivots in the Hispanic migration system. **Key Words: Hispanic concentration, Hispanic migration, population gateway, spatial redistributor.**

Kevin E. McHugh

KEVIN E. McHUGH (Ph.D., University of Illinois) is Assistant Professor of Geography at Arizona State University, Tempe, AZ 85287. His primary research interests are in migration and residential mobility.

The Hispanic population in the United States has grown rapidly over the last two decades, increasing from 9.1 million in 1970 to an estimated 18.8 million in 1987 (US Bureau of the Census 1988). Hispanics now represent the fastest growing minority in the nation. Between 1980 and 1987, the Hispanic population increased 30% while the non-Hispanic population grew less than 6%. Projections of the Hispanic population for the year 2000 range from 23 to 31 million (US Bureau of the Census 1986). According to the middle series projections, Hispanics will account for one-fourth of total US population growth between 1983 and 2000.

"Hispanic" is an umbrella term that refers to US residents whose cultural heritage traces back to a Spanish-speaking country (Valdivieso and Davis 1988). Other than having common ancestral ties to Latin America or Spain, peoples of Spanish origin in the US are highly diverse (Bean and Tienda 1988). Mexican-Americans, the largest group, account for 63% of US Hispanics in 1987; Puerto Ricans account for 12% and Cuban-Americans 5%. Hispanics with origins in Central America (excluding Mexico) and South America comprise 11%, and the residual category "other Hispanics" makeup the remaining 8% of US Hispanics (US Bureau of the Census 1988).

Hispanic immigration has received considerable scholarly attention, but Hispanic migration and population redistribution *within* the United States is seldom investigated (Garcia 1981). Some recent studies examine the geographical distribution of particular Hispanic groups, such as Boswell's (1984, 1985a, 1985b) work on Cuban-Americans and Puerto Ricans, Arreola's (1985) examination of Mexican-Americans, and Portes and Bach's (1985) longitudinal study of Cuban and Mexican immigrants in the United States. Bean et al. (1988) recently reviewed the geographical distribution and interregional migration of Hispanic groups. There has not been a comprehensive examination of place-to-place migration flows of Hispanics in the United States, partly due to the historical lack of information on Hispanic migration within the United States. Hispanic interstate migration for the period 1975–80 are the first place-to-place Hispanic migration data published by the Bureau of the Census (1985).

The purpose of this paper is to identify patterns of Hispanic migration and population redistribution within the United States. The paper focuses on whether Hispanics are becoming more or less geographically concentrated in the United States and on identifying recent migration patterns that are contributing to Hispanic population redistribution.

Hispanic population redistribution is examined in two ways. First, changes in the geographical distribution of Hispanics over recent decades are examined through state percentage shares of the total Hispanic population and state percentage shares of four major Hispanic groups: Mexican-Americans, Puerto Ricans, Cuban-Americans, and Central/South Americans. Shifts in the state shares over time indicate trends in the geographical concentration and deconcentration of Hispanic groups in the United States.

Second, I examine the role of immigration from abroad and internal migration within the United States in contributing to Hispanic population redistribution. These analyses provide insights into the relative importance of immigration versus internal migration in contributing to Hispanic population change at the state level. I also identify large net interstate migration streams within the US and compute the effectiveness of these streams in redistributing the Hispanic population. This shows the interstate connections most instrumental in redistributing Hispanics within the United States.

The paper draws upon the concept of spatial redistributors in contributing to a geographical understanding of the US Hispanic migration system. Spatial redistributors are places that exhibit asymmetry between patterns of in- and out-migration and thus serve as pivots in systems of population redistribution (Roseman 1977; Morrison 1977; Roseman and McHugh 1982). Key states should

* This research was supported by a Faculty Grant-in-Aid, Office of the Vice President for Research, Arizona State University. I thank Barbara Trapido for drafting the map.

TABLE 1
TOP NINE STATES IN HISPANIC POPULATION

State	1970 Population	% Dist.	1980 Population	% Dist.	1987[a] Population	% Dist.
California	2,369,292	26.1	4,544,331	31.1	6,249,000	33.3
Texas	1,840,648	20.3	2,985,824	20.4	4,207,000	22.4
New York	1,351,982	14.9	1,659,300	11.4	2,182,000	11.6
Florida	405,036	4.5	858,158	5.9	1,256,000	6.7
New Jersey	288,488	3.2	491,883	3.4	737,000	3.9
Illinois	393,204	4.3	635,602	4.4	692,000	3.7
Arizona	264,770	2.9	440,701	3.0	664,000	3.5
New Mexico	308,340	3.4	477,222	3.3	535,000	2.9
Colorado	225,506	2.5	339,717	2.3	347,000	1.9
Total for nine states	7,447,266	82.1	12,432,738	85.1	16,869,000	89.9
United States	9,072,602	100.0	14,608,673	100.0	18,790,000	100.0

[a] 1987 figures are estimates of the civilian, noninstitutional Hispanic population from the March Current Population Survey. They are not directly comparable to the 1970 and 1980 census populations.
Sources: US Bureau of the Census, 1982; US Bureau of the Census, Current Population Survey, March 1987, Public Use File.

serve as Hispanic spatial redistributors at the international and national scales. International redistributors are states that attract large numbers of Hispanics from abroad and redistribute Hispanics within the United States, thus serving as population gateways. Key states should also serve as internal redistributors of Hispanics, as indicated by large net interstate migration streams. The redistributor concept is particularly relevant to the geographic concentration and deconcentration of Hispanics in the United States.

I first summarize shifts in the geographical distribution of Hispanic groups in the United States. The second section examines Hispanic immigration from abroad and internal migration within the United States, emphasizing the role of key states as Hispanic population gateways. Geographical patterns of migration that contribute to Hispanic population redistribution within the United States are identified in the third section. The final section is a discussion of three key issues: (1) whether Hispanic groups are becoming more or less geographically concentrated, (2) determinants of Hispanic migration within the United States, and (3) linkages between Hispanic immigration from abroad and internal migration within the United States.

Geographical Distribution of Hispanics

The US Hispanic population is concentrated geographically. Nine states accounted for 82% of the total Hispanic population in 1970 (Table 1). This percentage rose to 85% in 1980 and an estimated 90% in 1987. The following states had 1987 Hispanic populations greater than 300,000: New York and New Jersey in the Northeast, Illinois in the Midwest, Florida in the Southeast, and California, Texas, Arizona, New Mexico, and Colorado in the Southwest.

There has been some redistribution between these nine states as measured by

their share of the total US Hispanic population, most notably a seven-point rise in California's share, so that California now accounts for one-third of all Hispanics in the country. Texas, Florida, and Arizona have also increased their share of the Hispanic population over the last two decades. New York's declining share between 1970 and 1980 is noteworthy. Illinois, New Mexico, and Colorado also posted small declines in their share of the Hispanic population.

The US Hispanic population is diverse in nationality and cultural heritage. Disaggregating Hispanics by national origin and showing state percentage shares in 1960 and 1980 indicate trends in the geographic concentration and deconcentration of individual Hispanic groups (Table 2).

Hispanics of Mexican origin dominate in the southwestern states of California, Texas, Arizona, New Mexico, and Colorado, and also in Illinois. California and Texas in 1980 accounted for three-fourths of the Mexican-origin population in the United States. The most important shifts in the distribution of Mexican-Americans are the increase in California's share coupled with a declining share for Texas. This long-term trend began early in the twentieth century. In 1910, 60% of persons of Mexican stock in the United States resided in Texas, and California accounted for only 13% (Grebler et al. 1970). At that time, Texas had greater employment opportunities for the Mexican population, particularly in agriculture. Throughout the twentieth century, California's share of the Mexican-origin population steadily increased as job opportunities shifted to California, initially in agriculture and later through urban expansion (Jaffee et al. 1980).

Illinois is the only state outside the southwest with a large Mexican-American population. The Mexican-origin population in Illinois grew from less than 2% of the national total in 1960 to nearly 5%

in 1980. The development of the Mexican-origin population in Illinois resulted from their "settling out" from midwestern migratory labor streams as well as from direct migration to Chicago in response to employment opportunities in railroad maintenance, steelmaking, meatpacking, and other manufacturing sectors (Grebler et al. 1970). In 1980, Chicago ranked third among metropolitan areas in Mexican origin population, behind Los Angeles and Houston (Bean et al. 1988).

Puerto Ricans are the largest Hispanic group in New York and New Jersey. The most important redistribution of Puerto Ricans has been away from New York to nearby states in the Northeast, in addition to Florida and California. New York's share of the Puerto Rican population dropped from 72% in 1960 to 49% in 1980. Conversely, New Jersey, Massachusetts, Connecticut, Pennsylvania, Illinois, Florida, and California increased their share of Puerto Ricans. The deconcentration of Puerto Ricans away from New York resulted from declining employment opportunities, poor housing, and crime problems in New York City (Boswell 1984).

Cuban-Americans represent the largest Hispanic group in Florida, where they have become increasingly concentrated in south Florida, partly in response to the Cuban Refugee Resettlement Program (Boswell and Curtis 1984). Cuban-Americans resettled outside south Florida under this government-sponsored program began returning to Miami in the late 1960s. By the mid-1970s this return flow increased to a major migration stream. A survey commissioned by *The Miami Herald* in 1978 found that 40% of the population of Cuban origin in Dade County were returnees to Miami from elsewhere in the United States (Boswell and Curtis 1984).

The increased concentration of the Cuban-origin population in south Florida continues in the 1980s. The 1980 census figures do not include the estimated 125,000 Cuban "Marielito" refugees who arrived in Miami shortly after the 1980 enumeration. In addition, Cuban return movement to south Florida has continued in the 1980s. Boswell and Curtis (1984) cite estimates prepared by the Cuban National Planning Council that between 65 and 70% of Cuban-Americans reside in Florida.

Outside Florida, sizable numbers of Cuban-Americans reside in New York, New Jersey, and California. Before the Castro revolution in 1959, New York City was the primary destination of Cuban immigrants. New York's share of Cuban-Americans declined from 45% in 1950 to 10% in 1980. Cuban-Americans in New Jersey are highly concentrated in the area of Union City–West New York, across the

Hudson River from New York City. This concentration of Cuban-Americans is the largest outside Miami. California accounts for 8% of the 1980 Cuban-origin population. Los Angeles ranks fourth among urban areas in Cuban-American population (Boswell and Curtis 1984).

Hispanics with origins in Central America (excluding Mexico) and South America are concentrated in New York and California, with smaller concentrations in Florida and New Jersey. New York has a greater number of Hispanics with origins in South America and the Dominican Republic, while California has larger numbers of Central Americans (Allen and Turner 1988). Although fewer in number than Hispanics of Mexican and Puerto Rican origin, persons with origins in Central and South America represent the fastest growing Hispanic group in the United States, increasing an estimated 40% between 1980 and 1987 (US Bureau of the Census 1988). Much of the recent influx of Central Americans is a response to political turmoil in El Salvador, Nicaragua, and Guatemala (Allen and Turner 1988).

Hispanics are a highly urban population. In 1980, 81% of Mexican-Americans resided in metropolitan areas. Other Hispanic groups show greater levels of metropolitan concentration: 96% for Puerto Ricans, 94% for Cuban-Americans, and 96% for' Central/South Americans. In comparison, 73% of non-Hispanic whites resided in metropolitan areas in 1980 (Bean et al. 1988). Twenty-nine metropolitan areas in 1980 had more than 100,000 Hispanics. Los Angeles and New York alone accounted for nearly one-quarter of the US Hispanic population (Davis et al. 1983).

Hispanics also show a propensity to concentrate in central cities within metropolitan areas. In 1980, 65% of Mexican-Americans, 81% of Puerto Ricans, 45% of Cuban-Americans, and 67% of Central/ South Americans living in metropolitan areas were in central cities. The comparable figure for non-Hispanic whites is only 35%. Cuban-Americans have shown the greatest suburbanization, indicating their higher socioeconomic status relative to other Hispanic groups (Bean et al. 1988).

Hispanic Population Gateways

Immigration from abroad has contributed greatly to Hispanic population growth in the United States. Data on Hispanic immigration from abroad coupled with internal migration within the United States show that key states serve as Hispanic population gateways (Table 3). These data include immigration from abroad and internal migration within the United States, 1975-80, for 15 states with Hispanic populations over 100,000 in 1980. These migration data are not available for

Hispanic groups defined by national origin and are based on Hispanics enumerated in the 1980 Census of Population. The actual number of Hispanic migrants, particularly from abroad, is greater because a significant share of undocumented immigrants were not enumerated in the 1980 census (Warren and Passel 1987; Bean and Tienda 1988). Migrants from abroad refer to persons of Hispanic origin residing outside the United States in 1975 and in the designated state in 1980. This calculation includes foreign immigrants as well as US citizens returning from abroad. The vast majority of Hispanic movers from abroad are immigrants.

As expected, California, Texas, and New York attract very large numbers of Hispanics from abroad; Florida, Illinois, and New Jersey also receive sizable numbers

of Hispanic immigrants. These six states are the primary Hispanic gateways to the United States. Immigration from abroad more than offsets internal net migration losses for four of these states: New York, New Jersey, Illinois, and California. Despite internal migration away from these core states, they maintain or strengthen their Hispanic population concentrations through immigration. Florida and Texas, on the other hand, experienced both substantial immigration as well as net gains from elsewhere in the United States for the period 1975-80.

Geographical Patterns of Hispanic Migration

Geographical patterns of Hispanic migration within the United States can be seen by mapping the 25 largest net inter-

TABLE 2
GEOGRAPHICAL DISTRIBUTION OF HISPANICS BY NATIONAL ORIGIN, 1960 AND 1980[a]

	Percent of US total							
	Mexican		Puerto Rican		Cuban		Central/South American	
State	1960	1980	1960	1980	1960	1980	1960	1980
Massachusetts	—	0.1	0.3	3.8	0.7	0.8	1.6	1.8
Connecticut	—	—	1.6	4.5	2.1	0.7	1.9	1.1
New York	0.1	0.4	72.2	49.2	31.9	10.1	29.5	34.9
New Jersey	0.1	0.1	6.5	12.2	6.9	10.7	5.1	8.4
Pennsylvania	0.1	0.2	2.1	4.4	1.5	0.6	1.9	0.8
Ohio	0.2	0.5	1.3	1.7	0.8	0.4	1.7	0.5
Illinois	1.7	4.7	3.9	6.7	1.9	2.3	3.7	3.2
Michigan	0.6	1.2	0.3	0.5	0.2	0.4	1.7	0.6
Florida	0.1	0.7	2.1	4.8	43.0	58.4	5.2	9.1
Texas	38.7	32.4	0.7	1.0	1.1	1.7	3.3	2.8
Colorado	4.2	2.4	0.1	0.2	0.1	0.2	0.7	0.4
New Mexico	7.1	2.7	—	0.1	0.1	0.1	0.1	0.2
Arizona	5.7	4.7	0.2	0.2	—	0.1	0.6	0.3
Washington	0.4	0.9	0.1	0.2	0.7	0.2	0.8	0.4
California	38.7	42.1	3.1	4.4	3.2	8.0	24.9	25.4
Total for 15 states	97.7	93.1	94.5	93.9	94.2	94.7	82.7	89.9

[a] States listed have 100,000 or more persons of Hispanic origin, 1980.
Source: Bean et al. 1988.

TABLE 3
INTERNAL HISPANIC MIGRATION AND HISPANIC MOVERS FROM ABROAD, 1975-80[a]

	Internal migration			Number from abroad
State	Inmigration	Outmigration	Net migration	
Massachusetts	13,848	12,619	1229	20,118
Connecticut	11,148	10,005	1179	15,937
New York	27,552	133,061	−105,509	139,961
New Jersey	41,478	43,917	−2439	51,198
Pennsylvania	14,118	13,739	379	16,279
Ohio	9928	14,571	−4643	6355
Illinois	25,882	48,105	−22,133	66,124
Michigan	10,595	15,582	−4987	5958
Florida[b]	106,042	40,406	65,636	96,273
Texas	120,749	97,702	23,047	155,851
Colorado	28,578	27,137	1441	8596
New Mexico	32,485	31,036	1449	7535
Arizona	30,567	29,440	1127	15,229
Washington	24,051	11,826	12,225	8890
California	132,948	139,357	−6409	412,958

[a] States listed are those with 100,000 or more persons of Hispanic origin, 1980.
[b] Number of Hispanic movers to Florida from abroad, 1975-80, does not include the estimated 125,000 Cuban "Marielito" refugees who arrived shortly after the 1980 census enumeration.
Source: U.S. Bureau of the Census, 1985.

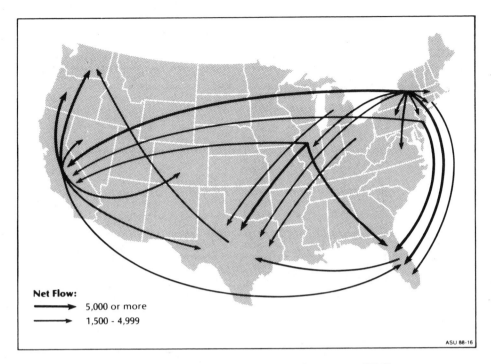

Figure 1. Large hispanic net interstate migration streams, 1975–80.

Net Flow:
→ 5,000 or more
→ 1,500 - 4,999

ASU 88-16

state migration streams for 1975–80, the most recent internal Hispanic migration data available (Fig. 1). In addition to the 25 net migration streams, the underlying gross migration flows and a percent effectiveness value for each interstate connection are reported (Table 4).

Percent effectiveness (Eij) indicates the level of net migration exchange between a pair of states relative to the size of the underlying gross migration flows. It is computed by dividing net migration by the sum of the gross migration flows in both directions, and multiplying the resulting ratio by 100:

$$Eij = [Nij/Mij + Mji] \times 100 \quad (1)$$

where

Eij = percent effectiveness of migration from state i to state j

Nij = net migration exchange between state i and state j (Mij − Mji)

Mij = gross migration flow from state i to state j

Mji = gross migration flow from state j to state i.

In absolute terms, Eij varies from 0 to 100%. A 0% effectiveness indicates that equal numbers of migrants are moving in both directions resulting in no population redistribution between the pair of states. Conversely, an effectiveness value of 100% would mean that *all* movement is unidirectional (either Mij or Mji = 0). Thus, effectiveness values indicate strong currents in a migration system (Plane 1984).

Several patterns of Hispanic migration are evident (Fig. 1). Net movement from northeastern and midwestern states to the three large Sunbelt states—Florida, Texas, and California—is conspicuous. Northeastern states are linked most strongly with Florida. Net flows from New York and New Jersey to Florida are very large and highly effective in redistributing Hispanics (Eij = 83.6% and 75.6%).

New York stands out for registering highly effective Hispanic migration losses to Florida, Texas, and California (Table 4). This result parallels the overall trend of large migration losses for New York during the 1970s. In fact, currents of migration from New York to Florida and California, as measured by percent effectiveness values, were stronger among Hispanics than non-Hispanics.

Bean et al's (1988) breakdown of Hispanic migration by national origin between New York and Florida, 1975–80, indicates why this stream is highly effective. Cuban-Americans constitute the greatest number of Hispanics in the New York-to-Florida stream. More than ten times as many Cuban-Americans migrated from New York to Florida as moved in the opposite direction. Many Cuban immigrants had been resettled from Miami to New York in the Cuban Refugee Resettlement Program (Boswell and Curtis 1984), so it is very likely that a substantial share of Florida-bound Cuban-Americans were returning to south Florida. Most Cuban-Americans returning to Dade County from outside Florida cite climate and a desire to be near family

and friends as reasons for their return (Boswell and Curtis 1984). Net movement of people of Puerto Rican and Central/South American origin from New York to Florida is also significant, although these streams are not as effective as Cuban-origin movement to Florida (Bean et al. 1988).

The midwestern states of Illinois, Michigan, and Ohio are linked most strongly with Texas (Fig. 1). These Hispanic migration streams are overwhelmingly Mexican-American, and are moderately effective in redistributing Mexican-Americans from the Midwest to Texas, with effectiveness values of 38.8% for Illinois, 25.2% for Michigan, and 39.1% for Ohio (Table 4). Significant numbers of Hispanics from Illinois move to Texas, Florida, and California, although the connection to Florida is most effective.

New York and California serve as spatial redistributors of the Hispanic population within the United States (Fig. 1). In addition to sending large numbers of Hispanics to Florida and California, New York is a redistributor of Hispanics within the Northeast. Large numbers of Hispanics move from New York to nearby states, including New Jersey, Pennsylvania, Connecticut, and Massachusetts. These four net migration streams from New York have moderately high effectiveness values (Table 4). Puerto Ricans are the dominant group in these streams, although Cuban-Americans and Central/South Americans are also likely to be present, especially in the stream to New Jersey. Puerto Rican migration away from New York relates to declining manufacturing employment, particularly in the textile and garment industries (Bean and Tienda 1988). Boswell (1984) also cites poor housing and crime as additional push factors in Puerto Rican migration from New York.

California is also a spatial redistributor of the Hispanic population. California gains Hispanics from New York, New Jersey, and Illinois, but loses Hispanics to western states, including Washington, Oregon, Nevada, and Colorado (Fig. 1). Thus, California is emerging as an interregional Hispanic redistributor, just as it has served as an interregional redistributor among Anglos since the late 1960s (US Bureau of the Census 1973). Overall, California recorded modest net out-migration of Hispanics within the United States, 1975–80, as losses to western states more than offset gains from northeastern and midwestern states.

Examining place of birth for Hispanics in western states also provides evidence that California is a Hispanic redistributor. In 1980, 50% of Hispanics in California were native to the state and 40% were foreign born. Only 12% of California Hispanics were born elsewhere in the United States. On the other hand, 40 to 50% of Hispanics in Washington, Oregon, and Nevada were born elsewhere in the United States, most likely California (US Bureau of the Census 1985).

For 1975–80, Texas gained Hispanics from northern states as well as from Florida and California (Fig. 1). Net flows to Texas from California and Florida, however, are small relative to large gross migration exchanges. Effectiveness values for the net streams to Texas are only 7.9% for California, and 18.0% for Florida (Table 4). In fact, Texas may be losing Hispanics to Florida and California since the recent decline in the energy-based Texas economy.

Discussion

Are Hispanics becoming more or less geographically concentrated in the United States? There has been some redistribution of the Hispanic population through internal migration as a result of (1) movement from northeastern and midwestern states to Florida, Texas, and California; (2) net movement from New York to nearby states in the Northeast; and (3) net migration from California to other western states. Immigration from abroad, however, continues to traditional areas of Hispanic concentration. For several states, heavy immigration among Hispanics has more than offset migration losses to other places within the country. A complete understanding of Hispanic population redistribution will require examination of both immigration and internal migration, as well as consideration of differentials in natural population increase among Hispanic groups.

The Hispanic population remains geographically concentrated in nine states, but this overall view masks differences among individual Hispanic groups. Cuban-Americans and Hispanics of Central/South American origin are becoming more concentrated, the former in Florida and the latter in California and New York. The increasing concentration of Cuban-Americans in south Florida is partly attributable to return migration following the Cuban Refugee Resettlement Program and to the strength of the Cuban-American community in Dade County. It is not surprising that most Central/South Americans concentrate in California and New York, given their recent arrival in the United States.

Hispanics of Mexican origin show some deconcentration away from core states. Bean et al. (1988) reached similar conclusions through their analysis of dissimilarity indexes that compared the geographic distribution of Hispanic groups and the overall US population. They found that the concentration of the Mexican-or-

TABLE 4
LARGE HISPANIC NET INTERSTATE MIGRATION STREAMS
AND PERCENT EFFECTIVENESS OF STREAMS, 1975–80

State i	State j	Gross flow i to j	Gross flow j to i	Net gain state j	Percent effect.
New York	Florida	38,398	3431	34,967	83.6
New York	New Jersey	28,080	7098	20,982	59.7
New Jersey	Florida	18,071	2515	15,556	75.6
New York	California	16,960	3369	13,591	66.9
Illinois	Texas	12,949	5710	7239	38.8
Illinois	Florida	7806	1754	6052	63.3
California	Oregon	8344	2459	5885	54.5
California	Washington	10,005	4178	5827	41.1
California	Texas	32,234	27,494	4740	7.9
California	Nevada	6916	2234	4682	51.2
New York	Texas	5702	1187	4515	65.5
New York	Pennsylvania	5626	1141	4485	66.3
New York	Connecticut	5644	1617	4027	55.5
New York	Massachusetts	5701	1833	3868	51.3
Illinois	California	9387	5707	3680	24.4
Florida	Texas	8499	5910	2589	18.0
California	Colorado	7888	5476	2412	18.0
New Jersey	California	3906	1510	2396	44.2
Michigan	Texas	4925	2942	1983	25.2
Ohio	Texas	3411	1492	1919	39.1
Connecticut	Florida	2385	635	1750	57.9
New York	Virginia	2288	539	1749	61.9
New York	Illinois	2664	1009	1655	45.1
California	Florida	7160	5531	1629	12.8
Texas	Washington	3037	1512	1525	33.5

Source: U.S. Bureau of the Census, 1985.

igin population has become less pronounced from 1960 to 1980. Bean et al. (1988) also found that Puerto Ricans exhibit some deconcentration away from core states, especially New York, over the 20-year period. Recent interstate migration has played the dominant role in the deconcentration of Puerto Ricans away from their New York core.

Hispanic migration patterns are broadly similar to overall patterns of migration within the United States. Hispanic migration from northern states to Florida, California, and Texas is part of the larger population redistribution to the Sunbelt (Biggar 1979; Long 1988). Currents of migration to Florida, California, and Texas tend to be stronger among Hispanics than non-Hispanics, perhaps because of the greater concentration of the Hispanic population and opportunities in the three large Sunbelt states. Social networks defined on the basis of ethnicity probably serve to channelize Hispanic migration flows to Florida, California, and Texas.

New York and California have emerged as spatial redistributors of Hispanics, just as they have redistributed the Anglo population. California's emergence as an interregional redistributor in the late 1970s—attracting Hispanics from states in the Northeast/Midwest and losing Hispanics to states in the West—follows a similar trend among Anglos. New York and California are likely to continue as central pivots in the Hispanic migration system: both have large Hispanic populations and serve as gateways for large

numbers of new immigrants to the United States.

Comparisons of immigration from abroad and internal migration of Hispanics within the United States should show that Hispanics born in the United States, or those residing in the United States for a number of years, are more likely to migrate than are recent immigrants. Recent immigrants are typically less familiar with the United States, know less English, and tend to be of lower socioeconomic status than longer-term residents. Recent immigrants tend to concentrate in ethnic enclaves for social and economic support. Grebler et al. (1970) found that Hispanics of Mexican origin showed greater rates of intercounty migration, 1955–60, the further they were removed from the immigrant generation.

Portes and Bach (1985) studied the link between immigration and internal migration through a six-year residential history of a sample of Mexican and Cuban immigrants who entered the United States in 1973. They found that Mexican immigrants were more likely to change residences after living in the United States for three years, and that less than 25% remained at the same residence over the six-year period, 1973–79. Slightly more than one-half of the Mexican immigrants remained in Texas (state of entry), one-fourth moved to other states in the Southwest, and 16% moved northward to Chicago. Return immigrants (those who had been to the United States previously) were more likely to move and showed a more dispersed pattern of settlement than first-time Mexican immigrants. This study demonstrates that experience in the United States as well as social and economic ties can be developed through circular migration between Mexico and the United States (Massey 1985). In contrast to the rather dispersed settlement pattern of Mexican immigrants, Portes and Bach (1985) found that the Cubans concentrated in Miami and remained there over the six-year period.

As the US Hispanic population grows, questions and issues relating to migration and population redistribution will be of increasing concern to social scientists and policy-makers at local, state, and federal levels. At the microlevel, there is a need for household level research that examines linkages between migration and socioeconomic and demographic status of Hispanics. Relationships between immigrant generation/length of residence in the United States, internal migration, and socioeconomic and demographic status will contribute to a broader theory of migration, adjustment, and assimilation.

At the aggregate level, Hispanic migration and population redistribution impacts labor markets and has important implications for the provision of educational and social services. The issue of geographic impacts is particularly important given the growth of the Hispanic population and uncertainties surrounding consequences of the Immigration Reform and Control Act of 1986.

Literature Cited

Allen, J. P., and E. J. Turner. 1988. *We the People: An Atlas of America's Ethnic Diversity*. New York: Macmillan Publishing.

Arreola, D. D. 1985. Mexican Americans. In *Ethnicity in Contemporary America: A Geographical Appraisal*, ed. J. O. McKee, 77–94. Dubuque, IA: Kendall/Hunt Publishing.

Bean, F. D., and M. Tienda. 1988. *The Hispanic Population of the United States*. New York: Russell Sage Foundation.

Bean, F. D., M. Tienda, and D. S. Massey. 1988. Geographical distribution, internal migration, and residential segregation. In *The Hispanic Population of the United States*, ed. F. D. Bean and M. Tienda, 137–77. New York: Russell Sage Foundation.

Biggar, J. C. 1979. The sunning of America: Migration to the Sunbelt. *Population Bulletin* 34:1–42.

Boswell, T. D. 1984. The migration and distribution of Cubans and Puerto Ricans living in the United States. *Journal of Geography* 83:65–72.

Boswell, T. D. 1985a. The Cuban-Americans. In *Ethnicity in Contemporary America: A Geographical Appraisal*, ed. J. O. McKee, 95–116. Dubuque, IA: Kendall/Hunt Publishing.

Boswell, T. D. 1985b. Puerto Ricans living in the United States. In *Ethnicity in Contemporary America: A Geographical Appraisal*, ed. J. O. McKee, 117–144. Dubuque, IA: Kendall/Hunt Publishing.

Boswell, T. D., and J. R. Curtis. 1984. *The Cuban-American Experience*. Totowa, NJ: Rowman and Allanheld Publishers.

Davis, C., C. Haub, and J. Willette. 1983. U.S. Hispanics: Changing the face of America. *Population Bulletin* 38:1–44.

Garcia, J. A. 1981. Hispanic migration: Where they are moving and why. *Agenda: A Journal of Hispanic Issues* 2:14–17.

Grebler, L., J. W. Moore, and R. C. Guzman. 1970. *The Mexican-American People*. New York: The Free Press.

Jaffee, A. J., R. M. Cullen, and T. D. Boswell. 1980. *The Changing Demography of Spanish Americans*. New York: Academic Press.

Long, L. 1988. *Migration and Residential Mobility in the United States*. New York: Russell Sage Foundation.

Massey, D. S. 1985. The settlement process among Mexican migrants to the United States: New methods and findings. In *Immigration Statistics: A Story of Neglect*, ed. D. B. Levine, K. Hill, and R. Warren, 255–92. Washington, DC: National Academy Press.

Morrison, P. A. 1977. Urban growth and decline in the United States: A study of migration's effect in two cities. In *Internal Migration: A Comparative Perspective*, ed. A. A. Brown and E. Neuberger, 235–54. New York: Academic Press.

Plane, D. A. 1984. A systematic demographic efficiency analysis of US interstate population exchange, 1935–80. *Economic Geography* 60:294–312.

Portes, A., and R. L. Bach. 1985. *Latin Journey: Cuban and Mexican Immigrants in the United States*. Berkeley, CA: University of California Press.

Roseman, C. C. 1977. *Changing Migration Patterns within the United States*, Resource Paper No. 77-2. Washington, DC: Association of American Geographers.

Roseman, C. C., and K. E. McHugh. 1982. Metropolitan areas as redistributors of population. *Urban Geography* 3:22–33.

US Bureau of the Census. 1973. Census of population: 1970. *Mobility for States and the Nation*, Final Report PC(2)-2B. Washington, DC: Government Printing Office.

US Bureau of the Census. 1982. Census of population: 1980. *Persons of Spanish Origin by State: 1980*, Supplementary Report PC80-S1-7. Washington, DC: Government Printing Office.

US Bureau of the Census. 1985. Census of population: 1980. *Geographical Mobility for States and the Nation*, PC80-2-2A. Washington, DC: Government Printing Office.

US Bureau of the Census. 1986. Projections of the Hispanic population: 1983 to 2080. *Current Population Reports*, Series P-25, No. 995. Washington, DC: Government Printing Office.

US Bureau of the Census. 1988. The Hispanic population in the United States: March 1986 and 1987. *Current Population Reports*, Series P-20, No. 434. Washington, DC: Government Printing Office.

Valdivieso, R., and C. Davis. 1988. US Hispanics: Challenging issues for the 1990s. *Population Trends and Public Policy* 17:1–16.

Warren, R., and J. S. Passel. 1987. A count of the uncountable: Estimates of undocumented aliens counted in the 1980 United States census. *Demography* 24:375–93.

RHINE-MAIN-DANUBE CANAL

Connecting the Rivers at Europe's Heart

Kenneth C. Danforth

Kenneth C. Danforth, based in Washington, DC has written extensively on Central and Eastern Europe since 1970.

A flotilla bearing thousands of European dignitaries progressed southward across an unlikely landscape—the Franconian Uplands of Bavaria.

The captains of the vessels cut back to dead slow as they approached the continental water divide between the villages of Hilpoltstein and Bachhausen. Roaring cascades shot out from the water cannons of a fire brigade and formed a curtain of water.

The date was September 25, 1992. A dream of a thousand years was about to be realized.

Standing in the lead ship, President Richard von Weizäcker, Prime Minister Max Streibl of Bavaria, and Federal Minister of Transport Günter Krause, all together turned a steering wheel connected with the ship's siren. The fire brigade shut off its cascade. At that signal, ships waiting on both sides of the divide started blowing their sirens. Then the convoy of passenger ships started southward.

Not all of the activity was ceremonial. Also waiting their turn to go through the locks in both directions were 15 barges, most of them laden with freight. They came from almost every nation washed by the Danube and Rhine. All were decorated with banners.

"I am one-hundred trucks," boasted one barge.

"I am the most environmentally friendly mode of transportation," proclaimed another.

Twenty other working ships stood by, waiting for the traffic to clear so they could proceed with their cargo.

Thus did business begin on the mighty Rhine-Main-Danube Canal, bringing to a successful conclusion 71 years of beaverlike industry interrupted only by World War II. At long last, huge barges would be able to go all the way from the North Sea to the Black Sea—2,174 miles—without ever encountering a wave. The long and perilous circuit from Belgium and the Netherlands to Romania and the Crimea would no longer be necessary.

The effort to link the seas dates all the way back to the reign of Charlemagne. The emperor, who'd traveled widely across the continent, knew that the Danube was only a few miles from the Main, which flowed into the Rhine. His engineers advised him that if his army could dig a canal to join a tributary of the Danube with a tributary of the Main, his ships could move from one end of his realm to the other.

Digging started in earnest in A.D. 793. The plan was a good one, but it was thwarted by torrential rains. As fast as his thousands of workers could shovel, the mud poured back down on them. The ditch became a grave for many of them. Charlemagne soon gave up.

More than 10 centuries passed before anyone tried again. This time the monarch in charge was King Ludwig I of Bavaria. He started his canal in 1837, and actually completed it in 1845. The Ludwig-Danube-Main Canal remained in use until late in World War II. But its peak year was 1850. In more than a century, it showed a profit for only 12 years. Eventually, its 101 small wooden lock-gates and its dependence on horses for towing made it a picturesque relic. But it was

doomed from the beginning by the onrushing phenomenon of the railroad.

The gargantuan effort to build a modern canal began as a cooperative charter between Germany and the Free State of Bavaria in 1921. After years of constructing locks and barrages on the Main and the German portion of the Danube, the builders turned at last to the centerpiece, the canal itself.

The completed canal immediately becomes not only the crown jewel in the vast inland-waterway system of Europe, but at 1,340 feet above sea level it becomes the crown itself—the highest of all the continent's navigable waterways.

It may be the greatest public-works project in history.

The opening of the canal marks an achievement of epic geopolitical proportions. Perhaps just as notable, its cost of 6.5 billion deutschmarks ($4.1 billion) has not cost the German taxpayers a single pfennig. Rhein-Main-Donau AG—the company that built the canal—pays off its interest-free government loans with money generated by hydroelectricity. It has already turned the canal over to the government. In 2050, when the last of the loan is paid off, the 55 power stations themselves will become government property.

From north to south, the canal runs from Bamberg on the Main, through Nuremberg, to Kelheim on the Danube. It is 107 miles long, more than twice the length of the Panama Canal.

The official opening of the canal was performed in the middle of the *Scheitelhaltung*, the summit-level reach, a 10-mile gouge through the very backbone of the continental divide. In the surrounding farmland, raindrops that fall only inches apart might flow either to the Rhine

➤ T w o   D i s t a n t   P o r t s   i n   E u r o p e ◄

once visited ports at both ends of the great new Rhine-Main-Danube Waterway—2,174 miles from each other—in less time than it takes a barge to chug from Düsseldorf to Strasbourg.

One afternoon I was exploring primitive Sulina, Romania. Sulina languishes at the mouth of one of the Danube Delta's three navigable channels, the central and most direct shot to the Black Sea. The next evening, shocking even to me, I was in the Netherlands ready to visit Rotterdam, the largest and most modern seaport in the world.

Here are brief impressions of the two farthest terminals of the new waterway created by the opening of Germany's RMD Canal.

Sulina and The Black Sea

Mihai Macarencu's 135-horsepower open boat sped us down the Danube from Tulcea, past the taut anchor chains of a dozen ocean-going vessels. Most of them were Romanian, many were rusty derelicts. They lay at anchor, bows against the current that had come all the way from the Black Forest.

Pairs of pelicans fluttered atop light poles, and formations flew overhead. Rows of planted poplars grew along stretches of the shore nearest isolated villages. Black-headed gulls skimmed the water.

Mihai, his face pinched scarlet by the sun, his voice loath to fight the wind, is a man of few words. In his longest communication in three days, he said, "The delta freezes over in winter and we don't have any roads. Nothing can be brought in when there's ice, and that's two months some years, so we have to plan ahead. We store up as much grain and canned food and dried meat as we can or people will starve."

I'd hoped that when we reached Sulina we'd be able to speed right on out to the Black Sea. Permission had been sought and denied—military security. An icebreaker, dredgers, and six destroyers were docked along the left bank. On the right bank stood a few big wooden houses left from the time when this was a Greek fishing settlement.

On a Sunday morning, both commercial and religious activities proceed at a lethargic pace. In the old Orthodox church, the aging faithful come to stand in lonely and forlorn reverence. Nearby, men walk up to the window of an outdoor beer stand, count out sticky currency, then sit musing over tepid brew at rickety metal tables. At the quay, a ferry from Tulcea ties up and a few passengers step quietly to shore. One man grins, holding up a jar of gray pickles to show his family the prize he's brought down the Danube.

The whole city of Sulina has only one stretch of pavement, the riverside promenade. It has no automobiles. It does possess a tractor, which can be hooked to an iron cart and used to haul produce along the city's rutted dirt lanes, billowing thick dust into every house.

At the Black Sea beach, which the people reach by trudging five miles across a barren lowland where hogs wallow in ditches, there are no hotels, cafés, or refreshment stands. An old woman in a little cart, switching a donkey's rump, brings two wooden kegs of water to sell to sand-encrusted vacationers who lie basting in their own sweat. They are among the elect; they have escaped Bucharest for a few days.

Rotterdam and The North Sea

The tall modern hotels of Scheveningen rose behind me. The cold North Sea taunted a thousand tentative swimmers. Twenty thousand more huddled behind glass-screened outdoor cafes, consuming genever, Heineken, and herring. Bikini-clad Dutch girls, mottled with goosebumps, toyed with the surf. Electronic amusements beyond number crowded the strand as far as I could see in either direction. Scheveningen is so close to Rotterdam that people pop up to the beach for the odd afternoon.

I got my best views and orientation of the teeming waterways of Rotterdam from the decks of De Maze, a gleaming white 60-foot yacht that the mayor and port authority sometimes provide for journalists. The first mate served little shrimp sandwiches and highballs as the captain maneuvered and talked about the harbor.

The Port of Rotterdam passed New York in trade volume 30 years ago, and now it's twice as big. The port receives more than 32,000 ocean-going vessels and 180,000 inland-waterway barges every year and handles more than 250 million tons of freight. As many as one hundred ships a day seek their places along 40,000 yards of quayside. The complexity of the traffic would rattle any safety engineer, but in Rotterdam it all goes smoothly.

Twenty-six radar stations monitor traffic around the clock. The radar screens identify each ship's profile and send it to a computer for confirmation of its name, registry, cargo, and assigned wharf.

The port stays in business in every kind of weather, including thick fog. Operators also know where every ship is every minute, not only in port but as far as 37 miles out to sea. The city requires every ship to take on one of Rotterdam's 360 pilots. A river pilot boards the ship out in the North Sea, brings it up the New Waterway, and hands it off to a harbor pilot for berthing.

Rotterdam had to start from worse than scratch after World War II, for the Luftwaffe had destroyed the city on May 14, 1940. Today, ironically, more imported goods reach Germany through Rotterdam than through all German ports combined.

No one can imagine Sulina ever catching up with Rotterdam. Yet in a way the two dramatically different ports are now sister cities—linked by the grand new transcontinental waterway. Contrasting them illuminates the tremendous potential for the growth of waterborne commerce in the Danubian lands of the former Soviet bloc.

—Kenneth C. Danforth

The canal links the Danube and the Main rivers, allowing ships to travel from the North Sea (via the port at Rotterdam, Netherlands) to the Black Sea (via the port at Sulina, Romania).

basin (and on to the North Sea) or to the watershed of the Danube (and thus to the Black Sea).

With the rolling green hills of the Franconian Uplands all around them, punctuated by a church steeple in the village of Sulzkirchen, the waterborne guests and a crowd on the shore listened. Speakers praised the engineers and statesmen who'd made it all possible, and applause rippled from ship to ship to shore. Then the beribboned craft, packed from bow to stern with some 3,500 honored guests, churned forward toward Bachhausen.

There they descended through an enormous lock—40 feet wide, nearly 100 feet deep, and 625 feet long—to the pond that would lead them to Berching, a little town that had just become a port. Outside Berching's intact medieval walls stood a gigantic beer tent. There, beneath the glow of illuminated canvas and food fellowship, the builders and the backers of this epic undertaking rejoiced at its completion. Nevertheless, the German press in general covered the inauguration of the canal as if it were an environmental catastrophe.

For several years, environmentalists and SPD ministers had almost killed the project. The company made enlightened compromises, and the last stretch of the canal to be built is a model for accommodating a major public project to ecological sensitivities. It spent $40 million to protect the ecology, including preservation of backwaters and marshes. While many environmentalists demonstrated against the canal, others worked quietly

to see that the project—which was too far along to be stopped—was carried out with maximum concern for the landscape and the plants and creatures that dwelt upon it.

Only a few years ago, Rhine shippers still feared that the canal was going to bring a predatory hoard of East Bloc price-cutters. Soviet, Romanian, and Bulgarian fleets were used to competing unfairly on the Danube. Communist finance ministers always went into a feeding frenzy whenever the whiff of hard currency was in the air. Red ink was a patriotic color to a manager who could sell his national-firm services for marks or francs at less than his free-market competitors could afford. The horrid word—which reverberated on the docks of Basel, Strasbourg, Mannheim, Frankfurt, Duisburg, Arnhem, Rotterdam, and Antwerp—was dumping.

Eastern boatmen were mainly military personnel, who were paid abysmal wages. A single ship did not need to show a profit; it was part of a larger economic scheme. But now everybody in Europe has to make a profit—or perish.

The Germans (back when they were "West Germans") warded off the threat by declaring the canal a "national waterway," not subject to the freedoms of either the Danube or the Rhine, both internationalized by formal conventions. They vexed most of the East European transportation officials by holding to a policy of bilateral agreements instead of letting them negotiate as a more powerful bloc. It was a classic case of commercial divide-and-conquer.

The canal was seen as a way to help open up Eastern Europe to trade and, thus, to greater exchanges of all kinds. And then the Soviet empire fell, the satellite nations went their various paths toward free markets, Germany became one nation, and the Soviet Union became many. The mission of the canal switched from one of opening the East Bloc to facilitating its economic development.

One of the first ships to sail from the Danube to the Rhine was operated by Mahart of Hungary. It arrived in Duisburg in early October.

Millions of tons of bulk goods—the mainstay of inland shipping—can now start moving quietly, cheaply, and cleanly by water instead of over the Danube basin's already choked highway and railway system. Coal, ore, pig iron, sand, gravel, scrap metal, fertilizer, and grains will be among the most important cargoes. Huge electrical transformers and printing presses, which are extremely unwieldy to ship overland, are finished products that Eastern Europe sorely needs.

An entire freight train would be needed to transport all the cargo that can be carried on just one big barge.

Company officials estimate that more than five million tons of freight will move through the canal in each of the first two years. After eight years, they believe the figure will climb to 10 million tons annually. Without a doubt, the canal will carry a large percentage of a projected 90 percent increase in East-West freight traffic between now and 2010.

The Chunnel
The Missing Link

Travel for millions of Europeans will never be the same again with the opening of the Channel Tunnel.

D a v i d   L e n n o n

David Lennon is the managing editor of the syndications department of the Financial Times *in London and a correspondent for* EUROPE.

THE CHUNNEL, as it is known in the UK, is the first dry-land link between the UK and France since the ice age 12,000 years ago. It will run for 23 miles below the seaway called the English Channel by the British and *La Manche* (The Sleeve) by the French.

Fierce enemies for two-thirds of the last 1,000 years, the UK and France have put aside their lingering linguistic and cultural differences in order to build the link which had been missing in the European transport system.

The dream of a fixed-link between this island and mainland Europe goes back almost two centuries; yet, implementation of the project took less than a decade once the governments of the UK and France decided to invite businessmen to submit their proposals.

Now, thanks to one of the largest European engineering projects this century, the road and rail networks of the UK and continental Europe will soon be joined together. Before we know it, trains will be whizzing people, cars, and freight back and forth under the channel in ever increasing numbers.

Reprinted with permission of *Europe: Magazine of the European Community,* March 1993, pp. 8-10. © 1993 by *Europe.* Subscriptions available at $19.95/year from *Europe,* 2100 M Street, NW, 7th Floor, Washington, DC 20037.

The Chunnel will close one of the crucial gaps in the European transport network and by the end of the century will provide rapid access to and from the UK and various centers all over the continent. A single transport network will serve the newly unified single market of Europe.

The Channel Tunnel is a rail link, so you still cannot actually drive between southern England and northern Europe. But drivers, whether of private cars, buses, or trucks will drive their vehicles onto specially constructed trains called "Le Shuttle," which will zip through the tunnel in 35 minutes.

No more than an hour after leaving the motorway to enter the terminal at one end, the vehicle should be out on the road on the other side—remembering, we hope, to switch from driving on the left to driving on the right, or vice versa.

The shuttle will operate on a "turn up and go" basis, with no advance booking required, just like a toll motorway. The shuttles will operate 24-hours a day and every 15 minutes during peak periods. The cost of using the service is expected to be similar to the price of the traditional cross-channel ferries.

The first shuttle trains are due through the tunnel at the end of [1993] with regular passenger and freight trains being added by the middle of [1994]. There could be as many as 400 trains rushing daily in each direction.

The national railway companies of the UK (BR), France (SNCF), and Belgium (SNCR) will operate high-speed passenger services linking London with Paris in three hours, while Brussels will take just ten minutes longer. As the rail networks on both sides of the channel are improved travel times will be considerably reduced and the speed of the service between city centers will rival intercity flights.

The project comprises three parallel tunnels: two single-track rail tunnels running in one direction only and a central service tunnel. The tunnels are entered at specially constructed terminals at Folkestone in southern England and Calais in northern France.

Work on digging the tunnel began in December 1987 and was completed three and a half years later, in June 1991. Since then one of the 11 huge, 1,000-ton Tunnel Boring Machines has

been sitting near the motorway at Folkstone with a big sign proclaiming "For Sale—One Careful Owner."

The tunnels are 31 miles in total length with 24 miles running beneath the Channel sea bed at an average depth of 128 feet. The longest rail tunnel in the world, Japan's Seikan tunnel linking the islands of Hokkaido and Honshu, is nearly two and a half miles longer than the Channel Tunnel, but its underseas section is only 14.3 miles.

Considerable economic and employment benefits should be generated for the areas where the terminals are located, and many hope the new complementary transport links will attract industries wishing to take advantage of the creation of this rapid transport system between the UK and the continent.

North Kent in Southeast England has suffered from severe industrial decline and infrastructure problems in recent times. Its commercial and civic leaders welcome the building between London and the Chunnel of the high-speed rail link, which they believe will offer enhanced development opportunities.

In France's depressed northern Pas-de-Calais industrial region the locals yearn for the operation of the Chunnel. Every town mayor in the region is hoping the new transport infrastructure will bring more investment and employment.

The governments of the UK and France decided from the outset that the Chunnel must be funded by private enterprise, not public funds. As a result, Sir Alastair Morton, Chief Executive of the joint Anglo-French concessionaire Eurotunnel, said he was introduced in New York as "the man who raised $10 billion for a hole in the ground."

Like most visionary projects, this one too has suffered a massive overrun in costs. The current estimate for the tunnel and its system of trains and terminals is $12.8 billion, almost double the original forecast. The final price tag is likely to be closer to $13.6 billion.

But neither the price nor the six-month delay in opening the link can compare to the breathtaking vision of the project and the revolutionary impact its operation will have on European transport.

Despite the growth in costs, the latest forecasts for shuttle traffic in the first full year of operation—12.8 million passengers and 8.5 million tons of

CHANNEL TUNNEL—A Brief History

The idea of a tunnel under the Channel originally surfaced in 1802 and the first attempt to bore an undersea route to France was made in 1880.

A 2,000-meter trial bore was made at Folkestone and a similar tunnel was also started in France from a shaft sunk near Sangatte. But the project was personally vetoed by the Duke of Wellington, who feared it could be used as an invasion route by the French cavalry.

The route of the present tunnel dates back to the end of the 1950s when the company that had owned the Suez Canal was looking for a new role. Nothing came of this.

Then, in 1974, work began on a large tunnel similar in size and configuration to the one that has now been built. A 400-meter length of service tunnel was bored before work was aborted by a British government concerned by the costs.

The financial basis for the decision to go ahead with today's tunnel was provided by a 1982 banker's report that showed the project could be economically feasible.

This report led to the formation of a group of British and French companies and a decision by the governments of the two countries that a fixed link should be built across the Channel.

In April 1985 bids were invited for the financing, construction, and operation of a fixed link. The Eurotunnel plan was selected in January 1986 and construction of the Channel Tunnel began in December 1987.

—David Lennon

freight—indicate that those who bought shares in Eurotunnel may be onto a winner. And the current estimate is that traffic will double in the next 10-15 years.

So, when the first huge train emerges gleaming from beneath the sea at Folkestone, it will no longer be possible for a British newspaper to repeat the headline of earlier this century "Fog in the Channel, Europe cut off."

AN ALTERNATIVE ROUTE TO

MAPPING HISTORY

J. BRIAN HARLEY AND
DAVID WOODWARD

J. Brian Harley is professor of geography at the University of Wisconsin-Milwaukee and director of the Office for Map History, American Geographical Society Collection. David Woodward is professor of geography at the University of Wisconsin-Madison.

THE SIXTEENTH-CENTURY GEOGRAPHER Hulsius once wrote, "Maps may be called the light or eye of history." It is in this role, as the primary documents for locating historical events, that early maps are still most used by the historians of the late-twentieth century. Nowhere are these uses more visible than in the study of the maps of what is sometimes still called the first great age of European exploration. J.A. Williamson, the distinguished historian of the Cabots, wrote discerningly in 1937 that "old maps are slippery witnesses." But for many researchers maps still hold out the prospect of locating the elusive landfalls of a Columbus or a Drake, of reconstructing the routes of navigators or tracks of the explorers, or perhaps of identifying a place described ambiguously in written texts. Maps are the coordinates of history.

During our researches in connection with the History of Cartography project and the Maps and the Columbian Encounter Exhibition program, we have come to understand the place of maps in early American history in some alternative ways. Maps can still be analyzed as credible and articulate witnesses to some aspects of the European voyages and explorations to America and in that sense we continue to interrogate them as a traditional record of events. But we are also discovering—as we re-read them as a visual language to uncover new meanings—that they have yet more to contribute to a richer history of the Columbian encounter.

First, we are seeking a new perspective that focuses on the use of maps and on the social conse-

quences of their making. The key question is "What happened when particular maps were made?" The research has shifted from a theoretical consideration of what maps were designed to do to what they actually did in society.

Inasmuch as Renaissance maps straddle a major transition in the history of European cartography—from the medieval to the modern—this question is especially fascinating. The dramatic shift in ways of thinking about the world, and in the way that vision was constructed, can be seen by comparing two of the dominant traditions of fifteenth-century cartography. The first tradition, the medieval *mappaemundi*, was allegorical, historical, and literary—a representation of the space of

---

From the *mappaemundi* to the Ptolemaic grid, maps have been both mirrors and catalysts of their times

---

Christianity. Often centered on Jerusalem, these maps were introverted to the interior of the classical and medieval world by a circumscribing ocean sea. Beyond the pillars of Hercules there was nothing: *Ne plus ultra* (Nothing more beyond). The second tradition—represented by the rediscovered world maps of Ptolemy—by popularizing coordinates of latitude and longitude, led to a major conceptual shift in ways of fixing geographical positions and hence in visualizing and controlling the world. From the early fifteenth century onwards, after Ptolemy's *Geography* had been translated into Latin, the system of latitude and longitude—the symbol of modern cartography—began its ascendancy.

From *América*, Vol. 3, Nos. 5/6, November 5, 1991, pp. 6-13. Reprinted with permission of América Press, Inc., 106 West 56th Street, New York, NY 10019. © 1991. All rights reserved.

4. SPATIAL INTERACTION AND MAPPING

The coordinate system derived from Ptolemy was quickly appropriated as an instrument of the first great age of European expansion into the overseas world. Whether or not such a seemingly humble innovation was a necessary condition for that expansion we cannot be sure. But by reversing the introspection of the *mappaemundi*, the Ptolemaic world map projected an image of extroversion. The numbered sequence of latitude and longitude values, known as the graticule, explicitly recognized the other half of the world. Even if it was not an accurate prediction of what was there, it was a rhetorical visualization of the unknown, an invitation to fill the blank space, and to explore that previously inauspicious West beyond the ocean sea.

The importance of the Ptolemaic grid in the visualization of the world is more difficult to assess at the level of individual historical actors. With Columbus—who, as nearly everyone now knows, began from the assumption that the world was round rather than proving that this was the case—we should be especially cautious. We know that he made careful comparisons of maps and first-hand observations to promote his voyage to America yet the Bible took its place as his authority along with the Classical geographers. In a moment of self-denial, he wrote that "reason, mathematics, and *mappaemundi* were of no use to me in the execution of the enterprise of the Indies," but this re-flects his doubts about the meaning of his mission as well as the geography revealed by his voyages. Certain questions obsessed Columbus: the size of the earth; the longitudinal width of Asia and its relationship to Japan; the corresponding width of the Western Ocean.

The common denominator of the key maps available to Columbus is that they bore a graticule either drawn or implied. So it was with the "navigation chart" drawn by the Florentine physician and astronomer, Paolo Toscanelli in 1474, and a copy of which was later sent to Columbus. Toscanelli describes his map thus: "Although I know that the world can be shown in the form of a sphere, I have determined . . . to show the same route by a chart similar to those made for navigation . . . the straight lines shown lengthwise on the said chart show the distance from west to east; those across show the distance from north to south . . . " We are looking here at a map graduated with parallels and meridians, a grid that enabled the user to measure the distance across the Atlantic between Lisbon and Quinsay near the city of Cathay in China (26 spaces of 250 leagues each). It may not be an exaggeration to assert that Ptolemy's grid was a critical preconception for the events leading to the meeting of two worlds in 1492. The world maps by Henricus Martellus Germanus, of which the large manuscript map he made in 1489 is the most dramatic

World Map,

from the Ulm 1482

edition of Ptolemy's

Geography, *reveals*

a world of the late

Greco-Roman times

with a closed

Indian Ocean and

a Mediterranean

Sea twenty degrees

too long

example, are crucial for such an understanding. Its graduation in latitude and longitude implies that only 90 degrees of longitude—a quarter of the world—existed between the Canary Islands and Japan.

Second, we find that the often unintended results of making maps are as important as those that were intended. Once again Ptolemy's grid, rectilinear, abstract and uniform is a stimulus for other geographical actions. We do not know if Ptolemy in the second century AD could have foreseen what the eventual historical effects of his invention would be, but the potential of a system of spatial reference as a source of power and inventory was quickly grasped in the Renaissance. Roger Bacon had already stressed the proselytizing power of a global coordinate system to his patron Pope Clement IV in the thirteenth century. Likewise, Jacopo d'Angelo, in presenting to Pope Alexander V the first Latin translation of Ptolemy's *Geography* in the early fifteenth century, wrote in the dedication that he hoped the book would serve as "an announcement of his coming rule . . . so that he may know what vast power over the world he will soon achieve." And later in his book *The Cross*, John Donne recognized the religious symbolism of a mathematical coordinate system: "Looke downe, thou spiest out crosses in small things;/ Look up, thou seest byrds rais'd on crossed wings;/ All the

Globes frame and spheres, is nothing else/ But the Meridians crossing Parallels." As Samuel Edgerton has put it, "the cartographic grid of the Renaissance was believed to exude moral power"; it expressed "nothing less than the will of the Almighty to bring all human beings to the worship of Christ under European cultural domination." The globe could now be grasped as a knowable totality and though finite, an expression of God's infinite wisdom.

The maps that carried the grid were also a super icon of this power to dominate. Initially, following Ptolemy, the "world map" had covered only the "inhabited" hemisphere of the Old World. The first map on which it extended to the whole world (globes excepted) was the *Cosmographia Universalis* made circa 1508 by Francesco Rosselli, a commercial printmaker in Florence. This small world map thus has an importance extending far beyond its modest appearance. Graduated with 360 degrees of longitude and 180 degrees of latitude, it is thus the earliest extant map of the world in the modern sense of "map" and "world." It takes on a special significance as drawn on an oval projection into which every point on earth could be theoretically plotted and upon which every potential route for exploration could be shown. The map had truly become a "totalizing device": it was a geographical idea of elegant simplicity.

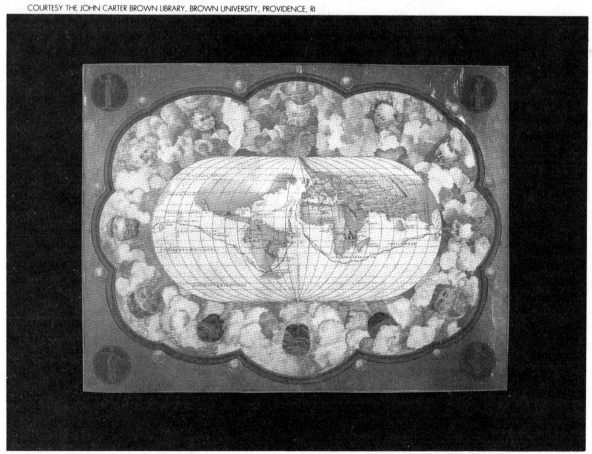

World Map, 1543–1545, from a portolan atlas by Battista Agnese, a Genoese cartographer active in Venice. Using the Ptolemaic grid, the map clearly shows Magellan's route around the world

The pre-conquest Nuttall screenfold from the Mixtec culture depicts the story of an ancestral migration. Researchers have reconstructed the geographic elements identifying place signs and relating them to modern locations

To the monarchs of early modern Europe bent on imperial and spiritual conquest, the map became a menu for colonization. As one chart maker put it in 1534, "With these charts, the reader may inform himself about all this new world, place by place, as though he himself had been there." The world could be carved-up on paper. By its abstracted simplicity and apparent objectivity, placing the viewer above the world even before the age of the astronaut, it also homogenized space and offered to the explorer and his patron a clean slate for exploitation. The Ptolemaic grid was the perfect cartographic instrument for collecting, collating and correcting geographical knowledge. Robert Thorne could boast in 1527, "there is no sea unnavigable, no land unhabitable." The circumnavigation of the world in 1522 had made everything possible. The world could now be viewed as a marked out stage, the *Theatrum Orbis Terrarum* of Ortelius' famous atlas of 1570 and the actors—the Europeans and the "Others"—could be assigned their place. Turn over its pages, and every map has a gridded backdrop on which the Europeanization of the world could be played out. The atlases of the sixteenth

and seventeenth century increasingly became the symbols as much as the instruments of European power. Spatial imagery had acquired a new meaning and a new strength.

Our third point addresses the need to consider the maps of peoples on both sides of the Columbian encounter. It highlights the extent to which mapping, rather than being primarily a European preserve, has a significant native American dimension. In the early colonial period, the fullest evidence for the coexistence of Native American and European traditions of mapping relates to Mexico. What the Europeans found in Central America after 1520, though it was not recognized as such, was an independent cradle for the invention of cartography. The claim of pre-Columbian peoples—no less than those of Europe, the Middle East, or China—to have created their own tradition of cartography is confirmed not only by eye-witness reports, as when Cortés was given a "cloth [map] with all the coasts painted on it," but also by the survival of picture maps from both the pre- and post-conquest eras. Somewhat later, it is also confirmed by some of the local maps and plans

COLOR FACSIMILE FROM THE SPECIAL COLLECTIONS GOLDA MEIR LIBRARY, UNIVERSITY OF WISCONSIN—MILWAUKEE

Maps to the nth degree

J. Brian Harley and David Woodward have edited a multi-disciplinary world history of cartography being published by the University of Chicago Press in six volumes. The series, entitled *History of Cartography,* is supported by The National Science Foundation, and private foundations and individuals. Volume 1, *Cartography in Prehistoric, Ancient, and Medieval Europe and the Mediterranean* was published in 1987; Volume 2, Book 1, *Cartography in the Traditional East and Southeast Asian Societies* is scheduled for publication in 1994.

The "Maps and the Columbian Encounter" supported by The National Endowment for the Humanities, has generated a traveling exhibition with a narrative about maps and their meaning from both sides of the encounter. It is described in an illustrated guide by J. Brian Harley, *Maps and the Columbian Encounter* (Milwaukee: Golda Meir Library, 1990), $12.95.

(*pinturas*) in the *Relaciones Geográficas* of New Spain in the 1580s. Though often drawn and painted in a European style, some maps perpetuate purely native or hybrid traditions of representation.

Only relatively recently has the maplike character of Mixtec ritual and genealogical manuscripts been fully recognized. For instance, the pre-conquest Nuttall screenfold and comparable codices are now interpreted as spatial histories showing episodes of an ancestral migration and conquest of the Valley of Mexico. Among the place signs identified in the Nuttall Codex are many of those for hills (shaped like a bell and painted brown), those for buildings, fortified sites, and historical events, and the strings of footprints for routes taken by the conquering lords. Such pictorializations of the land are no less a map than the Peutinger map deriving from the classical Roman world or a fifteenth-century map of a European estate. When the place signs are deciphered and related to a modern map, using archaeological, oral and pictorial evidence, we can reconstruct some of the historical places within the pre-conquest landscape: rivers, lakes,

and cities as seen through the lens of an indigenous culture. Mapping as geography and cosmography was by no means alien to the cultures of ancient America and for them, too, maps were the coordinates of history and an ideological perception of territory.

A striking contribution of Native Americans to the mapping of the Encounter—also revealed by recent research—lies in their often unacknowledged choreographing of European exploration. Beginning with Juan de la Cosa's world map of 1500, most European maps showing the New World conceal a hidden stratum of Indian geographical knowledge. It is difficult to imagine what some of the early European maps of America would have looked like had there been no Indian guides when the Europeans arrived in the New World. The extent to which Columbus and his contemporaries relied on Indian knowledge was considerable. As early as October 13, 1492, Columbus reported taking prisoners and on other occasions—alike by the Spanish, English, and French—Indians were kidnapped and taken back to Europe for interrogation. On their return they became the eyes of the mapmaker. A remarkable exchange of geographical information was in progress throughout the sixteenth century and much of it eventually given coordinates in European maps. Among those who had maps drawn for them by Indians are Cortés, Coronado, Escobar, Farfan, Captain John Smith, and Samuel de Champlain. Only a few European maps acknowledge these sources so that their existence has to be inferred from features such as large lakes, artificially straightened rivers running into the west, or the fabled "golden" cities such as Cibola, El Dorado, and Norumbega.

Nor should we picture Native American peoples as solely passive agents for European mapmakers. The mapping of America was not a one-way process in which colonized peoples offered no challenge to external surveys of their territory. It is true that in some regions native voices were quickly silenced and their maps were superseded by those of the politically dominant culture. But in other regions maps were used to help resist the imposition of colonial rule. A tradition of native mapping in Mexico—as part of native art in general—survived the Conquest. We glimpse the power articulated in European maps being partly resisted on native maps which offer an alternative cartography stressing the legitimacy of their own territory.

Among the manifestations of resistance are the maps used by Indians in land claims, where they became documents for litigation in colonial courts. Elsewhere, the genealogy of ruling Indian families was inserted on maps as a way of seeking to restore land or to reduce onerous taxes. The Codex Xolotl—a narrative of the original conquest of the Valley of Mexico by the ruling family of Texcoco—is also a cartography. As the Indians sought to remake their own history through a map, it links time and space, seeking to prove ancient nobility rooted in territory. These "alternative cartographies" stood as a challenge to Spanish power.

In many of the European colonial maps we witness a process of silencing the indigenous culture.

The imposition of a western way of seeing as the only valid form of representation excludes not only the Indians from the perception of their own land but also the Europeans from the perception of the Indians. If for the colonists cartography was psychologically enabling, for the colonized it let down a curtain on a familiar world. By insisting that the land will only appear in a certain way—and by affirming the "correctness" and "naturalness" of what it shows—European maps excluded the American way of seeing the land. The induction of this cartographic amnesia was a social projection of a future geography. Just as Indian maps were often destroyed, so too were the traces of an Indian geography frequently eradicated from the space of European maps. The maps increasingly convey, through the articulation of the blank spaces and *terra incognita*, the assumption of an uninscribed earth. Later, in North America, they would contribute to the myth of an empty frontier.

As we approach the 500th anniversary of the first voyage of Columbus in 1992, we are reaching toward an alternative cartographic history. Its tone is more reflective than celebratory. Although the maps of the Columbian era have been traditionally viewed as eloquent witnesses of European geographical discoveries, they are much more than that. They also reveal a fundamental shift in thinking by the geographers of Renaissance Europe about the shape and size of their material world. The talisman of this enlarged world was the Ptolemaic grid. But as well as inventorying places discovered by Europeans and identifying a terrain for evangelization, the coordinated space of the new maps was instrumental in the symbolic appropriation of Native American territory. By recognizing Indian peoples as the victims of a European cartography we have also begun to reinstate their contribution into the cartographic record of American history. By reading between the lines of the map, a start has been made in writing a different account of the mapping of a New World that was already old when Columbus arrived.

Cultural Commitments and World Maps

William E. Phipps

William E. Phipps, whose Ph.D. is from Scotland's St. Andrews University, is Professor of Religion and Philosophy at David and Elkins College, Elkins, West Virginia. As displayed by his articles in Natural History *(8/85),* Bios *(12/83), and* Journal of Ecumenical Studies *(7/79), he is specially interested in relating the history of religions to issues in the natural and social sciences. His latest book is* Genesis and Gender *(Praeger Press).*

We assume that world maps accurately represent reality, and we usually remain unaware that a map often reveals the cartographer's world view. Distortions in maps belonging to other civilizations are easily discerned, but misrepresentations by cartographers of one's own culture are more difficult to detect. A close examination of some historical maps reveals the universal tendency to pass off religious and national biases as matter-of-fact reality.

Oldest world map is a clay tablet excavated in Iraq.

The oldest extant world map is on a clay tablet excavated in Iraq. Dating from the sixth century B.C.E., it shows the earth to be virtually coterminous with Mesopotamia, the region of southwest Asia between the Tigris and Euphrates rivers. The Euphrates bisects the circular map, and the walled city of Babylon is prominently displayed along that river. Triangular islands protrude at the edge of a river encircling the earth. This circumfluent river reflects the Babylonian myth of the world emerging from the body of Tiamat, the salt water goddess.

This ancient Babylonian map expresses the outlook of the person who composed the heroic story of Gilgamesh. That epic tells of a man who is instructed by a god to construct a ship near the Euphrates in order to escape a forecasted deluge on the surrounding land — which is presumed to be the entire earth. The Genesis story of the devastating Noachian flood also assumes that the Mesopotamian valley is the whole world.

The Greek and Chinese worldviews

During the same sixth century, Anaximander became the first Greek to draw a map of the world. Crosscultural influence between the Babylonians and the Greeks is likely, because Anaximander's map also features the earth as a disk around which the *Okeanos* flows. Although all copies of Anaximander's map have perished, its outlines can be reconstructed by later descriptions. Europe is as large as all other regions combined. Greece is in the center and the waters of the Mediterranean Sea, the Black Sea, and the Nile River radiate out from the core country.

The early Greeks audaciously pinpointed the world's center. According to their mythology, Zeus released two eagles at the opposite extremities of the earth and they met at Delphi. Pious Greeks went to the oracle located there to receive messages from sun god Apollo. Plato writes: "This god is doubtless for all humans the interpreter of the ancestral religion; he sits in the middle of the earth at its navel and delivers his interpretation." [Republic 427] A decorated conical stone *omphalos* (navel) marking this sacred center was found in the ruins of the Temple of Apollo and is now in the Delphi museum.

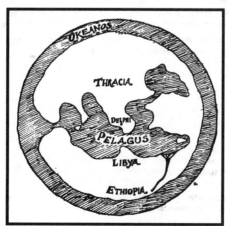

Reconstruction of Anaximander's map.

Through most of their history, the Chinese have unintentionally illustrated by their world maps the ethnocentric predicament of humans. The Chinese have used characters meaning "center kingdom" to refer to themselves. Accordingly, their maps are like wheels with Beijing's Temple of Heaven at the hub. The Chinese may also have been influenced by the

Beijing and the Great Wall are at the center of this (simplified) map of China. The rest of the world is consigned to the outer ring.

Babylonians because both feature a ring of water around their respective homelands. Also, little islands designate the habitations of all other peoples and animals — mythical and real. Awareness of this entrenched sinocentricity is helpful for interpreting political happenings in every period of Chinese history.

Christian Maps: Jerusalem as a "bellybutton"

The earliest Christian cartographer was a Byzantine monk named Cosmas who lived in the sixth century. In his book entitled *Christian Topography* he presupposes that a true map must follow biblical guidelines. The eighth chapter of Hebrews was interpreted to mean that the earth has the same rectangular shape as the tabernacle built by Moses. This fits with biblical references to "the four corners of the earth." [Isa. 11:12; Rev. 7:1] There is one body of water surrounding the earth because God declared on the third day of creation: "Let the waters under the heaven be gathered together into one place." [Gen. 1:9] This ocean is supplied by the Indus, Nile, Tigris, and Euphrates rivers which flow out of the Garden of Eden. [Gen. 2:10-14] The water around the rim of the world was probably also suggested by earlier Greek maps.

In the seventh century, Isidore of Seville found in the Bible a basis for drawing a circular world map. This Spanish bishop started with Ezekiel, who announces as God's spokesman: "This is Jerusalem; I have set her in the center of the nations." [Ezekiel 5:5] Also, another Israelite prophet pictured the Creator drawing the boundaries of the world with a compass.

From *Focus*, Summer 1991, pp. 7-9. Reprinted by permission of The American Geographical Society.

[Isa. 40:22] The Garden of Eden is above India at the top of the map. *Oriens* is inscribed above *Paradis* because humanity presumably originated (<*oriri*, to rise) in the direction of the rising sun. Thus, to orient maps has meant for much of European history to place the Orient at the most prominent place. Since the eye first notices what is at the top, Eden is give a place of honor. Isidore also has a single ocean surrounding the earth.

In the thirteenth century, Richard of Haldingham made the map of the world which was on display at the altar of the Hereford Cathedral. Jesus is enthroned at the top, above Paradise in the East, presiding over the Last Judgment. "The holy land," as designated by a biblical prophet, [Zechariah 2:12] fills an area between Mesopotamia and Egypt which is larger than Africa. Jerusalem, looking like a belly-button, is at the center of this English map. The Latin Vulgate, on which Richard relied, designates Jerusalem as "the navel of the earth" (*umbilicus terrae*). [Ezek. 38:12]

Based on the Psalter world map of the 13th century, this sketch has Jerusalem as its "bellybutton." Note "T and O" configuration.

The Greco-Roman division of the earth into Europe, Asia, and Africa was continued because it could be harmonized with the Genesis account of the earth being divided among three sons of Noah after the great flood. The continents were divided by three seas. The late Roman name *Mare Mediterraneus* suggests this, because "Mediterranean" is a combination of the words *medius* (middle) and *terra* (earth). This type of map was called by poet Leonardo Dati *un T dentre a uno O* ("a T within an O") because the Mediterranean formed the upright of the "T" and is intersected at the top by the Black and Red seas to form the horizontal ends. The ocean forms the O of this most popular world map of the Middle Ages.

These highly subjective T and O representations of reality probably encouraged European Christians to join crusades which aimed at recapturing the center of the world from the "infidels". It was humiliating for Western Europeans to feel as though they lived at the periphery of lands God gave to the faithful. Those original Zionists who succeeded in fighting their way to Jerusalem were pleased to see a particular pillar in the Church of the Holy Sepulcher. Erected there — and still standing — is a marker giving the precise location of the world's center!

From Muslim to Mercator

Muslims map-makers followed the ancient proclivity to represent the world as a flat disc, with religious convictions determining the heartland. The influential twelfth century cartographer Idrisi placed the Arabian peninsula at the center. He was familiar with the circular Christian maps from his extensive travels in Europe. Also, Isidore's map had been included in an Arabic translation of his *Etymologies*. Idrisi reoriented the widespread Christian map so as to give Islam the propaganda advantage. Christendom is demoted to the bottom of the map and the darker skinned people are at the top. Africa — Idrisi's home continent — is larger than all the other land areas combined. At that time Islam had become the dominant force in some kingdoms of West Africa and the Nile Valley. Europeans were certain that Idrisi's map was ludicrous and preferred St. Isidore's T and O map. Thus, at Augsburg in 1472, a woodcut of that map became the first to be printed after the press was invented.

For the past four centuries the most widely circulated map of the world has been some modification of Mercator's projection. Flemish cartographer Gerhard Mercator lived on the continent in which Christianity was centered and, like Cosmas, he attempted to portray the earth on a rectangular map. Mercator gave Europe and the Northern Hemisphere a form that greatly distorts reality.

To Mercator's credit it should be pointed out that he intended the map to be primarily used for navigational purposes. However, it quickly became and remains the most popular map for those interested in land areas. Europeans have looked at the way Mercator placed all of Asia to the east of his heartland and have designated areas of that continent the "Near East" and the "Far East." In order to place North America in the center, Americans have modified this map by dividing the largest continent and placing part of it at each of the longitudinal ends.

As compared with the globe, the Mercator projection resembles the distorting convex and concave mirrors at a fairground. Only on the equator is the map fully accurate. The Atlantic Ocean appears to be larger than the Pacific Ocean. Lands beyond the 40th latitude are greatly distorted, and this includes most of Europe. Britain, although approximately the same size as Ghana, appears to be twice as large. Europe is about the size of South America, even though the latter has several million more square miles. Scan-

dinavia looks much larger than India even though India is actually several times larger.

Africa and the Nile dominate, with the Arabian Peninsula and Mecca at the center of this sketch based on al-Idrisi's map.

Colonial exploitation was more easily rationalized by those who shared Mercator's view of the world. Starting with the pernicious equating of upper with superior, those in the Northern hemisphere have looked down on the underlings to the South. They are on top of Africa, South America, and Australia — "The Land Down Under." The Southern hemisphere is also made to look inferior by placing the equator two-thirds the way down the map. Its latitude is not extended to the full ninety degrees because an absurdity of the Mercator projection would be obvious to all: other continents would be dwarfed by the gigantic size of Antarctica! If the Northern hemisphere were cut off at the 60th parallel, as is the Southern, then most of Scandinavia would be omitted.

Alternatives and correctives

A number of alternatives to the Mercator projection have been made in an effort to counter its imperialistic representations. The Australian map of the world humiliates most earthlings by inverting the hemispheres and showing Australia in a dominating position. Likewise, the Turnabout Map of the Americas published in 1982 reverses the arbitrary but stereotyped placement of the Americas. It is advertised as providing a "corrective perspective" by eliminating the mischief bred by the Mercator projection. These consciousness-raising efforts are fitting retribution for centuries of Eurocentric cartography.

German historian Arno Peters published a map in 1973 that he claimed had overcome the Mercator limitations by projecting areas in proportion to their relative sizes. Peters' map is not essentially new but is a modification of projections made by Johann Lambert in the eighteenth century and by James Gall in the nineteenth. The Northern hemisphere remains at the top but the equator is located in the middle of the map. South America and Africa are several times as long as Europe. Many millions of copies of the map have been produced in a number of languages

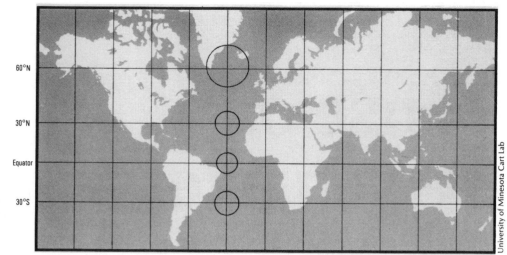

Gerhard Mercator's projection (map first used in the Summer 1986 issue of FOCUS)

with the support of the United Nations Development Programme. The publisher, Friendship Press, points out that the Mercator map is skewed to the advantage of the hemisphere where whites have traditionally lived. "The Soviet Union appears to be more than double the size of Africa, in spite of the fact that Africa is actually much larger," it states. Consequently, "the traditional map is not compatible with objectivity, which is required in a scientific age." By contrast, the Peters map shows "fairness to all people by setting forth all countries in their true size and location."

However, these attempted corrections may be as dangerous as the Mercator projection which they hope to replace. Third World regions near the equator are exaggerated at the expense of regions that have dominated them. By simplistically maintaining that accuracy has at last been achieved, reverse prejudices may emerge. Africa is at the center of the Peters map both horizontally and vertically.

Portrait or photograph?

Cartographers produce what is more like a portrait than a photograph, so it is unwise for any to claim that any world map shows the earth as it is. A whale might conceivably criticize all human maps, for they exaggerate places where land creatures live. A whale's eye view of the world would center on the South Pole because such a map would feature the hemisphere which is eighty-eight percent water and the single ocean which covers three-fourths of the globe's surface.

A projection (literally, "something thrown off") is an apt term for a map because it is a sphere trying the impossible stunt of acting like a plane. Even if persons were to confine themselves to the use of globes, prejudices would not necessarily be eliminated. Astronauts who photographed the earth on their moon trek convinced many that there is no real up or down to our terrestrial ball. Yet our globes are generally mounted so that the North Pole is at the top. Those who belong to the Northern hemisphere can then imaginatively look down on the small minority of the global population in the lower half. It makes us dizzy to look at the globe with the South Pole at the top and we are disoriented because the

descriptive words on it are upside down.

In a widely circulated Christmas advertisement by an American corporation, an artist depicts our globe held so that the United States sparkles in the sight of God. Obliterating the view of much of the Southern hemisphere are the divine hands holding the globe. Above the picture is this caption: "He holds the world, its beginning and its end." The advertisers presumed that their customers are devoted to a planetary deity who has "the whole wide world in his hands" and who is mainly concerned about Americans.

The history of cosmic maps reveals a parallel struggle with geocentrism and heliocentrism. Prior to the Renaissance virtually all people accepted without question that our planet was stationary and at the center of the universe. When that conviction was painfully destroyed, humans retained some sense of cosmic security by believing that our sun was stationary and at the hub of the universe. Isaac Newton then showed by his laws of gravitation that there is no center. Our solar system is not even at the center of our local galaxy, the Milky Way. Our earth

is one of the smaller satellites of one of the smaller stars of one of an unlimited number of galaxies. The words of Pascal, a contemporary of Newton, are apropos: "The universe is an infinite sphere whose center is everywhere and whose circumference is nowhere."

World maps often function in the opposite way from what is commonly expected. Whereas they are viewed as horizon broadening, they may fail to aid us in venturing forth to other cultures. Maps can be culturally confining and can reinforce the ethnocentricism that is native to all humans. Ethnocentricity is even more subtle than egocentricity and therefore more difficult to eliminate. Most people are probably unaware that they have this social disease. New Englander Oliver Wendell Holmes wrote: "The axis of the earth sticks out visibly through the center of each and every town or city." However, that falsely located axis is not visible by eyesight but can be perceived by educational insight. After exposure to a variety of cartographic perspectives, we can fully grasp that maps tend to center in their creators. Then we can begin to surmount cultural conditioning and extricate pivotal orientation.

Further Readings

Bagrow, Leo and Raleigh Skelton, 1964. *History of Cartography*. Cambridge: Harvard University Press.

Thrower, Norman, 1972. *Maps and Man.* Englewood Cliffs: Prentice Hall.

America's business-eye-view: "He's got the whole world in his hands."

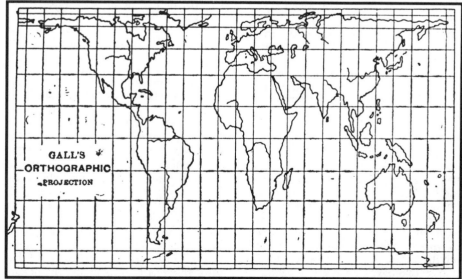

GALL'S ORTHOGRAPHIC PROJECTION

The 1973 Peters projection is derived from James Gall's 1885 Orthographic Equal Area projection. Reproduced with permission of the Royal Scottish Geographical Society. For more on projection, see Phil Porter and Phil Voxland, 1986," Distortion in maps" FOCUS V. 36 no. 2.

A sense of where you are

Powerful computerized maps are helping make order out of an increasingly complicated world

A map is more than lines and figures on a sheet of paper. It is a state of mind. From the days of the earliest explorers, map makers have drawn the terra incognita of the Earth into closer and closer focus, shaping world views as they sketched ever-more-accurate images of the planet. Artist Tom Van Sant of the GeoSphere Project in Santa Monica, Calif., recently took the process to its ultimate end with a bit of cartographic history: the first map of the world that is not drawn at all. Rather, it is pieced together entirely from satellite photos.

Van Sant's new view of the world is part of a cartographic revolution that is transforming flat and static maps into dynamic displays of information about the earth and its inhabitants. Using powerful computers, high-resolution video monitors and vast memory banks, a new generation of map makers is merging the crisp photo imagery from satellites and aircraft with volumes of data to create maps that are as vital a tool for navigating in the modern world as sea maps were for ship captains centuries ago.

Highlighted in a six-part television series called "The Shape of the World" airing on PBS this month, computerized maps are rapidly becoming an integral part of scientific research, business marketing and urban planning. The series is funded by IBM, one of many companies racing to become part of the rapidly growing business of supplying software and hardware for the new electronic map technology, known as "geographic information systems" or GIS.

Pioneered by such companies as the Environmental Systems Research Insti-

tute of Redlands, Calif., and Intergraph Corp. of Huntsville, Ala., GIS systems are already being used by city governments as the geographic equivalent of a "crystal ball." Municipal maps that combine data on water tables, residential housing and air-pollution levels help officials find an optimum site for a landfill, for instance, and computerized maps that show updated traffic information provide the fastest routing for ambulance and police service. The military is using highly detailed GIS maps of Kuwait City made prior to the gulf war to assess bomb damage and help restore power and phone lines, and the U.S. Fish and Wildlife Service used the new map technology to analyze the impact of an Alaskan oil pipeline on caribou migrations. In the future, scientists hope to use GIS to combine data from various earth monitoring satellites to keep track of global climate changes.

Boardroom mapping. By allowing map makers to easily merge almost any kind of data with a spatial map, GIS technology has become a key tool in corporate boardrooms as well. Petroleum companies use GIS technology to analyze geological formations and locate the best spots for drilling; the telephone company U.S. West is targeting its marketing of "call waiting" and other services with a GIS program that merges customer profiles with phone information; and a movie distributor uses GIS to analyze what kinds of films draw the best crowds in various neighborhoods. The new mapping technology is even being downsized to work on a typical personal computer: One sys-

tem offered by MapInfo of New York is used by banks, for instance, to track the geographical distribution of their mortgages.

Merging maps with other data to create analytic tools has a proud tradition: During the Revolutionary War a French cartographer used hinged overlays to display troop movements during the Battle of Yorktown. In 1855, a British physician plotted the location of victims of cholera on a map of London to trace the source of the epidemic to a public water pump.

But it has only been with the advent of computers that map makers have had the power to merge vast reservoirs of data with highly accurate geographic displays. Typically, map makers start with a photograph from a satellite or an airplane. The geographical data are either scanned into the computer or entered by means of a handheld device that a map maker uses to "trace" a coastline, for instance, or follow the contours of a mountain range. Once in the computer, the geographic data can be merged with almost any existing database—basic census information, rainfall patterns, income distribution, the populations of various insect species, disease prevalence or a country's mineral resources.

For all their potential, the new map making techniques still come up short of a common map maker's dream: to be able to give a map gazer instant access to a wealth of information—past and future—about any spot on the globe. The sheer mass of data required for such a task would require highly sophisticated computer programs that do not

 From *U.S. News & World Report*, April 15, 1991, pp. 58-60. © 1991 by U.S. News & World Report.

yet exist, says Terry Smith of the University of California at Santa Barbara. Smith is an associate director of the National Center for Geographic Information and Analysis, a consortium of universities funded by the National Science Foundation that is developing computer programs and other technologies to handle the huge amounts of data needed for modern mapping. This task will become even more challenging in the near future, Smith says. Several Earth-monitoring satellites, planned for launch in the mid-'90s, will begin beaming down data on everything from weather to vegetation patterns to geological formations. Making sense of it all will be a herculean task, says Smith, requiring programmers to give computers what in essence amounts to a "map of the map" to help navigate through the data.

More honest maps? The sophistication of GIS notwithstanding, cartographers are quick to point out that the burgeoning mounds of data provide no guarantee that the new maps are any more accurate, or any less deceptive, than their predecessors. By their very nature, maps are selective presentations of information, with sprawling cities shown as round dots, hilly terrain smoothed to flat sheets of pastels and statistical mi-

---

"MENTAL MAPS"

The politics of cartography

All maps distort reality. Flattening a spherical world without misrepresenting some part of it is simply not possible. Map makers have long appreciated this fundamental paradox of cartography, but in recent years the distortions in maps have become increasingly political. Indeed, map making appears to be the most recent target of the political correctness movement that is currently sweeping through American education and publishing.

The political debate centers mostly on a 400-year-old map, the Mercator projection, the image of the world most familiar to American schoolchildren. Originally designed as a navigational tool, it still works fine for that purpose, but it gets the relative size of land masses dead wrong. The Soviet Union doesn't dwarf Africa, as the Mercator makes it appear. Nor is Europe larger than South America.

Maps of the mind. Such distortions are of more than passing significance, critics of the Mercator insist, because "mental maps" are crucial to cultural perceptions. "In our society," argues Salvatore Natoli of the National Council for Social Studies, "we unconsciously equate size with importance and even with power, and if the Third World countries are misrepresented, they are likely to be valued less."

The most provocative challenger to the Mercator is the Peters projection, named after its creator, German historian and cartographer Arno Peters. Although the map is nearly 20 years old, it is being heavily promoted today by the World Council of Churches and various United Nations organizations, which distribute the projection as "a map for our day" and as a politically sensitive replacement for the outdated Mercator. The Peters map is an "equal area" projection that claims to represent relative sizes of land masses more accurately. In publicizing his map, which noticeably elongates the Third World continents of Africa and South America, Peters has bitterly denounced the "European arrogance" of the Mercator map, saying that the cartographic profession has

FRIENDSHIP PRESS

Stretching truths. *Arno Peters has campaigned for the adoption of a map that displays the world's continents at their true size, though not their true shape.*

intentionally used it to foster European imperialist attitudes.

The Peters map has been roundly attacked by the cartographic establishment on technical and philosophical grounds, but publishers and educators appear to find the map's politics appealing. Despite an intense lobbying effort by leading cartographers that dissuaded Prentice-Hall from publishing the map, in October, HarperCollins Publishers in New York released the first American edition of a world atlas based on the Peters map.

The schools, too, are yielding to the tide of cartographic political correctness. More and more are using the Peters map not only to raise consciousness about the inevitability of map distortions but also to counter the strong European emphasis of the Mercator and other maps. The Texas Education Agency has required textbook publishers who sell books to Texan children to explain that the world does not really resemble the traditional Mercator image. Publishers of new texts must also include comparisons of different map projections. In Paramus, N.J., educators are using the Peters projection with other maps to get students to look at the world differently.

It is human nature for map makers to put their own interests at the center of the world and keep what is foreign peripheral. The enthusiastic reception that the Peters map is currently receiving suggests that the Third World's point of view is, if nothing else, politically fashionable.

BY SCOTT MINERBROOK

nutiae lumped into coarse generalizations. "It's not only easy to lie with maps," notes geographer Mark Monmonier of Syracuse University, whose book *How to Lie With Maps* has just been published. "It's essential." With their fleet computers and access to vast stores of information, map makers have more rather than less power to skew their presentations, whether their goal is to show the "ideal" location to build a house or to demonstrate that a development project will have a negligible impact on the environment. The fact that these distortions can subtly shape public perception has not been lost on developers, advertisers and propagandists who promote maps with a view of the world that fits their particular agenda (see preceding page).

Nor is electronic cartography likely to supplant paper maps. Indeed, the result of the new technology will most likely be more paper maps, not fewer. Not only does GIS make it easier to update maps more frequently, it also makes it possible to produce customized, "one time" maps designed only for specific tasks. Paper maps will stay around for a more basic reason as well. They are a convenient method for bringing information to the most sophisticated computer on earth, the human brain. "The human visual system is extremely powerful," says Smith, and the brain's ability to organize the dots, lines and colors of a traditional map into meaning is unparalleled by any machine.

From the dazzling displays of electronic cartography to the intricately wrought scrolls of the ancient explorers, maps have long been a source of pleasure as well as information for a species whose most sophisticated sense is vision. By capturing the map maker's view of the world, maps touch the core of human curiosity that drives both art and science. Whether a cosmological theory of the universe or a simple block diagram in a shopping mall, maps provide a fundamental message that seems essential to the human condition: *You are here.*
WILLIAM F. ALLMAN

THE NEW CARTOGRAPHERS

Gregory T. Pope

In a windowless chamber, scrubbed clean by hissing climate control, stand towers, racks, and cabinets of computing power. Here, in the U.S. Geological Survey's Reston, Virginia, headquarters, a nineteen-inch monitor winks on, revealing a three-dimensional relief map of the Loma Prieta area south of San Francisco. At the punch of a button the image wheels into motion, taking the viewer around the region's peaks and valleys on a ten second flying-carpet ride.

In a similar room at the SPOT Image Corporation, about a mile down Reston's tidy Sunrise Valley Highway, technicians are poring through data beamed down several days ago from *SPOT 2*, a commercial satellite that circles the globe every twenty-six days. These days, any mapmaker with as little as $500 to spare can freshen up an outdated map. By simply summoning a map on a computer screen, and superimposing a SPOT image over it, the computer whips through an automatic change-detection routine, shifting borders and other features to match the satellite picture.

It's fitting that both the Geological Survey and the SPOT Imaging Corporation—two driving forces behind today's cartographic revolution—call Reston home, a city of 32,000 that wouldn't have shown up on most maps twenty-five years ago. The landscape—and the art of cartography—change quickly now. Gone are the pens and sapphire-tipped gravers of yesteryear. More precise than ever before, thanks to computerization and satellite imagery, modern maps no longer stop at simply guiding the user from point to point. Like crystal balls, they must let the user gaze into the future to answer thorny questions: where to site a landfill, how to route a fleet of ambulances, how the terrain will look after an earthquake has rippled through.

New demands have compelled mapmakers to package geographical information in innovative ways. The new cartographers use software that translates spatial features—houses, street intersections, rivers, city blocks—into mathematical points, lines, and polygons. These geometric elements are identified and tagged as street names, zip codes, soil types, population figures, whatever, and are downloaded into a computer's memory. The result isn't a map in the traditional sense but rather an information pool where every feature is digitized and stored in a vast database. Such a database, called a geographic information system, or GIS, is the crux of the new cartography. Using a GIS, cartographers can analyze geographic information without ever looking at a map. Or they can unleash snazzy graphics routines to display the information in a more familiar maplike form.

The venerable mapmaking houses, whose profits hinge on the exactitude of their products, have wholeheartedly embraced GIS technology, largely because keying corrected digital data into a computer takes a lot less time than the traditional chore of scribing a whole new map onto a sheet of coated film. More than likely, a GIS database spawned the road map or world atlas you just bought.

The elite among the new cartographers sharpen their view of the world by employing a satellite capability known as remote sensing. *Landsat 1*, lofted by NASA in 1972, showcased the first glimpse of remote sensing's power, offering views of the earth unprecedented in their accuracy. Currently, *Landsats* 4 and 5 can pick out objects the size of a baseball diamond from almost 700 miles above the planet.

But these days *SPOT 2*, launched and managed by the French space agency, boasts the keenest eyes in the sky available to civilian mappers, training high-resolution scanners on a continuous forty-mile ribbon around the earth's surface. Operating in multispectral mode (capturing red, green, and near-infrared wavelengths), SPOT's twin sensors can distinguish features sixty-five feet across; in black-and-white mode, they can resolve landmarks as small as thirty-two feet. The images are

179

"rubber sheeted" by ground-based computers, a technique that stretches and doctors the data to compensate for the planet's curvature, the spacecraft's viewing angle, and other distortions. With aggressive marketing (Maps are lies, SPOT is truth, goes one unofficial corporate slogan), SPOT Image Corporation peddles its data in a variety of formats.

Among popular SPOT products are remotely sensed duplications of the U.S. Geological Survey's famous topographic maps. These break up the nation into standard rectangular map chunks known as quadrangles, or quads. SPOT quads provide an instant update of the Survey's quadrangles, some of which have fallen out-of-date. "You can't say they all reflect today's landscape," says University of South Carolina geographer David Cowen of the Survey maps. "It took the Survey a hundred years to map the country. SPOT quads help with fine-tuning." Aided by Cowen's department, South Carolina's government, for example, has mustered up-to-date SPOT coverage of the entire state, identifying new roads in rural regions to help plot economic development.

The one-two punch of high resolution and multispectral imaging has made SPOT the instrument of choice for mapping changes in land cover, especially encroachment on wilderness areas. Trees and tar, cornfields and concrete reflect the sun's radiation in distinctive spectral signatures that SPOT can measure. The U.S. Fish and Wildlife Service used a SPOT image to draw a vegetation base map of the Great Dismal Swamp National Wildlife Refuge, which straddles the Virginia-North Carolina border, in order to trace the shifting industrial and agricultural areas around the refuge. Meanwhile, planning new roads around Jacksonville, the Florida Department of Transportation employed SPOT data to distinguish sensitive wetlands from resilient uplands.

Yet even SPOT's acuity falls short of what some cartographers would like. Squint hard at a SPOT image and the geometry of streets and rivers stands out but you can't single out fence posts, electric pylons, manholes, and other small objects. In addition, objects viewed from space must be pinned to a matrix of longitude and latitude before they can be translated onto a map. Ground-based surveying teams, toting plane tables, theodolites, and trigonometric handbooks have traditionally fixed small features onto maps and continue to do so. Until remote-sensing satellites become more eagle eyed, their jobs are safe.

Just as mapmakers are turning to space-based tools, their counterparts in the surveying business also look skyward to refine their art. Many surveyors now rely on a fleet of military satellites called the Global Positioning System (GPS). These spacecraft broadcast meticulously timed digital pulses; a ground-based receiver, collecting signals from at least four satellites, can use these pulses to reckon its own longitude, latitude, and altitude.

The Pentagon designed GPS satellites to position soldiers and guide smart weapons to their targets. But GPS receivers were soon commercially available, alarming defense strategists: Anyone on the planet could tap GPS signals to steer a missile right into the men's room in the White House East Wing, for example. So last year the Department of Defense scrambled the signals so that only military receivers could use GPS to its full potential. (Ironically, the Pentagon had to lift GPS encryption during the Persian Gulf War because the military ran out of receivers, forcing them to equip soldiers with commercial GPS receivers.)

But surveyors have devised a neat trick to overcome the signal encryption. They place a GPS receiver on a landmark whose longitude and latitude is already known. They then walk around with another receiver to calibrate the degree of signal distortion. After adjusting for signal distortion, the receiver can establish its whereabouts to within three feet; stationary, to within inches.

The technique isn't cheap but for fast-growing areas in the Sunbelt it's worth the cost. Local communities have traditionally maintained a plethora of separate charts for tax assessment, property lines, police precincts, water and sewer lines, and so on. Many now want to scan the collection into a computer, coalescing the maps into a unified GIS to help plot growth. But there are problems. "The maps are supposed to fit together but sometimes old maps don't fit well because of the cartographic license they took back then, perhaps moving a stream's location or fudging a hill's placement so it would fit on the map," says Bill Daly, manager of mapping services for Huntsville, Alabama.

To resolve these inconsistencies, Daly's department turned to GPS. Surveyors laid out a GPS-derived network of control points using landmarks that stand out on different, but overlapping, aerial photographs. Guided by these landmarks, they could piece the different photos together. The result: a seamless, accurate base map of the area, detailed enough to show manholes, tiny streams, and other features. Cartographers then whisked the photographic map into a computer to generate custom renderings such as three-dimensional maps. Beyond giving the city the exact locations of its infrastructure, Daly's group has used the technique to identify slopes too steep or too aesthetically sensitive for proposed development projects.

Similar projects are underway at the U.S. Geological Survey's GIS Research Lab, where a staff of eight is developing maps that peer into the future. "There are a lot more uses for our data than just paper maps," says Nick Van Driel, chief of the GIS lab. For example, four years ago the Phelps-Dodge Mining Company offered to trade land with the Forest Service, hoping to sink an open-pit copper mine into terrain covered by Arizona's Prescott National Forest. The Forest Service called on the GIS Lab, where researchers loaded a computer with the Survey's own elevation data, a *Landsat* image, paper maps showing geology and water tables, land ownership boundaries, and the proposed mine plan. The computer responded with a three-dimensional view of the terrain, showing the mining operation's pits, ponds, and piles and spotlighting changes in drainage caused by the relandscaping. The image then swiveled to reveal how the proposed mine would stand out from different points along a nearby scenic road. Faced with this analysis, the land-swap deal collapsed.

The ability of GIS to model the future has given land-use planners powerful new capabilities. "There's a major transition going on right now," explains Douglas Gerull, executive vice president of the mapping sciences division of Intergraph Corporation, a leading vendor of computer-based mapping equipment. "First people used computers to make maps better, faster, and cheaper. Now they're using the data to manage the areas they mapped."

Managing data is a pet project of University of Wisconsin professor of landscape architecture Ben Niemann and his colleagues. Merging Landsat imagery with property-line maps and soil charts, they developed a GIS database that pinpointed sources of erosion in rural Wisconsin. Next they coupled the system with a water-quality assessment to model the impact of a proposed corporate headquarters on a nearby lake. Most recently they created a three-dimensional terrain map and trickled pesticides across the computerized landscape to illustrate how the chemicals would seep into bodies of water. "Before you implement a new policy," says Niemann, "you can predict its consequences on the landscape—you can ask, 'What if?' "

While local planners are benefiting from the new cartography, accurate geographic information is buoying the fortunes of modern commerce, just as improved navigational charts led Renaissance-era merchants to the

riches of new worlds. Oil companies like Shell and Amoco siphon data from three-dimensional geological maps to keep track of exploration. For Federal Express and UPS, computerized maps deliver the fastest possible routes for package trucks. U.S. West, the phone company, calls on GIS to guide the marketing of premium services and the placement of new switching facilities.

The private sector's love affair with the technology has helped boost GIS into a big business. Sales of cartographic computers and software totaled $1.4 billion last year worldwide. But the explosive growth has lead to new concerns and disturbing trends.

True, the precision demanded by computer systems has prodded many cartographers to smarten up their data with remote sensing and other tools, but it's still possible, with some creative fudging, to shoehorn incompatible data into a single digital map. And when digital mapmakers have concocted a GIS from old maps without considering their original purpose, problems can multiply. "What happens when census maps are used for routing emergency vehicles? Who is liable when an ambulance runs into a dead end street?" asks Michael Goodchild, codirector of the National Center for Geographic Information and Analysis in Santa Barbara, California. "Maps are being held accountable relative to reality in ways they never should be." Other concerns are even more alarming. With we-know-where-you-live information so readily available, will the new cartographers turn from solving social problems to impinging on individual freedom? "The downside of the new cartography is that this is the technology of George Orwell," says Intergraph's Gerull. "If anything makes the Big Brother society possible, this will."

Easy access to geographic data has inspired a different perspective in

the mind of Jack Dangermond, president of Environmental Systems Research Institute, or ESRI, a top seller of mapping software. Dangermond, considered a guru among the new cartographers, sees GIS as a means of opening windows for public scrutiny of governmental behavior. In short, as the technology becomes cheaper, the new cartographers will no longer just be bureaucrats—they'll be you and me. "We are the consumers of government," he argues. "Why not be a well-informed consumer?"

The current battle over congressional redistricting may furnish a proving ground for Dangermond's vision. With widespread use of GIS technology, state legislatures and governors' offices will no longer monopolize the tools to analyze the results of the 1990 census. As of April 1, 1991, the census data, which chart a block-by-block analysis of population, voting-age population, and racial makeup, has been publicly available on CD-ROM.

"Democrats, Republicans, caucuses, ethnic groups, anyone, can come up with their own proposal for reapportionment because the database is defined by public law," says Cowen of the University of South Carolina. "Equal access to the data elevates what was once a closed-door back-room deal to a level of democracy we've never seen before." In the case of redistricting, the new cartography stands to unlock a Pandora's box. But ESRI's Dangermond says that's just the point. He believes that digital mapping will allow people to grasp the connections between cultural, physical and geographic patterns. "We're reaching the ragged edge of sensitivity on this planet," he says. "Things are getting more complex, and there's not enough time to focus on them. GIS promises to interrelate things. Even if we can't solve our problems, we'll understand them."

Population, Resources, and Socioeconomic Development

The final section in this anthology includes articles on several important problems facing humankind. Geographers are keenly aware of regional and global difficulties. It is hoped their work with researchers from other academic disciplines and representatives of business and government will bring about solutions to these serious problems.

Probably no single phenomenon has received as much attention in recent years as the so-called population explosion. World population continues to increase at unacceptably high rates. The problem is most severe in the less developed countries where, in some cases, populations are doubling in less than 20 years.

The first article in this section addresses the serious implications of an increasingly aging global population and the impact of this change on social, medical, and economic systems. The human population of the world passed the 5 billion mark in 1987, and it is anticipated that population increase will continue well into the twenty-first century, despite a slowing in the rate of population growth globally since the 1960s. By 2080 the human population of Earth could reach 14 billion.

In the next article, the implications of population growth and food supply are discussed. As the human population increases, many more agriculturalists will be forced to depend on marginal land for growing food. Martha Farnsworth Riche's article follows, which discusses the radically changing ethnic composition of the U.S. population. The next two articles deal with socioeconomic problems in sub-Saharan Africa and South Africa.

Daniel Okun reviews another dilemma in Third World development: the widely inadequate supplies of fresh water, a commodity we all take for granted. In the last article, two futurists list 50 trends facing the world.

Looking Ahead: Challenge Questions

Are you affected by the population explosion? How?

Can you give examples of how economic development adversely affects the environment? Need this always happen?

What are natural resources? Give some examples.

How do you personally feel about the occurrence of starvation in Third World regions?

What might it be like to be a refugee?

Is colonialism a historic entity, or is it present today in certain forms?

How is Earth a system?

Are world systems sustainable? For how long?

What is your scenario of the world in the year 2000?

Unit 5

The Aging
of the Human Species

*Our species has modified the evolutionary forces that have always
limited life expectancy. Policymakers must consequently prepare
to meet the needs of a population that will soon be much older*

**S. Jay Olshansky, Bruce A.
Carnes, and Christine K. Cassel**

*S. Jay Olshansky, Bruce A. Carnes and
Christine K. Cassel have worked extensively on estimating the upper limits to
human longevity. Olshansky is a research
associate at the department of medicine,
the Center on Aging, Health and Society
and the Population Research Center of the
University of Chicago. In 1984 he received
his Ph.D. in sociology from that institution. Carnes, a scientist in the division of
biological and medical research at Ar-
gonne National Laboratory, received his
Ph.D. in statistical ecology from the University of Kansas in 1980. Cassel is chief of
general internal medicine, director of the
Center on Aging, Health and Society and
professor of medicine and public policy at
Chicago. She received her M.D. from the
University of Massachusetts Medical Center in Worcester in 1976.*

For the first time, humanity as a whole is growing older. The demographic aging of the population began early in this century with improvements in the survival of infants, children and women of childbearing age.

It will end near the middle of the next century when the age composition of the population stabilizes and the practical limits to human longevity are approached. No other species has ever exerted such control over the evolutionary selection pressures acting on it—or has had to face the resulting consequences.

Already the impact of the demographic transformation is making itself felt. In 1900 there were 10 million to 17 million people aged 65 or older, constituting less than 1 percent of the total population. By 1992 there were 342 million people in that age group, making up 6.2 percent of the population. By 2050 the number of people 65 years or older will expand to at least 2.5 billion people—about one fifth of the world's projected population. Barring catastrophes that raise death rates or huge inflations in birth rates, the human population will achieve a unique age composition in less than 100 years.

Demographers, medical scientists and other workers have anticipated the general aging of the human species for several decades, yet their attention has been focused almost exclusively on the concurrent problem of explosive population growth. We believe, however, that population aging will soon replace growth as the most important phenomenon from a policy standpoint. In a more aged population, the patterns of disease and disability are radically different. Many economic and social institutions that were conceived to meet the needs of a young population will not survive without major rethinking. Attitudes toward aging and the aged will have to be modified to address the demands of a much larger and more diverse older population.

Age structure is a characteristic of populations that reflects the historical trends in birth and death rates. Until recently, the shape of the human age structure was fairly constant.

Before the mid-19th century the annual death rates for humans fluctuated but remained high, between 30 and more than 50 deaths per 1,000 individuals. Those elevated, unstable rates were primarily caused by infectious and parasitic diseases. The toll from disease among the young was especially high. Often almost one third of the children born in any year died before their first birthday; in some subgroups, half died. Because childbirth was very hazardous, mortality among pregnant women was also high. Only a small segment of the population ever lived long enough to

From *Scientific American*, April 1993, pp. 46-52. © 1993 by Scientific American, Inc. All rights reserved. Reprinted by
permission.

face the physiological decrements and diseases that accompany old age.

The only reason *Homo sapiens* survived such terrible early attrition was that the number of births more than compensated for the deaths. It was common for women to give birth to seven or more children in a lifetime. The higher birth rates were part of a successful survival pattern that reflected an array of favorable evolutionary adaptations made by humans.

Together the evolutionary constraints and adaptations produced a long-term average growth rate for the human species that, at least before the mid-19th century, hovered just above zero. The age structure of the population had the shape of a pyramid in which a large number of young children made up the broad base. At the apex were the few people who lived past their reproductive adulthood. The mean age of the population was low.

Clearly, much has changed since then. During the 20th century, the disparity between high birth rates and low death rates led to population growth rates that approached 2 to 3 percent and a population doubling time of only about 25 years. In the U.S. today, people aged 65 and older make up 12.5 percent of the population; by 2050 they will constitute 20 to 25 percent. This change is the result of declining mortality during the early and middle years. It was initially brought forth by improvements in sanitation and was later assisted by other public health measures and medical interventions. Collectively, they asserted control over the death rates from infectious and parasitic diseases and from maternal mortality.

The series of steps by which a population ages has been the subject of considerable research. Indeed, the patterns of this demographic transformation and the speed with which they occur are central to understanding the social problems now on the horizon.

Initially, declines in infant, child and maternal death rates make the population younger by expanding the base of the age pyramid. Yet that improvement in survival, along with social and economic development, leads to a drop in birth rates and the beginning of population aging. Fewer births produce a narrowing of the pyramid's base and a relative increase in the number of people who are older.

As risk of death from infectious and parasitic diseases diminishes, the degenerative diseases associated with aging, such as heart disease, stroke and cancer, become much more important. Whereas infectious and parasitic diseases usually occur in cyclic epidemics, the age-related diseases are stable and chronic throughout an extended life. Consequently, the annual death rates fall from high, unstable levels to low, steady ones of eight to 10 persons per 1,000. Abdel R. Omran, when at the University of North Carolina at Chapel Hill, was the first to describe this change as an "epidemiologic transition." The rate of change and underlying causes of the transition differ among subgroups of the population.

In the final stage of the epidemiologic transition, mortality at advanced ages decreases as medical and public health measures postpone the age at which degenerative diseases tend to kill. For example, heart disease, stroke and cancer remain the primary causes of death, but healthier ways of life and therapeutic interventions permit people with those diseases to live longer. Disease onset and progression can also be delayed.

Once the birth and death rates in a population have been in equilibrium at low levels for one average life span—approximately 85 to 100 years—the age structure becomes almost permanently rectilinear: differences in the number of persons at various ages almost disappear. Thereafter, more than 90 percent of the people born in any year will live past the age of 65. About two thirds of the population could survive past 85, after which death rates would remain high and the surviving population will die rapidly. Such age structures have been observed in laboratory mice and other animals raised in controlled environments.

A crucial feature of the rectilinear age structure is its stability. If birth rates increase and temporarily widen its base, its rectilinear shape will gradually reassert itself because nearly all the members of the large birth generation will survive to older ages. Conversely, if the birth rate falls, the aging of the population will temporarily accelerate because the young become proportionally less numerous. The rectilinear age structure persists as long as early and middle-age mortality remain low.

The trend toward stable, low death rates has already been observed for a substantial segment of the world's population. Nevertheless, no nation has yet achieved a truly rectilinear age structure. Countries such as Sweden and Switzerland are much further along in the demographic transformation to population equilibrium than are other developed nations.

In the developed nations, two major phenomena have had a particularly noteworthy influence on the transformation of the age structure. The first is the post–World War II baby boom, the rise in birth rates that occurred during the middle of the century. Although 100 years is usually enough time for an age structure to become stable, the high birth rates of the baby boom postponed the aging of the population by widening the base of the age structure again. As the baby boomers grow older, however, the average age of the population will increase much faster. The stabilization process will probably take about 150 years for the developed nations, in which rectilinear age structures should become common by 2050.

The second factor that influenced population aging in developed nations was the unexpected decline in old-age mortality that began in the late 1960s. Few scientists had anticipated that death rates from vascular disease could substantially be reduced at older ages. A fall in old-age mortality accelerates population aging by raising the age at which death becomes more frequent and the age structure begins to narrow. Death has become an event that occurs almost exclusively at older ages for some populations.

In many developing countries and in some groups within developed nations, human populations still face intense selection pressures. Consequently, some developing nations are not likely to reach equilibrium even by the middle of the 21st century. Nevertheless, the pace at which the population ages will accelerate throughout the developing world for the next 60 years.

For example, in China, which has both the largest population and the largest number of elderly people, the population aged 65 and older will increase from 6.4 percent (71 million people) to about 20 percent (270 million people) by 2050. China will then contain more people over 65 than the U.S. now has at all ages. India, which has the second largest elderly population, should experience even greater proportional increases.

We must emphasize that the demographic momentum for both population growth and population aging is already built into the age structures of all nations: the people who will become old in the next half century have, of course, already been born. These demographic forces will present a formidable set of social, economic and health problems in the coming decades—many of which are as yet unforeseen by policymakers and are beyond the capacity of developing countries to handle.

5. POPULATION, RESOURCES, AND SOCIOECONOMIC DEVELOPMENT

By the middle of the 21st century the transformation to an aged population should be complete for much of humanity. No one yet knows whether medical science will thereafter succeed in postponing the age at which rapid increases in the death rate begin. Will the apex of the age distribution retain its shape but shift to older ages, or will mortality be compressed into a shorter time span? The answer, which could profoundly affect economic and health issues, depends on whether there is an upper limit to longevity and a lower limit to the death rate.

For decades, the question of how low death rates can go has puzzled researchers. In 1978 demographer Jean Bourgeois-Pichat of Paris calculated that the average human life expectancy would not exceed 77 years. He arrived at that figure by theoretically eliminating all deaths from accidents, homicides, suicides and other causes unrelated to senescence. He then estimated the lowest death rates possible for cardiovascular disease, cancer and other diseases associated with aging. In effect, he eliminated all causes of death except those that seemed intrinsic to human biology. Yet shortly after its publication, Bourgeois-Pichat's life expectancy limit had already been exceeded in several nations. Other demographers have speculated that life expectancy will soon approach 100 years, but their theoretical estimates require unrealistic changes in human behavior and mortality.

In 1990 we took a more practical approach to the question of longevity. Rather than predicting the lower limits to mortality, we asked what mortality schedules, or age-specific death rates, would be required to raise life expectancy from its current levels to various target ages between 80 and 120 years. To determine the plausibility of reaching the targets, we compared those mortality schedules with hypothetical ones reflecting the elimination of cancer, vascular problems and other major fatal diseases. We demonstrated that as the actuarial estimate of life expectancy approaches 80 years, ever greater reductions in death rates are needed to produce even marginal increases in life expectancy.

Our conclusion was that life expectancy at birth is no longer a useful demographic tool for detecting declines in death rates in countries where mortality rates are already low. Furthermore, we suggested that the average life expectancy is unlikely to exceed 85 years in the absence of scientific breakthroughs that modify the basic rate of aging. Like others before us, we demonstrated that even if declines in death rates at older ages accelerate, the gains in life expectancy will be small.

Why is the metric of life expectancy so insensitive to declining old-age mortality in low-mortality countries? First, for as long as reliable mortality statistics have been collected, the risk of death has always doubled about every eight years past the age of 30. That characteristic of human mortality has not changed despite the rapid declines in death rates at all ages during this century. A 38-year-old man today has a longer life expectancy than one from a century ago, but he is still twice as likely to die as a 30-year-old man.

Moreover, there is no indication that humans are capable of living much past the age of 110 regardless of declines in death rates from major fatal diseases. Thus, as death becomes ever more confined to older ages, the decline in death rates will inevitably stop. The point of deceleration occurs as life expectancy approaches 80 years.

Finally, in low-mortality countries, cardiovascular disease and cancer account for three of every four deaths after age 65. Those diseases are, in effect, competing for the lives of individuals, particularly at advanced ages. If the risk of dying from any single disease were reduced to zero, the saved population would simply be subject to high mortality risks from other causes—yielding a surprisingly small net gain in life expectancy. As deaths become concentrated into older ages, the competition among causes of mortality grows more pronounced.

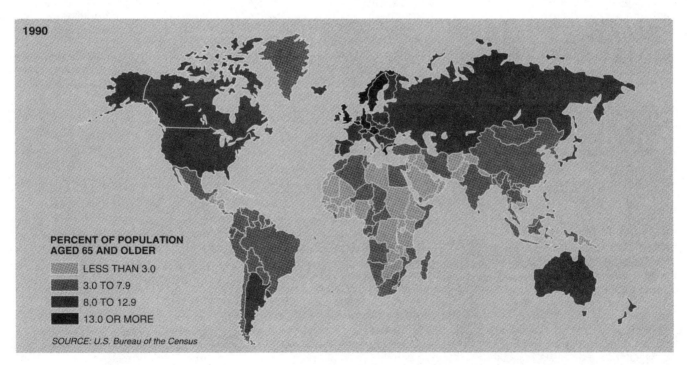

1990

PERCENT OF POPULATION AGED 65 AND OLDER

LESS THAN 3.0
3.0 TO 7.9
8.0 TO 12.9
13.0 OR MORE

SOURCE: U.S. Bureau of the Census

AGING OF THE WORLD POPULATION will become much more apparent during the 21st century. The trend is already pronounced in the industrialized countries. Within just a few decades, much of the population in the developing world will

Conceivably, however, medical researchers may learn how to slow the rate of senescence itself, thereby postponing the onset of degenerative diseases and the causes of old-age mortality. Toward that goal, many scientists working in the fields of evolutionary and molecular biology are now trying to learn why organisms become senescent.

In an influential paper written in 1957, evolutionary biologist George C. Williams, who was then at Michigan State University, proposed a mechanism for the evolution of senescence. His theory and subsequent predictions rested on two arguments. First, individual genes are involved in multiple biological processes—a widely accepted concept known as pleiotropy. Second, he proposed that certain genes conferred survival advantages early in life but had deleterious physiological effects later. He then linked those assumptions to the prevailing concept that an individual's evolutionary fitness is measured by the genetic contribution that he or she makes to subsequent generations.

Williams then argued that an individual's odds of reproducing successfully would inevitably diminish over time because he or she would eventually die from an accident or some other uncontrollable cause. As individuals fulfill their reproductive potential, selection pressures should diminish, and any genes that had damaging effects later in life could not be eliminated by natural selection. Williams argued that this process, called antagonistic pleiotropy, provided a genetic basis for aging.

Another theory, proposed in 1977 by biologist T.B.L. Kirkwood of the National Institute for Medical Research in London, is a special case of antagonistic pleiotropy. He assumed that organisms must always divide their physiological energy between sexual reproduction and maintenance of the soma, or body. The optimum fitness strategy for a species, he argued, involves an allocation of energy for somatic maintenance that is less than that required for perfect repair and immortality. Senescence is therefore the inevitable consequence of the accumulation of unrepaired defects in the cells and tissues. Under Kirkwood's disposable soma theory, senescence is the price paid for sexual reproduction.

The disregulation of genes may provide a mechanism that links the antagonistic pleiotropy and disposable soma theories into a unified concept of disease and senescence. Two concepts central to the modern paradigm of molecular biology are required: gene regulation and pleiotropy. It is assumed in molecular biology that genes are carefully regulated and that the proteins produced by gene activity are typically involved in multiple, often interacting processes. Over time, a gradual accumulation of random molecular damage could disrupt the normal regulation of gene activity, potentially triggering a cascade of injurious consequences. Richard G. Cutler, a gerontologist at the National Institute on Aging, has referred to this process as the dysdifferentiative hypothesis of aging.

The severity of the consequences will depend on how critical the affected processes are at the time of their disregulation and the ability of the organism either to compensate for or to repair the damage. If the damage disrupts the regulation of cell growth or differentiation, cancer could result. Antagonistic pleiotropy describes cases where the temporal expression of a gene becomes disregulated. For example, a gene that is essential early in life may be harmful if expressed later. Gene disregulation and pleiotropy also provide a biological mechanism for the disposable soma theory. Aging may occur when the normal repair and maintenance functions of cells become disregulated and gradually degrade physiological function.

The accumulating evidence suggests that sites of molecular damage may not be entirely random. Some regions of the genome appear to be inherently unstable and may therefore be more susceptible to the disruption of gene regulation. When the damage occurs in somatic cells, disease or senescence, or both, may occur. The consequences of damage to the germ cells (eggs and sperm) run the gamut from immediate

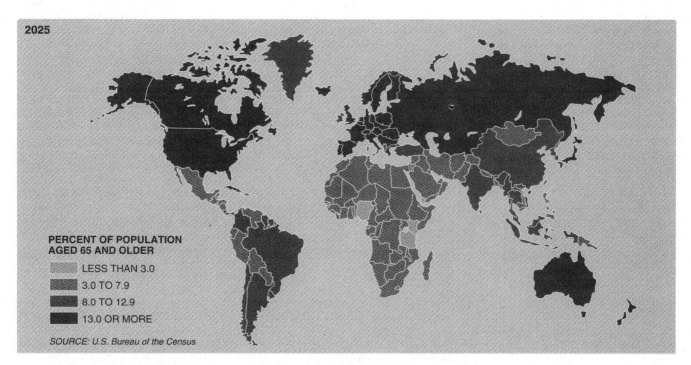

2025

PERCENT OF POPULATION AGED 65 AND OLDER

LESS THAN 3.0

3.0 TO 7.9

8.0 TO 12.9

13.0 OR MORE

SOURCE: U.S. Bureau of the Census

also be dramatically older. This demographic transformation is occurring because mortality at young ages has diminished.

The social, medical and economic changes that accompany the aging of the population will pose significant problems.

5. POPULATION, RESOURCES, AND SOCIOECONOMIC DEVELOPMENT

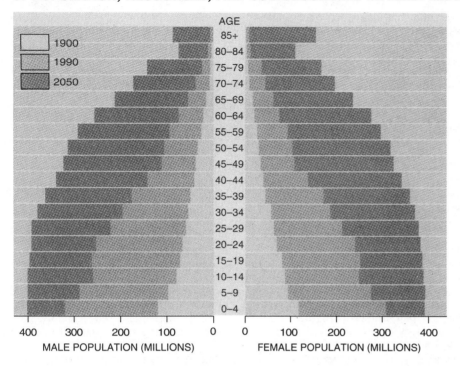

AGE STRUCTURE of the population is changing dramatically. For the past 100,000 years, the human age structure had the shape of a narrow pyramid. Since 1900, it has become wider and more rectilinear because relatively larger numbers of people in the growing population are surviving to older ages. By the middle of the 21st century it will be very nearly rectangular.

cell death to genetic changes that can be passed to the next generation. Propensities for disease and competency of somatic maintenance and repair are probably inheritable traits.

If there is a biological clock that begins ticking when a sperm fertilizes an egg, it probably does not go off at some predetermined date of death encoded in the genes. Rather the breakdown in gene regulation is a product of purely random events acting over a lifetime on a genome that contains inherited instabilities. As our understanding of biomolecular mechanisms grows, it may eventually become possible to manipulate disease processes and to slow the rate of senescence, thereby extending the average life span.

Although its link to molecular mechanisms is uncertain, one method of lengthening life span is known: dietary restriction. Early in the 20th century, researchers found that laboratory rats fed a low-calorie diet lived longer than those allowed to consume food at will. Those findings have been repeated for several species, including mice, flies and fish. Work by Richard Weindruch and his colleagues at the National Institute on Aging and by Roy L. Walford and his colleagues at the University of California at Los Angeles has suggested that dietary restriction may slow some parameters of aging in nonhuman primates.

These studies suggest life span can be extended by postponing—without eliminating—the onset of fatal diseases. Caloric restriction does not alter the rate of physiological decline in the experimental animals, nor does it change the doubling time for their death rate. Instead the animals appear to live longer because the age at which their death rates begin to increase exponentially is delayed. Dietary restriction seems to help preserve somatic maintenance for a longer time. Although it is not practical to expect enough people to adopt a calorically restricted diet to increase the average human life span, research may be able to identify the mechanisms at work and thereby extend longevity by other means.

Few observers had imagined that the demographic evolution of the human age structure would reveal a new set of diseases and causes of death. Will future reductions in old-age mortality reveal even more, new senescent diseases? Or will the prevalence of existing senescent diseases simply increase? Given the health care industry's focus on further reducing the impact of fatal diseases and postponing death, these issues will become critical to policy-makers attempting to evaluate the consequences—both medical and economic—of an aging population.

One of the most important issues is whether the trend toward declining old-age mortality will generally benefit or harm the health of the overall population. In a controversial paper published 12 years ago, physician James F. Fries of Stanford University hypothesized that the biological limit to human life is fixed at about 85 years. Better life-styles and advances in medical technology, he said, will merely compress mortality, morbidity and disability into a shorter period near that limit. His underlying premise was that changes in diet, exercise and daily routines will postpone the onset age both of the major fatal diseases (heart disease, cancer and stroke) and of the debilitating diseases of old age (including Alzheimer's disease, osteoporosis and sensory impairments).

Fries's compression-of-morbidity hypothesis has since been challenged by many scientists who posit an expansion of morbidity. They argue that the behavioral factors known to reduce the risks from fatal diseases do not change the onset or progression of most debilitating diseases associated with aging. Further reductions in old-age mortality could therefore extend the time during which the debilitating diseases of aging can be expressed. In effect, an inadvertent consequence of the decline in old-age mortality may be a proportional rise in the untreatable disabilities now common among the very old. This view has been referred to as trading off longer life for worsening health.

The expansion-of-morbidity hypothesis serves as a consequence and a corollary to the evolutionary theories of aging. As a larger and more heterogeneous population survives into more advanced ages, the opportunities increase for the known senescent diseases to become more prevalent. New diseases associated with age (possibly resulting from the pleiotropic effects of gene disregulation) may also have a greater opportunity to manifest themselves.

The ramifications of the expansion-of-morbidity hypothesis are so alarming that an international organization of scientists has been formed under the direction of demographer Jean-Marie Robine of INSERM in France to test its validity. The group's focus is the complex relation between declining old-age mortality and the relative duration of life spent healthy or disabled. Robine and his colleagues have demonstrated that women in Western societies can ex-

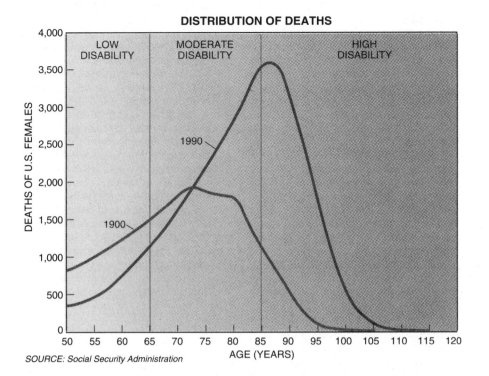

DISTRIBUTION OF DEATHS

LOW DISABILITY | MODERATE DISABILITY | HIGH DISABILITY

1990

1900

DEATHS OF U.S. FEMALES

AGE (YEARS)

SOURCE: Social Security Administration

PATTERNS OF DEATH AND DISABILITY are shifting as an epidemiologic transition occurs in the aging population. Because of healthier ways of life and medical interventions, people are surviving longer with heart disease, stroke and cancer. Yet because of their extended survival, they may suffer longer from the nonfatal but highly disabling illnesses associated with old age.

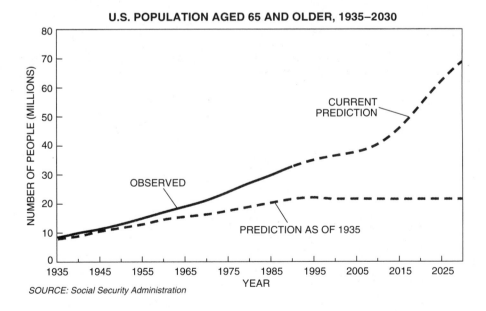

U.S. POPULATION AGED 65 AND OLDER, 1935–2030

NUMBER OF PEOPLE (MILLIONS)

CURRENT PREDICTION

OBSERVED

PREDICTION AS OF 1935

YEAR

SOURCE: Social Security Administration

STRAINS ON SOCIAL PROGRAMS, such as Social Security and Medicare, will continue to emerge as the population ages and life expectancy increases. The number of beneficiaries in the Social Security program, for example, is growing much faster than was anticipated when the program was first conceived decades ago.

pect to spend up to one quarter of their lives disabled and men up to one fifth. Wealthier people are more likely to live longer and be healthier than those who are less well-off.

The data also suggested that recently the average number of years that people spend disabled has grown faster than those that they spend healthy. In other words, although people are enjoying more healthy years while they are young and middle-aged, they may be paying the price for those improvements by spending more time disabled when they are older. Because of the known problems of data reliability and comparability and of the short periods observed, current trends in morbidity and disability must be interpreted with caution.

The dilemma we face as a society is that medical ethics oblige physicians and researchers to pursue new technologies and therapeutic interventions in efforts to postpone death. Yet that campaign will inadvertently accelerate the aging of the population. Without a parallel effort to improve the quality of life, it may also extend the frequency and duration of frailty and disability at older ages. Society will soon be forced to realize that death is no longer its major adversary. The rising threat from the disabling diseases that accompany most people into advanced old age is already evident.

There is every reason for optimism that breakthroughs in molecular biology will permit the average life span to be modified. Just how far life span could be extended by slowing the rate of senescence is the subject of much speculation and debate. No one has yet demonstrated that human senescence can be modified by any means.

It is also unclear how those breakthroughs might influence the quality of life. If slowing the rate of senescence postpones all the physiological parameters of aging, then youth could be prolonged and disability compressed into a short time before death. If only some parameters of aging are amenable to modification, however, then the added years may become an extension of disabled life in old age.

We can identify with certainty some of the social problems that an aging population will face. Two of the most difficult will be the financial integrity of age-based entitlement programs, such as Social Security and Medicare, and the funding of health care. Social security programs in the U.S. and other countries were created when the age structures were still pyramidal and life expectancies were less than 60 years. The populations receiving benefits from those programs are much larger—and living considerably longer—than was anticipated at their inception. Given that the demographic momentum for larger and longer-lived older populations already exists, it is inescapable that such programs cannot survive in their present form much beyond the second decade of the next century.

Because declining mortality allows most people to survive past the age of 65, Medicare will need to cover tens of millions of people in the U.S. Many of them will need coverage for several decades. Medicare has few effective restraints on the use of expensive acute care, which is critical for treating many fatal illnesses. Yet it covers almost none of the expense of chronic long-term care—the need for which will grow as rapidly as the population ages. As a result, the cost of the Medicare program (like that of health care in general) will escalate swiftly, eroding the political will for systemic reforms that include long-term care. Can we continue to invest in ever more costly health care programs that are not designed to handle the unique demands of a growing and longer-lived aging population?

If during the next century life expectancy increases even marginally above the current estimates, the size of the beneficiary populations for age-entitlement programs will be two to five times greater than is already anticipated. That change would result in extreme financial hardship.

In the developed nations the demographic evolution of the age structure is beneficial in the short run: the coffers of the entitlement programs are swelling with the tax dollars from an unusually large cohort of working-age people. It would nonetheless be unwise to let that temporary condition lull us into complacency. When the age structure in those nations becomes rectilinear, the ratio of beneficiaries to taxpayers will mushroom, and surpluses in entitlement programs will vanish.

The financial integrity of age-entitlement programs has already been jeopardized in some countries. The worst problems will arise globally just after the year 2010, when the generation of baby boomers reaches entitlement age. The certainty of the demographic evolution of population aging will soon force governments to restructure all their entitlement programs.

The demographic evolution of the age structure will have an impact on many aspects of human society, including the job market, housing and transportation, energy costs, patterns of retirement, and nursing home and hospice care, to mention only a few. For example, if current trends toward early retirement persist, future retirees will draw benefits from age-entitlement programs for 30 years or more and spend up to one third of their lives in retirement. Thus, the current patterns of work and retirement will not be financially supportable in the future. Social structures have simply not evolved with the same rapidity as age structures. The rise in life expectancy is therefore a triumph for society, but many policy experts view it as an impending disaster.

Although we have emphasized the dark side of aging—frailty and disability—it is also true that the demographic evolution of the age structure will generate a large healthy, older population. All older people, both the healthy and the sick, will need the chance to contribute meaningfully to society. Achieving that end will require an economy that provides ample, flexible opportunities for experienced and skilled older persons, as well as modifications in the physical infrastructures of society. Changes in attitudes about aging will be essential.

The medical establishment is continuing to wage war against death. Researchers in the field of molecular biology are still searching for ways to slow the basic rate of aging. Those efforts lead us to believe that the aging of the population will also continue and perhaps even accelerate. Everybody wants to live longer, and medicine has helped that dream come true. Only now is society beginning to comprehend what it has set in motion by modifying the natural selection forces that have shaped the evolution of human aging.

FURTHER READING

IN SEARCH OF METHUSELAH: ESTIMATING THE UPPER LIMITS TO HUMAN LONGEVITY. S. J. Olshansky, B. A. Carnes and C. Cassel in *Science,* Vol. 250, pages 634–640; November 2, 1990.

EVOLUTION OF SENESCENCE: LATE SURVIVAL SACRIFICED FOR REPRODUCTION. T. B. L. Kirkwood and M. R. Rose in *Philosophical Transactions of the Royal Society of London,* Series B, Vol. 332, No. 1262, pages 15–24; April 29, 1991.

LIVING LONGER AND DOING WORSE? PRESENT AND FUTURE TRENDS IN THE HEALTH OF THE ELDERLY. Special issue of *Journal of Aging and Health,* Vol. 3, No. 2; May 1991.

THE OLDEST OLD. Edited by Richard Suzman, David Willis and Kenneth Manton. Oxford University Press, 1992.

AN AGING WORLD II. K. Kinsella and C. M. Taeuber. Center for International Research, U.S. Bureau of the Census, 1993.

THE GLOBAL THREAT OF UNCHECKED POPULATION GROWTH

Garrett Hardin

Garrett Hardin is professor emeritus of human ecology at the University of California, Santa Barbara, and has long been concerned with population issues. His most recent book is Living within Limits: Ecology, Economics, and Population Taboos *(Oxford University Press, 1993).*

Although almost two centuries have passed since Malthus disturbed the slumbers of humanity with his dire predictions of over-population, the controversy has still not abated. In the final analysis, the only way to approach this problem beneficially is through the application of unremitting common sense and unwavering objectivity.

In addition, I would submit, in order to allow common sense and objectivity their free rein, we must avoid the trap of statistic slinging. In what we could take as a helpful warning, one public-spirited statistician, Darrell Huff, once penned a light-hearted book titled *How to Lie with Statistics.* And Benjamin Disraeli, Queen Victoria's favorite prime minister, said: "There are three kinds of lies: lies, damned lies, and statistics."

Scientists are inclined to argue with this, holding that statistics (properly used) are one of the glories of the scientific method. But since statistics are often *not* properly used, it must be admitted that Huff and Disraeli had a point. As used, statistics are often a sort of black magic accompanied by a disparagement of common sense. That won't do. As the logician Willard Van Orman Quine has said, "Science itself is a continuation of common sense."

Therefore, this essay will avoid statistics. The opaqueness of statistical arguments makes it easy for analysts to "get away with murder." Though often wonderfully useful, statistics can also serve as a substitute for thought. And empirical findings, the fodder of statistics, are so numerous and so ambiguous that almost any conclusion can be supported by a plausible argument. But empirical studies have great prestige in our science-sensitized society, particularly because they can be so selected and arranged as to seem to support faith in perpetual growth, the religion of the most powerful actors in a commercial society.

The literature on human population growth is enormous. Blessedly, most of it can be safely ignored. A handful of principles enables us to incorporate the meaning of a great mass of data in a few memorable images, which the following attempts to provide.

This article appeared in *The World & I,* June 1993, pp. 376-391. Reprinted with permission from *The World & I,* a publication of The Washington Times Corporation. © 1992.

THE DWINDLING OF CONCERN OVER POPULATION GROWTH

There is today, of course, a sharp division of opinion about the seriousness of human population growth. On the one hand, a legion of economists say, "Why worry? An increase in people doesn't matter, because an unfettered market economy can solve all problems of shortages." On the other hand, a brigade of environmentalists assert that shortages are real and ultimately decisive.

In 1970, the first Earth Day was celebrated with a great to-do about the perils of population growth. (Paul Ehrlich's *Population Bomb* had exploded into the public consciousness just two years earlier.) But when the second Earth Day came around twenty years later, scarcely a word was said about population. Why the change?

Had world population growth stopped? Hardly: An additional 1.7 billion souls had made their appearance—an increase of 47 percent. Had poverty and malnutrition been conquered? Exact figures are hard to come by, but about 2 billion people worldwide live below what Americans call the "poverty line." From time to time, it becomes fashionable for Americans to become upset by the poverty in one of these areas. There was Ethiopia a few years ago. Then Somalia … What next?

Many people harbor a vague hope that we will be able in the not-too-distant future to reduce population by shipping off excess people to other planets. Even to optimists, however, the economics of this proposal is forbidding. From 1969 to 1972, a dozen Americans made it as far as the moon for a few hours, at a cost of billions of dollars. During the same three years, the world's population increased by more than two hundred million.

Maybe the population problem will automatically solve itself? In a sense this is true: If human beings don't solve the problem, nature will. Starvation and disease can reduce the numbers drastically. But only a humanity hater would call that a solution. So what are we to do? Obviously, we must ask more questions first.

To begin with, why were the recent Earth Day celebrations so silent about population? The philanthropic foundations and business concerns that subsidized the 1990 celebrations called the tune. It was all right to talk about the environment, but population was a no-no. Needless to say, their spokespersons did not tell the public why they suppressed mention of population.

In the American economic environment, population growth has many positive benefits. More people means more things sold. More sales means more profits. More profits make it easier to collect the taxes needed to build up the infrastructure. Moreover, an expanding economy is an exhilarating one for the young to grow up in.

On the other hand, a nongrowing population discourages sales, makes taxes harder to bear, and just generally creates a depressing psychological atmosphere—especially for those who are just coming into the job market. Young people focus principally on the present and are hard to convince that no growth—or worse, negative growth—can be justified by a conceivable future benefit. To the nearsighted, there are more pluses than minuses to population growth.

But we must muster the courage to examine the likely effects of unchecked population growth in the next decade, the next century, and beyond. If population continues to increase at the present rate, in a mere six hundred years the earth would be standing-room-only, provided sustenance could temporarily be found for all the new people.

POPULATION GROWTH IS RELENTLESS, LIKE COMPOUND INTEREST

It was the *hypothetical* consequences of population growth that disturbed Thomas Robert Malthus in 1798. Every human population inherently has the potential of growing like compound interest. The population is now growing at about 1.7 percent per year. That may not sound like much, but, if it continues, the present 5.4 billion people will become more than 21 billion by the year 2074. Wherever you live, you should try to imagine what your life would be like with four times as many people living in your area. Imagine the highway congestion; imagine the pollution of air and water; imagine the disappearance of wilderness, wetlands, and wildlife. And imagine the increasing difficulty of getting the ear of overloaded congressmen. Democracy suffers.

Projections of present trends into the future are both logical and absurd. No doubt, it is their absurdity that turns off many people. Since an absurd conclusion is almost unimaginable, why waste time discussing it? Why not think about pleasant things, like sales and purchases, burgeoning profits, and the excitement of human activities? In the public arena, optimism sells; pessimism does not.

Our fundamental postulate is as follows: Population grows by compound interest, like money in a savings account. With money, a capital sum bears interest, which then becomes part of the capital and bears more interest. Similarly, a living population produces children, who (in time) produce more children. With money, no positive rate of interest, if continued, is too small to produce, ultimately, a claim on capital that is too great to be paid. Similarly, no positive rate of growth of a human population is too small to produce, ultimately, a population of descendants too great for the limited resources. The inescapable conclusion, then, is that eventually the limits of the earth will bring to an end the increase of the terrestrial population of *H o m o sapiens*.

Some rationalist critics of population growth see not only a problem but a *cancer* growing. Such critics have been castigated as "pessimists" and "misanthropes." But it has taken courage to be an outspoken rationalist in this area. Such courage was shown by the physician Alan Gregg (1890–1957), a vice president of the Rockefeller Foundation, which spent many millions of dollars in saving lives in distant and overpopulated countries. As he neared retirement age, Gregg, in an article in *Science* magazine titled "A Medical Aspect of the Population Problem," asked some hard questions of his organization's philanthropic policies. He wrote as follows:

I suggest, as a way of looking at the population problem, that there are some interesting analogies between the growth of the human population of the world and the increase of cells observable in neoplasms. To say that the world has cancer, and that the cancer cell is man, has neither experimental proof nor the validation of predictive accuracy; but I see no reason that instantly forbids such a speculation. ...

What are some of the characteristics of new growths? One of the simplest is that they commonly exert pressure on adjacent structures and, hence, displace them. ... The destruction of forests, the annihilation or near extinction of various animals, and the soil erosion consequent to overgrazing illustrate the cancerlike effect that man—in mounting numbers and heedless arrogance—has had on other forms of life on what we call "our" planet.

Metastasis is the word used to describe another phenomenon of malignant growth in which detached neoplastic cells carried by the lymphatics or the blood vessels lodge at a distance from the primary focus or point of origin. ... It is actually difficult to avoid using the word *colony* in describing this thing physicians call metastasis. Conversely, to what degree can colonization of the Western Hemisphere be thought of as metastasis of the white race?

Gregg proceeded to draw parallels between the physical wasting and the tainting of the blood caused by cancer and the "plundering" of the planet by industrial nations and the concomitant pollution of waterways.

Incurable optimists who hope that emigration to other celestial bodies could one day solve the earthly population problem do not appreciate the magnitude of the need. At the present time, we would have to ship off more than a quarter of a million people each day to keep the earth's population from increasing. And the irony is this: The passengers on each severely limited spaceship would have to submit themselves to the very restraint they were fleeing from, namely, having their reproduction rigidly controlled by their government. This paradoxical property of space migration rules it out as an escape from the necessity of population control. We are thus brought right back to earth again.

THE POLITICS AND ECONOMICS OF IGNORING POPULATION GROWTH

Does population growth necessarily create suffering and pain? Well-fed government officials usually either cover up the extent of such suffering or admit the suffering and pounce on other factors as scapegoats.

Such scapegoating took place at the first United Nations World Population Conference in Bucharest in 1974. The head of China's delegation used both denial and scapegoating to steer the conference away from thoughts of population control. "Population is not a problem under socialism," he said, and then went on to serve up some tempting scapegoats. "The primary way of solving the population problem," he said, "lies in combating the aggression and plunder of the imperialists, colonialists, and neocolonialists, and particularly the superpowers." His analysis was warmly welcomed by other delegates from the Third World.

The leader of the Indian delegation contributed the most memorable phrase of the conference: "Development is the best contraceptive." Operationally, this translates into: "Instead of demanding that we poor countries control our populations, you rich countries should give us money for erecting factories, building dams, and eliminating poverty." In retrospect, however, it looks as though the Indian delegation was just grandstanding at Bucharest because, two years later, the central government of India issued the following statement for internal consumption:

If the future of the nation is to be secured … the population problem will have to be treated as a top national priority. … It is clear that simply to wait for education and economic development to bring about a drop in fertility is not a practical solution. The very increase in population makes economic development slow and more difficult of

achievement. The time factor is so pressing, and the population growth so formidable, that we have to get out of the vicious circle through direct assault upon this problem as a national commitment.

It is fair to say that, in 1974, China, the most populous country in the world, made a shambles of the international population conference. Just ten years later, the United States, the richest country in the world, took over China's destructive role at the second UN conference on population in Mexico City, repeating India's slogan, "Development is the best contraceptive."

There's the old saying that politics makes strange bedfellows. So also do unconsciously shared ideologies. At first glance, the ideologies of China and the United States seemed (in 1984) to be very different: Marxism in China, capitalism in the United States. However, the common and unconsciously shared ideology of the two was (and is) a deep faith in technological progress. Holders of the reins of power in both nations believe that technology can solve all problems. Yet faith in technology is selective. Technology that attacks the demand end of the supply-and-demand equation is not generally approved of. Communists denigrated contraception in 1974, and capitalists rejected abortion in 1984. Both rejections stemmed in part from a childlike belief that technology can increase supply without limit. If there is no limit to supply, why risk squelching demand, that great engine of material growth?

Advancing technology expands the usable support base (though common sense always tells us that there are limits to how long the earth can sustain this base). How long can this expansion keep ahead of a growing population? Malthus fumbled this question; this was his greatest error. (His detractors have never let Malthusians forget it.) Malthus thought that *any* increase in the British population would diminish the per capita supply of subsistence (by which he principally meant food). So, what actually happened after Malthus published his essay? In the next quarter of a century, the British population increased by 25 percent, but real wages increased another 25 percent *per worker*. A most un-Malthusian result! In industrialized nations, this trend has continued to the present day, though the Third World has not been quite so fortunate.

How did this un-Malthusian result come about? For one thing, new geographic frontiers were opened up for exploitation by Europeans, principally in the Americas. For another, science and technology showed us how to get more human wealth out of the same old materials. We can hardly blame Malthus for not foreseeing this second development in 1798, when scientific engineering was just getting started.

For food, we are still almost entirely dependent on the farms, but we have learned how to get more food per acre. How? By genetic improvement of crops, of course; but that has limits (though they are incompletely known). Most of the gain has come from cheaper energy.

ENERGY AS THE CRUCIAL CONSTRAINT ON POPULATION GROWTH

A great change began in 1859, when Edwin Drake successfully drilled for oil in Pennsylvania. This led to the cheap oil that now runs our tractors, which produce food more cheaply than hand labor. Oil moves food to markets cheaper than coolies can. And oil is used to produce chemical fertilizers, which greatly augment agricultural productivity. Figuratively speaking, modern man eats oil.

Cheap energy has made our time a wonderful one to live in, but we know that the supply of oil will come to an end soon—probably in the lifetime of today's children. There will always be billions of barrels of oil left in the ground, but the remaining stocks will be at deeper and deeper levels. People will stop pumping up petroleum when it takes one thousand calories of energy to pump oil that yields only one thousand calories of energy when burned. The same logic holds for all the other energy-rich minerals—oil shale, natural gas, coal, and peat. The exhaustion horizon of these is farther off, but the end is only a few centuries away, a trifling segment of time as history is measured.

Many minerals are essential in a technological civilization like ours. Their use has been escalating faster than the growth of population, so one might expect their price to have risen faster still. But the record shows otherwise, as marketing expert Julian Simon delights in pointing out. For the past hundred years the average price of minerals has magically remained about the same, after correcting for inflation. How come?

The magic is in the market. In one way or another, every commodity has competitors. For the transmission of energy, aluminum wire competes with copper. When the price of copper rises too much, more aluminum is used. Moreover, reexamining our technology often shows us ways to use less of a given commodity. Economy replaces extravagance. When the cost of travel goes up, we sometimes find that a message can substitute for a trip. Information competes with transport.

Nevertheless, energy will ultimately be a major limiting factor of population size. In the nineteenth century, the central importance of the sun in supplying man's energy needs became clear. Burnable wood is really trapped solar energy. The energy resident in oil and coal was trapped by plants that grew millions of years ago. It can justifiably be called *fossil sunlight*. (Other sources of energy—tidal energy, for instance—are of trifling importance.) Such seemed to be the whole story until 1896, when radioactivity was discovered. Less than a decade later, Einstein wrote the equation that connected matter with energy. Because it takes only the tiniest amount of matter to produce tremendous amounts of energy, it seemed that we could forget about possible shortages of energy.

In 1954, financier Lewis Strauss, chairman of the Atomic Energy Commission, proudly predicted, "Our children will enjoy in their homes electrical energy too cheap to meter!" The slogan "too cheap to meter" was too good to forget. But it was not true. The financier was apparently not told that 50 percent of a homeowner's electric bill is attributable to the cost of transmitting the electricity. If production costs were zero, your electricity bill—without some sort of technological solution to the transmission prob-

Malthus on Population and Resources

Thomas Malthus

The following is an excerpt from Thomas Malthus' Essay on the Principle of Population (1803 edition). First published in 1798, his pessimistic essay was the original and most famous tocsin regarding the apparent threat of unchecked population growth to human quality of life.

According to a table of Euler, calculated on a mortality of 1 in 36, if the births be to the deaths in the proportion of 3 to 1, the period of doubling will be only 12 4/5 years. And these proportions are not only possible suppositions, but have actually occurred for short periods in more countries than one.

Sir William Petty supposes a doubling possible in so short a time as ten years.

But to be perfectly sure that we are far within the truth, we will take the slowest of these rates of increase; a rate, in which all concurring testimonies agree, and which has been repeatedly ascertained to be from procreation only.

It may safely be pronounced therefore, that population when unchecked goes on doubling itself every twenty-five years, or increases in a geometrical ratio.

The rate according to which the productions of the earth may be supposed

---

lem—could never be less than 50 percent of what it is now. And, of course, the cost of producing usable energy—again if technology failed to discover a really cheap production method—can never come close to zero.

THE HOLLOWNESS OF THE NUCLEAR ENERGY HOPE

And there are other reasons for believing that the "promise" of unlimited power from the atom is nothing more than a will-o'-the-wisp or even a dangerous siren song. In deciding what to do about nuclear energy, it is essential to remember two things. First, there is no "away" to throw to. Second, we cannot escape human nature.

Ecologists have always told us that throwing away is impossible. "Away" is not *no* place; "away" is *some* place. Nuclear power plants have made the "away" problem much worse. Energy is the principal thing we want from a nuclear plant, but we get also a diverse menu of dreadful by-products. (This will be true even with fusion energy, if we ever achieve it. Though the reaction itself may be "clean," it renders the surrounding reactor highly radioactive. How are we to get rid of that junk?)

For forty years, we have been accumulating radioactive trash at our reactor sites, and we still do not know what to do with it. Rivers are no "away"; neither is groundwater; neither are the oceans; neither is the atmosphere. It has been proposed that we load the wastes into spaceships to be blasted into the sun. Unfortunately, the blasting can miscarry—as everyone living near either of our space launch sites in Florida and California knows. Erratic missiles have to be destructed; their trash then rains on the earth. And God help us earthlings if that trash is radioactive!

This brings us to the second reason for distrusting nuclear power: human nato increase, it will not be so easy to determine. Of this, however, we may be perfectly certain, that the ratio of their increase must be of a totally different nature from the ratio of the increase of population. A thousand millions are just as easily doubled every twenty-five years by the power of population as a thousand. But the food to support the increase from the greater number will by no means be obtained with the same facility. Man is necessarily confined in room. When acre has been added to acre till all the fertile land is occupied, the yearly increase of food must depend upon the amelioration of the land already in possession. This is a stream which, from the nature of all soils, instead of increasing, must be gradually diminishing. But population, could it be supplied with food, would go on with unexhausted vigour; and the increase of one period would furnish the power of a greater increase the next, and this without any limit.

From the accounts we have of China and Japan, it may be fairly doubted whether the best-directed efforts of human industry could double the produce of these countries even once in any number of years. ...

In Europe [however] there is the fairest chance that human industry may receive its best direction. The science of agriculture has been much studied in England and Scotland; and there is still a great portion of uncultivated land in these countries. Let us consider at what rate the produce of this island might be supposed to increase under circumstances the most favourable to improvement.

If it be allowed, that by the best possible policy, and great encouragements to agriculture, the average produce of the island could be doubled in the first twenty-five years, it will be allowing probably a greater increase than could with reason

---

ture. Many academicians cringe at the term *human nature*. Why? Apparently because so many silly things were said about it in the past. But there *is* such a thing as human nature.

Language is treacherous. In official descriptions of control systems, the passive voice predominates. It is said, for instance, that the driveshaft *is connected* to a gear, the wheels *are moved* by impulses, or the system *is checked* for reliability. A filming of the total operation would show that a *human being* connects the driveshaft to a gear, a *human being* establishes the arrangement of the parts, a *human being* checks arrangements against plans. And some other human being audits the record to verify that the checking has been done. The passive voice is wrong because no machine arranges and runs itself: It is actively assembled, run, and checked by human beings, every one of whom is less than 100 percent reliable.

One seldom finds the slightest acknowledgment of the existence and importance of human beings in the descriptions of nuclear energy systems. The use of the passive voice converts active agents into *nonpersons*, to use George Orwell's term. Why worry about human nature when only nonpersons are involved?

A few commentators have dealt with the reliability issue. Alvin Weinberg, long the director of the Oak Ridge National Laboratory, even proposed a solution. We must, he said, create a sort of nuclear priesthood who will be even more devoted to the management of nuclear systems than the religious priesthood of the Middle Ages was to the care of sacred documents.

The shepherds of nuclear power plants must be competent. When a near catastrophe occurred at the Three Mile Island plant back in 1979, disaster was not averted until highly trained professionals replaced the regular staff, who had been bewildered by conflicting signals. In the Chernobyl disaster in Russia in 1986, hube expected.

In the next twenty-five years, it is impossible to suppose that the produce could be quadrupled. It would be contrary to all our knowledge of the properties of land. The improvement of the barren parts would be a work of time and labour; and it must be evident to those who have the slightest acquaintance with agricultural subjects that, in proportion as cultivation extended, the additions that could yearly be made to the former average produce must be gradually and regularly diminishing. That we may be the better able to compare the increase of population and food, let us make a supposition which, without pretending to accuracy, is clearly more favourable to the power of production in the earth than any experience that we have had of its qualities will warrant.

Let us suppose that the yearly additions which might be made to the former average produce, instead of decreasing, which they certainly would do, were to remain the same; and that the produce of this island might be increased every twenty-five years by a quantity equal to what it at present produces: the most enthusiastic speculator cannot suppose a greater increase than this. In a few centuries it would make every acre of land in the island like a garden.

If this supposition be applied to the whole earth, and if it be allowed that the subsistence for man which the earth affords might be increased every twenty-five years by a quantity equal to what it at present produces, this will be supposing a rate of increase much greater than we can imagine that any possible exertions of mankind could make it.

It may be fairly pronounced therefore, that, considering the present average state of the earth, the means of subsistence, under circumstances the most favourable to human industry, could not possibly be made to increase faster than in an arithmetical ratio.

---

man nature was a major contributing factor. And when the next catastrophe occurs—where? when?—we may be sure that human nature will be at fault.

From such experiences, one might conclude that all we need to do is put Ph.D. physicists in charge, 24 hours a day, 365 days a year. If you think that is an achievable solution, try to imagine what it is like to work in the control room of a nuclear plant. Imagine yourself sitting, day after day, watching the dials. In a well-designed system, built-in automatic responses take care of most of the situations that arise. Days go by, and you don't have to do a thing.

Weeks go by …

Months … perhaps even years, without incident.

Then … at some unforeseen moment … a horn blasts: *Something has happened!* Hastily, you try to remember what it is you are supposed to do. You take the manual out of a drawer; your fingers, made clammy with fear, get entangled with the pages. Seconds pass … more seconds. In a panic you shout: "Hey, Joe, what do I do now?"

On the nuclear time scale, seconds are an eternity. Corrections have to be made *now* … or NEVER. A nuclear power plant is no place for the poorly trained. It calls for people with Ph.D.s in physics. But how can we persuade such experts to accept such an utterly *boring* job? Would an Einstein accept it? Or an Edward Teller? Or an Alvin Weinberg? Need we ask?

Here's the paradox. The better a high-technology plant is designed, built, and inspected, the rarer will accidents be. The rarer the accidents, the more boring supervisory work becomes. The more boring the work, the more difficult it becomes to hire highly trained supervisors. Nationwide, thousands of supervisors are needed. Conflicting social forces will no doubt equilibrate at some level of unreliability, but even a low level of accidents is intolerable in the management of the atom.

Adam Smith would have recognized our problem. In 1776, long before anyone

The necessary effects of these two different rates of increase, when brought together, will be very striking. Let us call the population of this island eleven millions; and suppose the present produce equal to the easy support of such a number. In the first twenty-five years, the population would be twenty-two millions, and the food being also doubled, the means of subsistence would be equal to this increase. In the next twenty-five years, the population would be forty-four millions, and the means of subsistence only equal to the support of thirty-three millions. In the next period the population would be eighty-eight millions, and the means of subsistence just equal to the support of half of that number. And at the conclusion of the first century, the population would be a hundred and seventy-six millions, and the means of subsistence only equal to the support of fifty-five millions; leaving a population of a hundred and twenty-one millions totally unprovided for.

Taking the whole earth instead of this island, emigration would of course be excluded; and supposing the present population equal to a thousand millions, the human species would increase as the numbers 1, 2, 4, 8, 16, 32, 64, 128, 256, and subsistence as 1, 2, 3, 4, 5, 6, 7, 8, 9. In two centuries the population would be to the means of subsistence as 256 to 9; in three centuries as 4096 to 13, and in two thousand years the difference would be almost incalculable.

In this supposition no limits whatever are placed to the produce of the earth. It may increase for ever, and be greater than any assignable quantity; yet still the power of population being in every period so much superior, the increase of the human species can only be kept down to the level of the means of subsistence by the constant operation of the strong law of necessity acting as a check upon the greater power.

---

even dreamed of atomic energy, the great economist wrote: "The man whose whole life is spent in performing a few simple operations, of which the effects too are, perhaps, always the same, or very nearly the same, has no occasion to exert his understanding. ... He naturally loses, therefore, the habit of such exertion, and generally becomes as stupid and ignorant as it is possible for a human being to become."

If, as the supplies of fossil energy give out, we propose to base our civilization and our survival on nuclear energy, then—if Smith is right about human nature—we will unintentionally make the survival of civilization dependent on relatively stupid and ignorant people sitting in front of dials in nuclear power plants. That would be a stupid policy for a civilization to follow.

One thing is certain: When it comes to the design of complicated devices like nuclear power plants, we must never forget that the most reliable thing in the world is the ultimate unreliability of human beings.

POPULATION GROWTH *CAN* BE CONTROLLED

So, we begin to see where logic leads us. Though a few people may escape to other astronomical bodies, emigration from the earth cannot deal with a monstrously large population addicted to growth. A large, industrialized population requires a great deal of energy to live in the style to which it has become accustomed. Nuclear energy is ruled out by human nature. The human species must soon stop depending on the capital of stored fossil sunlight and adjust its budget to living on the income from today's sunlight. All talk of a "sustainable growth" policy becomes ludicrous when we realize that, once we are reduced to living on the daily energy income of sunlight, the population must be greatly reduced. Thus, it is virtually insane to sweep the population issue under the rug.

Of course, the difficulties of population control—particularly if we must actually *reduce* the size of the population—are staggering.

There are, however, some guiding principles that can serve to ease the way. First, we must not look for a population policy for the whole world—because there is no global government to enforce a global policy. Each nation must work on its own policy in the light of its own values. That means that immigration must be virtually stopped everywhere. No country can match its population size to its resources unless immigration is firmly controlled.

The second point, a corollary of the first principle, is: Globalism is dangerous. Kenneth Boulding wisely said: "There are catastrophes from which there is recovery, especially small catastrophes. What worries me is the irrevocable catastrophe. That is why I am worried about the globalization of the world. If you have only one system, then if anything goes wrong, everything goes wrong." In other words, don't put all your eggs in one basket. Given many sovereign nations, it is possible for humanity to carry out many experiments in population control. Each nation can observe the successes and failures of the others.

Third, this implies that nations should keep their noses out of other nations' business (as should America in regard to China's population control program), which means that principles of human rights ought to be nonuniversally applied. After all, whose rights were those that were adopted by the United Nations in its Universal Declaration of Human Rights? Were the selecting criteria universal? Or was some sort of moral imperialism at work?

Fourth, we shouldn't panic when we hear someone speak of "reducing the population." That doesn't mean killing anyone. All it means is that the chosen birthrate must be less than the normal death rate. Then attrition eventually brings the numbers down to a sustainable level. People can survive attrition with dignity.

Fifth, we ought also to challenge certain of our basic assumptions. For example, it is generally accepted that the infant mortality rate is a valid measure of the state of a civilization. (The United States has a low infant mortality rate, so we are civilized; China, economically poorer, has a higher rate.) *Mortality*—death—can be easily tallied, but *morbidity*—pain and suffering—is much harder to measure. Yet morbidity may be the more important measure of happiness.

Sixth, birth control clinics should be subsidized generously because it is far cheaper to prevent the birth of an unwanted child than it is to take care of it later. An abortion can be performed for less than a thousand dollars. Contraception services are even cheaper. Empirical studies show that raising a child to age eighteen, at what we regard as an average American standard of living (but without a college education), costs parents and taxpayers well in excess of $100,000.

Humanity is facing a great challenge, but *Homo sapiens* has done great things in the past. We have learned how to go around the world in hours instead of months or years. We have learned how to store an encyclopedia of knowledge in a tiny capsule. We've conquered most of the diseases. Recently we learned how to propel ourselves to the moon. These have all been splendid achievements. But they have been what we might call "outside jobs"—accomplishments outside the essential nature of the human species.

Controlling population, however, is a different sort of problem. Only when we understand human nature can we succeed in this task. This may be no more difficult than splitting the atom, but it calls for different sorts of abilities.

We must stop urging our young to train for outer space and start encouraging them to tackle the inner space of human nature. Only with an enlarged understanding of the possibilities and limitations of human nature can we hope to wrest the control of population from nature's heartless hands and bring it under more compassionate human control.

THE LANDSCAPE

OF

We have seen the victims of mass starvation. We have shuddered at the images of millions of people arriving at remote camps for aid. Many of the world's famines result from wars or civil strife. But we rarely see that hunger also grows from environmental ruin.

HUNGER

BRUCE STUTZ

Bruce Stutz is features editor at Audubon. *He is the author of* Natural Lives, Modern Times.

The numbers alone are staggering: In 1990, 550 million people worldwide were hungry, 56 million more than in the early 1980s. During that time, the number of malnourished children in the developing world increased from 167 million to 188 million. And experts predict that the situation will only get worse as food production in the poorest countries continues to decline.

Deforestation, desertification, and soil erosion have devastating effects on food production. Where forests are cut down, the soils are washed or blown away. Where land is planted too often or grazed too long, it can no longer support crops or cattle.

Ironically, modern agriculture—the science of growing food—has had the greatest impact on the decline in the food supply. In the 1960s governments began encouraging nomads to settle in one place, to raise one cash crop instead of several, to herd only one kind of livestock, and the soils quickly became exhausted.

This "green revolution"—intensified farming of "improved crop varieties" with irrigation, chemicals, and pesticides—at first raised productivity, but the long-term results were just the opposite. According to Mostafa K. Tolba, former executive director of the United Nations' Environment Programme, the process "made agroecosystems increasingly artificial, unstable, and prone to rapid degradation."

Population growth and refugee migration add to the problems of environmental degradation. When the land becomes too crowded and the soil too exhausted to support life, farmers move into forests, slashing and burning new farms. As the land gives out, the people move on, again and again. Eventually, they find their way to the refugee camps.

At the 1972 United Nations Conference on the Human Environment in Stockholm, environmental issues were considered secondary to issues of economic development. But the 1968–74 drought in Ethiopia and the Sahel made it evident that the environmental costs of traditional economic development might be too high.

The Ethiopian government estimated that its highlands were then losing ap-

From *Audubon*, Vol. 95, No. 2, March/April 1993, pp. 54-63. © 1993 by Bruce Stulz. Reprinted by permission.

proximately 1 billion tons of soil a year through water and wind erosion. So when development experts convened at the United Nations Conference on Environment and Development (UNCED) last June, the agenda had changed. The environment had to be protected, it was decided. The diplomats finally recognized the simplest truth: The land that produces the food must be preserved.

The food web inextricably connects plants, animals, and people with the water, soil, and atmosphere of the planet; the hunger web begins as those connections are severed. Air and water pollution contaminate and ruin food sources. The predictions are that global warming will also have an effect, changing planting times, growing seasons, even the ability of crops to survive in their present ranges.

On the following pages *Audubon* examines the environmental causes of mass hunger and some possible solutions. Most do not involve high-tech, grand-scale megaprojects; the best of them are low-tech, local, modest in scale. The solutions may not be as dramatic as scenes of armies massing to feed millions. But they are sustainable—which means that in the future hunger may be defeated without the help of armies.

DEFORESTATION

Slash, Burn, Plow, Plant, Abandon

Forests now cover 27.7 percent of all ice-free land in the world. In 1990 wood was the main energy source for 9 out of 10 Africans, providing more than half of their fuel. By the end of this decade, according to Mostafa Tolba, 2.4 billion people will be unable to satisfy their minimum energy requirements without consuming wood faster than it is being grown.

As human beings encroach on the world's remaining woodlands, deforestation will exacerbate the problems of hunger. For when hillsides are denuded, soil erosion sets in.

In Haiti, where forests once covered most of the land, 40 to 50 million trees are cut each year to supply firewood, cropland, and charcoal. At the current rate of deforestation, Haiti's forests will cease to exist within two or three years.

Already, loss of forests has caused massive soil erosion, and when drought strikes, the quality of the remaining soil will decline. When rain finally *does* fall, runoff will be too rapid and farmers will be forced to abandon cultivation. Since the 1970s, food aid to Haiti has risen sevenfold.

In Bangladesh and India, deforestation has caused another kind of problem, increasing the frequency and force of floods. Bangladesh used to suffer a catastrophic flood every 50 years or so; by the 1980s the country was being hit with major floods—which wash away farms and rice paddies—every four years. Between the late 1960s and late 1980s, India's flood-prone areas grew from approximately 25 million hectares to 59 million. (One hectare equals 2.4 acres.)

PROTECT AND PRESERVE

Although some countries, notably Brazil and Costa Rica, have preserved tracts of their forested land, less than 5 percent of the world's remaining tropical forests are protected as sanctuaries, parks, or reserves.

Regenerating woodlands by replanting them would provide some measure of relief. Over the past 10 years, China has reforested some 70 million hectares of endangered landscape. The U.N. Food and Agriculture Organization estimates that 1.1 million hectares of trees are successfully planted each year worldwide.

Modifying wood stoves to make them more efficient and increasing use of solar cooking would slow the decline of forests by decreasing reliance on wood for fuel.

Using the forests sustainably—by tapping trees for rubber, for example, or developing environmentally sound tourism—would provide more revenue than slash-and-burn agriculture.

DESERTIFICATION
The Spreading Barrens

Every year nearly 6 million hectares of previously productive land becomes desert, losing its capacity to produce food. The United Nations defines desertification as "land degradation in arid, semiarid, and dry subhumid areas [drylands] resulting mainly from adverse human impact." Translated into human suffering, that phrase means that by 1977, 57 million people had seen their lands dry up. By 1984 the number had risen to 135 million worldwide. Today, one-sixth of the total world population is threatened with desertification.

When drylands—which make up about 43 percent of the total land area of the world—revert to desert, hunger follows almost axiomatically. Most crops can't survive in the parched landscape, and harvests fail. Further, withered root systems can't hold the soil, and winds finally erode whatever topsoil remains.

Arid landscapes are so fragile that they break down quickly. A drought can mean catastrophe. In Mozambique, for instance, civil war combined with a worsening drought last year to leave 3.1 million people in need of food aid, 1.2 million more than in 1991.

But Africa is not the only place where food supplies are threatened by desertification. In Russia, annual desertification and sand encroachment northwest of the Caspian Sea were estimated to be as high as 10 percent. Around the drying Aral Sea, the desert has been growing at some 100,000 hectares per year for the last 25 years, an annual desertification rate of 4 percent.

STAVING OFF DISASTER

Agroforestry—planting trees as windbreaks and shade to protect pastureland—contributes to the maintenance of hard-used fields. In Kenya, for example, the Green Belt Movement has embarked on a large-scale tree-planting program.

Massive irrigation projects, such as those tried in Nigeria (see "Death of an Oasis," [Audubon] May-June 1992), are less practical, benefiting only a few at great cost to the environment. Small-scale projects, as low-tech and low-cost as collecting and managing rainwater, are often slow and laborious, but according to the Bread for the World Institute on Hunger and Development, they have succeeded in reclaiming hundreds of hectares of degraded land.

SOIL EROSION
A Worldwide Dust Bowl

Worldwide, erosion removes about 25.4 billion tons of soil each year. Deforestation and desertification both leave land open to erosion. In deforested areas, water washes down steep, naked slopes, taking the soil with it. In desertified regions, exposed soils, cleared for farming, building, or mining, or overgrazed by livestock, simply blow away. Wind erosion is most extensive in Africa and Asia. Blowing soil not only leaves a degraded area behind but can bury and kill vegetation where it settles. It will also fill drainage and irrigation ditches.

When high-tech farm practices are applied to poor lands, the result is often a combination of soil washing away and chemical pesticides and fertilizers polluting the runoff.

In Africa, soil erosion has reached critical levels, with farmers pushing farther onto deforested hillsides. In Ethiopia, for example, soil loss occurs at a rate of between 1.5 billion and 2 billion cubic meters a year, with some 4 million hectares of highlands considered "irreversibly degraded."

In Asia, in the eastern hills of Nepal, 38 percent of the land area consists of fields that have been abandoned because the topsoil has washed away. In the Western Hemisphere, Ecuador is losing soil at a rate 20 times what would be considered acceptable by the U.S. Soil and Conservation Service.

And even in the United States, 44 percent of cropland is affected by erosion.

DEFEATING THE ELEMENTS

According to the International Fund for Agricultural Development (IFAD), traditional labor-intensive, small-scale efforts at soil conservation—which combine maintenance of shrubs and trees with corp growing and cattle grazing—work best.

In the Barani area of Pakistan, a program begun by IFAD in 1980 to control rainfall runoff, erosion, and damage to rivers from siltation has resulted in a 20 to 30 percent increase in crop yields and livestock productivity.

POPULATION GROWTH
More Mouths to Feed

With a population growth rate of 1.7 percent, the world added almost 100 million people in 1992; an increase of some 3.7 billion is expected by 2030. Since 90 percent of the increase will occur in developing countries in Africa, Asia, and Latin America, the outlook is bleak: None of those countries can expect to produce enough food to feed a population increasing at such rates.

Population growth and environmental damage go hand in hand with poverty and hunger. In sub-Saharan Africa, for example, as colonial governments replaced pastoral lifestyles with sedentary farming, populations grew and farming and grazing intensified. Today, 80 percent of the region's pasture- and rangelands show signs of damage, and overall productivity is declining. Yet during the next 40 years the sub-Saharan population is expected to rise from 500 million to 1.5 billion.

Today only Bangladesh, South Korea, the Netherlands, and the island of Java have population densities greater than 400 people per square kilometer. (By comparison, the population density of the United States works out to 27 per square kilometer.)

By the middle of the next century, one-third of the world's population will probably live in overcrowded conditions. Bangladesh's population density could rise to 1,700 per square kilometer.

In Madagascar, population pressures have forced farmers to continuously clear new land. Virtually all the lowland forests in the country are gone. But cleared soil wears out quickly. Per capita calorie supply in Madagascar has fallen by 9 percent since the 1960s, probably the greatest decline anywhere in the world.

In Nepal, one of the world's poorest nations, increased population (700 people per square kilometer of cultivable land, the world's highest average) has forced villagers to expand their farm plots onto wooded hillsides. Marginal farmers rely on livestock, which they allow to graze in the remaining forests. Terraced soil once used for crop production has been abandoned for lack of nutrients, putting more pressure on ever-diminishing forest resources.

CHOOSING THE FUTURE
If a fertility rate of slightly more than two children per couple can be achieved by the year 2010, the world's population will stabilize at 7.7 billion by 2060. If that rate is not reached until 2065, world population will reach 14.2 billion by 2100.

According to a 1992 World Bank report, improving education for girls is an important long-term policy in the developing world. The more educated a woman is, the more likely she is to work outside the home and the smaller her family is likely to be. Choice also plays a role here: The United Nations' World Fertility Survey has found that women would have an average of 1.41 fewer children if they were able to choose the size of their family. Access to birth control methods could help lower the world's population by as many as 1.3 billion people over the next 35 years. During the Reagan-Bush years U.S. funding to programs offering such information was cut back; but President Bill Clinton has reversed that stand.

Somalia: An Ecopolitical Tragedy

Somalia's breakdown of law and order has created new waves of famine in recent months, but the groundwork for civil strife was laid by years of misuse of the nation's precious grazing lands and water resources.

Almost the entire country is categorized by the United Nations Environment Programme as susceptible to soil degradation, and most of the country is overgrazed.

Over the years the degradation was accelerated by the parceling out of communal grazing lands to private owners, which undermined traditional systems of land management. Private herds—which are generally larger than those owned communally—have stripped the hillsides bare, causing wind erosion during droughts and runoff during rains.

Building dams across valleys to halt water runoff in the north of the country has made matters worse by disrupting the natural drainage systems.

In the south the productivity of irrigated fields has been lessened by poor water management, which has created saline, waterlogged soils on the edge of the desert. There the salt will render the soil useless for food production in the decades to come.

—*Fred Pearce*

ENVIRONMENTAL REFUGEES
A Moveable Famine

The cycle of overpopulation, poverty, environmental ruin, and famine begins all over again when those trying to find a better life flee their ruined homelands. Land degradation is the largest cause of environmental-refugee movements.

According to Jodi L. Jacobson, a senior researcher at Worldwatch Institute, in Washington, D.C., 135 million people live in areas undergoing severe desertification.

But when those refugees move into areas that are already stressed by overpopulation or too intense agriculture, they place an added burden on the environment. Refugees need wood for fuel, water to drink, land on which to graze their livestock, and grain to eat—all of which are already scarce.

Jacobson estimates that some 10 million people worldwide are refugees from environmental ruin. "Competition for land and natural resources is driving more and more people to live in marginal, disaster-prone areas," she says, "leaving them more vulnerable to natural forces. Hence, millions of Bangladeshis live on *chars*, bars of silt and sand in the middle of the Bengal delta, some of which are washed away each year by ocean tides and monsoon floods."

CLIMATE CHANGE
Global Warning

Since 1800 atmospheric concentration of carbon dioxide (CO_2) has increased by about 25 percent and continues to rise each year. Over the same period atmospheric methane concentrations have doubled. Since the 1960s more than 100 separate studies have confirmed that a doubling of the CO_2 concentration would raise average surface temperatures by one to four degrees centigrade; three degrees is the figure used by the United Nations' Intergovernmental Panel on Climate Change.

Although a small number of scientists dispute these findings, weather and climate remain the biggest concern for farmers as warming begins to change growing seasons, irrigation needs, and land use. These changes will be especially serious in tropical regions, where farmland is already marginal and crops are growing near the limits of their temperature tolerance.

Those who live along shorelines will be hardest hit: The greenhouse effect could cause a global mean sea level rise of about six centimeters per decade. At that rate, many islands would become uninhabitable, and currently productive lowlands would be flooded. The developing countries that now experience the worst food shortages can expect to be hurt most by global warming. A study conducted by the University of Oxford and the Goddard Institute for Space Studies, funded by the U.S. Environmental Protection Agency, found that with an increase in average temperatures of three to four degrees centigrade, grain production in developing countries would decline by 9 to 11 percent by the year 2060, putting between 60 and 360 million people at risk from hunger—10 to 50 percent more than the currently predicted 640 million.

CLOSING THE GREENHOUSE

The use of renewable, or nonfossil, fuels—and more efficient use of all fuels—would go far to control the buildup of CO_2.

The World Bank recommends that governments remove energy subsidies and that they tax the use of carbon fuels. Maintenance of the world's large remaining forests would also help: Tropical deforestation accounts for 10 to 30 percent of the CO_2 released into the atmosphere.

The Oxford–Goddard Institute study found that slowing population increases could allow developing nations to cope more readily with the changing climate by changing land use and farming practices.

Hunger and the Ozone Hole

In early 1992 researchers from five U.S. marine-science institutes reported a drop of 6 to 12 percent in phytoplankton production under the Antarctic ozone hole. It appears that tiny marine organisms, which constitute more than half of all biomass on earth, may not be able to withstand harmful wavelengths of ultraviolet light—UV-B radiation—that penetrate the earth's thinning ozone layer.

When ozone holes form in the spring, most of the fish, shellfish, and crustaceans that humans harvest are in their larval, planktonic stages—floating in the topmost layer of the ocean. "Increasing intensities of UV-B radiation near the surface could negatively impact the reproductive potential of some of our most valuable marine resources, including tuna, pollock, cod, halibut, and flounder," wrote John Hardy, an associate professor at Huxley College, Western Washington University, in the November 1989 issue of *Oceanography*. Juvenile crabs, lobsters, shrimp, and anchovies are also vulnerable.

Just how such damage might move up the food chain is not known, but in Newport, Oregon, the Environmental Protection Agency's stratospheric ozone–depletion team found preliminary evidence of retarded growth in amphipods fed with phytoplankton that had been exposed to UV-B.

Although the leading industrial nations have agreed to halt production of the worst ozone-destroying compounds by 1996, ozone depletion is expected to continue for several decades as existing chemicals seep into the stratosphere. An ozone hole is likely to appear over the Northern Hemisphere, where ozone is dwindling at an estimated 1 percent per year. Each 1 percent decline in ozone is thought to increase exposure to biologically harmful ultraviolet light by at least 2 percent.

The United Nations Environment Programme warns that a 16 percent reduction in stratospheric ozone (which could occur in the next few years) would trigger a 6 to 9 percent drop in seafood production. Oceans now provide more than 30 percent of the animal protein eaten by humans.

—*Brad Warren*

We're All Minorities Now

SUMMARY Racial and ethnic diversity increases the differences between urban, rural, rich, and poor Americans. Children are most likely to be nonwhite or Hispanic, but the aging of diversity will have profound effects on consumer markets in the 1990s. Businesses can respond by using consumer information to unite diverse niches into profitable markets.

Martha Farnsworth Riche

Martha Farnsworth Riche is director of policy studies at the Population Reference Bureau in Washington, D.C.

The United States is undergoing a new demographic transition: it is becoming a multicultural society. During the 1990s, it will shift from a society dominated by whites and rooted in Western culture to a world society characterized by three large racial and ethnic minorities. All three minorities will grow both in size and share, while the still-significant white majority will continue its relative decline.

Whites represent eight in ten Americans, the 1990 census found, down from nine in ten as recently as 1960. Subtract white Hispanics, and you discover that only about three out of four Americans are non-Hispanic whites.

During the 1980s, the U.S. received 6 million legal immigrants, up from 4.2 million during the 1970s and 3.2 million during the 1960s. Few immigrants now are of European origin. Immigrants also tend to have more children than the non-Hispanic

All Americans are now members of at least one minority group.

white population, as do Hispanics and blacks. Together, these two factors are boosting the share of minorities in the population.

These trends are also creating diversity within the minority population. According to the Census Bureau, the 1990 census missed 1 in 20 blacks and Hispanics. Nevertheless, it gives an accurate picture of the rapid growth in their numbers. In 1990, 12 percent of Americans identified themselves as black, 9 percent as Hispanic origin (some of whom are also black), 3 percent as Asian or Pacific Islander, 1 percent Native Americans, and 4 percent "other." The first three groups will continue to grow faster than the white population. As each group grows, diversity within them will grow too.

These trends signal a transition to a multicultural society. If you count men and women as separate groups, all Americans are now members of at least one mi-

nority group. Without fully realizing it, we have left the time when the nonwhite, non-Western part of our population could be expected to assimilate to the dominant majority. In the future, the white Western majority will have to do some assimilation of its own.

Government will find that as minority groups grow in size relative to one another, and as the minority population gains on the dwindling majority, no single group will command the power to dictate solutions. The debate over almost any public issue is likely to become more confrontational. Reaching a consensus will require more cooperation than it has in the past.

The new demographic transition may be particularly difficult for business because it parallels an equally momentous economic transition. As the economy moves away from manufacturing and phys-ical skills and toward services and knowledge skills, a real danger emerges. The economic transition is increasing inequality in both incomes and opportunities. This inequality happens within and across racial and ethnic groups, and it has the potential to polarize both consumers and employees.

DIVERSITY DIFFERENCES

Immigration will add more Americans in the 1990s than it did in the 1980s, due to legislation enacted in 1990. The Immigration and Naturalization Service projects that legal immigration will exceed 700,000 per year starting in 1992. That compares with 600,000 immigrants per year as recently as the late 1980s. Illegal immigration will push the total even higher.

The 1990 law will also increase diversity among immigrants—notably at the upper end of the income scale. It allows people who have no family here to immigrate if they have highly prized work skills, or if they are ready to make a significant business investment. The law nearly tripled the number of visas (to 140,000 a year) for engineers and scientists, multinational executives and managers, and other people with skills in demand. This includes 10,000 visas a year for investor immigrants who will put at least $1 million into the economy and create ten jobs. (The entrance fee drops to $500,000 in rural areas and areas of high unemployment.)

Immigrants tend to join their peers, and their peers tend to live in large coastal cities. California, New York, Texas, Florida, Illinois, and New Jersey are expected to get three of every four new immigrants, who will be joining already-large minority populations in those states. In California, non-Hispanic whites will become a minority within the next two decades.

Central cities are still the front line for processing immigrants into society, and native-born minorities and older immigrants are also moving into suburban areas. Asians are most likely to integrate into white suburbs. Suburban blacks are still relatively segregated, according to research by Richard D. Alba and John R. Logan of the State University of New York at Albany. Hispanics fall somewhere in between.

These locational patterns ensure that multiculturalism will evolve unevenly across the country. As a result, many

YOUTH equals DIVERSITY

Most immigrants are young, and minorities generally have more children than whites. As a result, minorities are a progressively larger share of the population at younger ages.

(percent of population in each group by age)

	white	black	Hispanic	Asian/ Pacific Islander	American Indian	other races
0 to 9	74.8%	15.0%	12.6%	3.3%	1.1%	5.9%
10 to 19	75.1	15.1	11.6	3.3	1.1	5.4
20 to 29	77.3	13.1	11.5	3.3	0.8	5.5
30 to 39	79.9	12.0	8.9	3.3	0.8	4.0
40 to 49	82.9	10.4	7.1	3.1	0.7	2.9
50 to 59	84.4	10.1	6.4	2.6	0.6	2.3
60 to 69	87.4	8.8	4.8	1.9	0.5	1.5
70 to 79	89.3	7.9	3.5	1.4	0.4	0.9
80 or older	90.4	7.5	3.2	1.0	0.3	0.8
All ages	80.3	12.1	9.0	2.9	0.8	3.9

Note: Hispanics may be of any race; therefore, the percentages do not total to 100.

Source: 1990 census data

states and cities will become increasingly unlike the rest of the country.

Multiculturalism is not monolithic, either. The difference among Hispanic subgroups has been well documented; Cuban Americans are an economic and political dynasty in Miami, but no similar clout exists for Puerto Ricans in New York or Chicanos in Texas and California. One-quarter of the Hispanic population in 1990 was the product of immigration during the 1980s, if you include the children of immigrants. And 43 percent of Hispanics are immigrants from the 1970s and 1980s, according to Jeffrey Passel and Barry Edmonston of the Washington, D.C.-based Urban Institute.

Differences are even more pronounced in the fast-growing Asian American population. Passel and Edmonston report that 43 percent of the Asian American population in 1990 came from immigration during the 1980s, and 70 percent from immigration during the 1970s and 1980s. In 1970, the Asian American population was dominated by the Japanese. In 1980, the top group was the Chinese. Thanks to new immigration, the 1990 census found the Fili-

pino American population had grown almost as large as the Chinese American population, and both grew far beyond Americans of Japanese origin. Both the Asian Indian and the Korean populations now rival the Japanese population in size.

Different patterns of childbearing also play a role in creating a more diverse society. Fertility rates are still higher for minority groups than they are for non-Hispanic whites. In 1988, Hispanic women had the highest rate, with 96 children per 1,000 women aged 15 to 44. Black women had a rate of 87 per 1,000, compared with 63 per 1,000 for white women. As a result, two-thirds of minority families had children in 1990, compared with fewer than half of non-Hispanic white families.

Hispanic and nonwhite women will still have higher fertility rates in the 1990s, primarily because they come from younger populations, according to Juanita Tamayo Lott, president of a Washington, D.C. consulting firm. But these rates should diminish as these populations age. Nonwhite and Hispanic fertility rates should resemble white rates by the mid-21st century, she says.

The trend is clear. If current conditions continue, the United States will become a nation with no racial or ethnic majority during the 21st century. This may happen as early as 2060, according to demographer Leon Bouvier of the Center for Immigration Studies.

THE AGING OF DIVERSITY

The engine driving the diversity trend is the relative youth of minority populations. In 1988, non-Hispanic whites were older than any minority group, with a median age of 31.4 years. Hispanics were the youngest, with a median age of 24 years. Blacks were second youngest, at 25.6, while "other" races (mainly Asians) had a median age of 27.

The median age is increasing for all racial and ethnic groups, but Hispanics and blacks will remain younger than non-Hispanic whites. According to Census Bureau projections, non-Hispanic whites will have a median age of 41.4 years in 2010. That's ten years older than the median age for blacks in 2010 (31.4 years) and 12 years older than for Hispanics (29.3). "Other" races will have a median age of 35.6.

As a result, different age groups are becoming multicultural at different rates. In 2000, 72 percent of Americans will be non-

> **Immigration will add more Americans in the 1990s than it did in the 1980s.**

Hispanic white, according to Decision Demographics, a Washington, D.C. consulting firm. But fewer than two in three children will be non-Hispanic white. Non-Hispanic whites will account for 63 percent of children under age 8, 65 percent of children aged 8 to 13, and 66 percent of children aged 14 to 17. In contrast, nearly 80 percent of Americans aged 45 or older will be non-Hispanic white. Multicultural milestones show up first in the youngest ages.

These differences in the composition of age groups combine with differences in life expectancy to make the elderly population disproportionately white. However, with the notable exception of black men, the gap in life expectancy between whites and nonwhites has been narrowing. All these trends will eventually increase the multicultural character of the older population.

Multiculturalism is seeping into every aspect of American society, including language. The battle to make English the official language of the United States seems to have fizzled out, as Spanish-speaking Americans make it clear that they intend to retain their native language. As a result, many English-speaking Americans are discovering with a shock that they cannot communicate when visiting certain sections of California, Florida, or Texas. Bilingual signs and forms are becoming commonplace in many parts of the country.

Next spring, the Census Bureau will release the first data on "linguistically isolated" households. These are households in which no member aged 5 or older reported speaking English "very well." The numbers of such households did not merit a separate tabulation in previous censuses. But the 1980 census found 23 million households that spoke a language other than English at home. It also found 10 million households that had a less-than-adequate command of English. These numbers will be considerably larger in the 1990 count, thanks to immigration, according to Census Bureau demographer Paul Siegel.

Other factors are influencing the evolution of racial and ethnic identities. More and more Americans are of mixed parentage, and they are demanding to be recognized as multiracial.

Communications technologies are also changing the way people identify with their ethnic roots. For example, African films are gaining a significant audience here, particularly among African Americans. VCRs, fax machines, and other new technologies create important opportunities for cultural exchange in both directions. At the same time, it reduces the impetus for immigrants to assimilate into the "mainstream."

More than ever, the way for minorities to gain broader opportunities in American society is to get a college education. But relative to whites, college enrollment rates

> **In a multicultural society, businesses can thrive by finding common ground across racial and ethnic groups.**

actually declined for blacks and Hispanics during the 1980s.

As educational attainment becomes increasingly important to individual success, differences in educational attainment will produce sharply different socioeconomic profiles for different racial and ethnic groups. This trend could create a population polarized by both race and economic opportunity. Whites and Asians could increasingly dominate high-income high-status occupations, leaving blacks and Hispanics with low-income low-status occupations.

Even if employment discrimination suddenly ceased to exist, the lower educational attainment of minorities would keep many of them from entering newly opened doors. Poorly educated young black men are already shut out of the broader society; nearly one in four of those aged 20 to 29 is behind bars or on probation or parole.

As America participates increasingly in the world economy, business leaders could use a multicultural work force as a powerful competitive edge. But the opportunities will not be distributed equally among different racial and ethnic groups. The challenge is to maximize our comparative advantage in the world economy while still offering upward mobility to all Americans.

HOW BUSINESS CAN RESPOND

"The typical consumer-citizen of California in the late 1990s may be a 38-year-old professional who does Zen meditation. At home, she listens to Celtic folk music because her grandparents were Scottish. But she spends her vacations in northern Mexico to study Tarahumara culture, after picking up a taste for ranchero music," says Paul Saffo, who follows technology for the Institute for the Future in Menlo Park, California.

In a multicultural society, businesses

Recently, I was walking behind a proper-looking Asian man who seemed to be in his late 60s. He was carefully holding a small yellow book with both hands. As we reached a crosswalk, I moved closer to get a better look at the mysterious book. I assumed it was a sacred text full of ancient Eastern wisdom. As the light changed, the man carefully slipped the book into his jacket pocket. It was the *N.A.D.A. Official Used-Car Guide*.

Cultures mix like foods. The combinations are surprising and strange at first, but they can make sense once you get used to them. One example is the frozen pizza-flavored eggroll. Another is Korean-born country-and-western singer Kim Tsoy, the Eastern Cowboy. He is surprising—all six feet two inches of him—but his career choice also makes a lot of sense. Kim's parents came to America when he was four. His father worked for Voice of America; his mother studied opera at the Juilliard School. Music was Kim's best means of communicating in the New World. When he was thrust in front of his first-grade class during show and tell, he shared the English he knew best by performing a Mario Lanza song he had memorized from the movie *The Great Caruso*.

Kim grew up on a mix of Jim Reeves, Elvis Presley, and James Brown. He taught himself to play the guitar, and he hit the road at age 15. He has a strong silky voice and a three-octave range, so he can play to any audience. But Kim chose country music because it was the best medium for expressing his love of God, country, and family. "The country audience is gracious and loyal," he says. "When you get past the drinking and cheating songs, the traditional songs are about honor and family values. Country music most closely matches my traditional Korean background."

Kim and his band, Next of Kin, work in the country-and-western nightclubs out of Kissimmee, Florida, the Osceola County seat. Both Florida and Kissimmee are full of cultural surprises. For example, Florida has lots of cowboys and more cattle than Wyoming, Mexico, or Arizona. Local residents can remember when Kissimmee was just a cow camp on Lake Tohopekaliga, and they remain proud of the cowboy heritage. "The outlaws down here were meaner than the ones out West," said one guy I met in a bar. "They hid out in the swamps. A boat don't leave no trail."

Today, Kissimmee is the self-described "gateway to Walt Disney World" and other central Florida tourist attractions. In his book *Up For Grabs*, author John Rothchild calls the area Turnstile Valley. The huge attractions have captured Florida's leisure and labor market, he says, even though they lack the one natural asset that was thought to be the key to Florida's appeal: the beach. Kissimmee is 50 miles from the ocean.

Central Florida's attractions, the countless businesses that support them, and the umpteen-thousand hotel rooms have created an insatiable need for unskilled labor. As a result, the area's minority populations have skyrocketed. Osceola has the fastest-growing Hispanic population of any county in the United States. Blacks, whites, Hispanics, Asians, Native Americans—and native-born cowboys—are squeezed together in a world of concrete, swamps, and heat. Kissimmee, the old cow town, is now host to the world's largest tourist roundup. It's like an old western boom town, recast in neon lights and plastic.

Kim Tsoy likes it that way. "My record company wants me back in Nashville, but I still love playing to live audiences in small towns," he says. "People are all the same: they just want to relax and forget about their troubles. When I walk on to a country stage, I get plenty of strange looks. But once I open my mouth, they're mine."

His audience seems to agree. I met two rough-looking cowboys who were discussing something with the new owner of a local dance club. One of the cowboys was holding a can of baby powder. The owner, who had recently arrived from New Jersey, didn't know that whenever Kim Tsoy sings, these guys sprinkle baby powder on the floor so their boots can keep up with the music. I asked one of the cowboys what he thought of Kim. "We think he's fan-tastic," he said. "I just can't believe that sound is coming out of that head."

---

thrive by finding common ground across racial and ethnic groups. Businesses that try to target each group separately will be stunted by prohibitive marketing costs. Others will meet this challenge by helping multicultural consumers mix and match their lifestyles. Multicultural consumers will take discrete cultural pieces and mix them into custom-tailored wholes.

Another common need is information and entertainment that explains the world to multicultural consumers from their point of view. Last year, a widely publicized journalism study faulted young Americans for their ignorance of important news figures and news events. But

On every dollar bill is the phrase E pluribus unum, "from diversity comes unity."

given their increasingly multicultural nature, it's no surprise that today's youth had little interest or knowledge in what was going on in Eastern Europe, but were up-to-the-minute on developments in South Africa.* Consumer information and entertainment businesses are going to

*See "What's News with You," American Demographics, *November 1990, page 2.*

have to reposition both their content and their advertising to appeal to today's multicultural youth as they become tomorrow's multicultural adults.

Education is a major common need. The educational establishment has not adequately responded to the multicultural challenge, and that creates an opportunity for business.

Communications technology is building a new common ground for an increasingly multicultural population. We saw this during recent events in China and in Eastern Europe. We are going to see more of it as technological evolution lets our most recent arrivals keep close contact with their roots instead of cutting them off.

These developments create new opportunities for consumer businesses that can unlock culture from its origin and allow others to share in it. One example is the Japanese adoption of the Wild West, as Tokyo executives import log cabins from Montana and vacation on American dude ranches. As the world's first multicultural society, the United States is uniquely positioned to both understand and profit from the emerging global culture.

All this means that consumers are becoming simultaneously part of a global culture and a local community. It also means that these ties are based on common interests. Moreover, technology increasingly allows Americans to switch readily and frequently from one viewpoint to another. The marketer's new challenge is to find not only the right person with the right message but also to find them at the right moment.

Some of those moments will be global moments, as everyone in the world watches a soccer match, or a war. Some will be culturally specific moments, as Muslims or other groups share a moment that is invisible to everyone else. Some will be purely and simply local. But in every case, the common ground will be interests, concerns, and lifestyles.

Without necessarily realizing it, businesses have been preparing to meet this challenge by building detailed consumer information systems. Combined with attitude and behavior research, these systems can efficiently unite niches into markets. The systems' geographic specificity will extend marketing efficiency by allowing marketers to pay attention to the geographic variations in diversity.

For example, a Nissan television campaign featured a multicultural design team engineering cars "for the human race." The tag line made sense nationally, because it was broadly targeted image advertising. More directly targeted messages have to identify their audiences more closely. An ad that takes a multicultural society as a given is right for Los Angeles, but it might strike a strange note in rural Indiana.

Retailers can use locally based information systems to efficiently target specific demographic and market segments. Mark London is president and CEO of Equity Properties in Chicago, a firm that remodels and re-leases shopping centers whose trade areas have changed significantly. He recently analyzed an anchor store that was doing badly in a repositioned Miami mall. The store managers hadn't understood two crucial concepts. First, upscale Hispanic women don't have the same fashion preferences as other upscale women. Second, they don't have the same preferences as other Hispanic women. When the store learned to feature upscale Hispanic fashions, sales rebounded.

On every dollar bill is the phrase *E pluribus unum*, "from diversity comes unity." If this fundamental American belief can survive, our country will become a microcosm of an increasingly interdependent world. America can still offer hope to other countries, and to all of its citizens. But it can only work if we meet the multicultural challenge.

TAKING IT FURTHER

For more information about the changing makeup of minority populations, contact the Bureau of the Census's Data User Services Division at (301) 763-5820. A good general source of information on immigration is the *Statistical Yearbook* of the Immigration and Naturalization Service; telephone (202) 376-3066. For more on the impact of multiculturalism, contact the Institute for the Future at 2740 Sand Hill Road, Menlo Park, California, 94025; telephone (415) 854-6322. Reprints of this article may be purchased by calling (800) 828-1133.

Why Africa Stays Poor
And Why It Doesn't Have To

David Aronson

David Aronson grew up in Nigeria and the Ivory Coast and spent a year in Zaire on a fellowship. He is a graduate of Wesleyan University and the University of Florida and is currently working on a book about Zaire.

The images are so familiar that we have become all but inured to them: starving African children outlined against a broad expanse of empty sky; ragged, impoverished families huddled together on a stony steppe. They could be Biafrans in 1968, Sahelians in 1973, or Ethiopians in 1985. The most recent pictures are from Somalia, a barren stretch of East African coastland that juts into the Indian Ocean. Once a consolation prize in the Cold War (the real trophy in the Horn was Ethiopia, a richer and more populous nation), Somalia has since disintegrated into fiefdoms of grizzled warlords armed with Kalashnikovs and AK-47s. Now 2,000 Somalis die every day from hunger and its attendant diseases, and reports from elsewhere in Africa suggest that Somalia is only the beginning; according to the United Nations, 20 million to 60 million people are at risk of starvation throughout the eastern and southern parts of the continent.

The news out of Africa has been so grim for so long that the continent seems hopeless, its problems ineradicable. Yet this perception—though it has been reinforced by endless images of famine, disease, and warfare—is both untrue and unnecessarily fatalistic. There are some good reasons, I will argue, for hoping that Africa's next 30 years will be better than the last three decades.

It must first be acknowledged, however, that Africa is in a dismal state. Though statistics cannot convey the miseries of living in a shantytown outside Lagos, or the hardships of wresting a subsistence from the leeched landscape of the Sahel, they can suggest the dimensions of Africa's problems. In the 1980s, per capita African gross national product declined by nearly 25 percent. African farmers produce 20 percent less food today than they did in 1970, and there are twice as many mouths to feed. (At its current rate, Africa's population will reach nearly three billion by the year 2050.) Direct private investment in Africa constitutes 2 percent of the world's total, and experts predict that much of that miniscule figure will be siphoned off as Eastern Europe and Asia rebuild. One French diplomat has written: "Economically speaking, if the entire black Africa, with the exception of South Africa, were to disappear in a flood, the global cataclysm would be approximately nonexistent."

By the late 1980s, most of Africa's states were facing outright financial ruin. The 1990 World Bank annual development report listed 27 African countries among the world's 40 poorest. It seemed to matter very little what path to economic growth these countries were ostensibly pursuing; political scientist Crawford Young argued as much in his book *Ideology and Development in Africa.* "Does ideology matter?" he asked. "My reading of the evidence does not lead to a single unambiguous conclusion." "Socialist" Zambia, Tanzania, and Ghana; or "capitalist" Zaire, Kenya, and Nigeria; the still verdant Congo or the desertified Mali—by 1990, all were mired in economic and political stagnation.

Africa's troubles are not just economic. Civil war and ethnic strife have erupted in over a dozen African countries. Liberia and Somalia have followed Chad and Mozambique into Hobbesian anarchy. Even Kenya and the Ivory Coast—long touted as the two "success" stories of black Africa—have been jolted by a series of anti-government demonstrations. Meanwhile, the AIDS pandemic looms over Africa like a modern-day bubonic plague; the World Health Organization estimates that the disease will kill *20 percent* of Africa's working adults by 1996, including disproportionate numbers of the educated and successful. Perhaps the most sobering aspect of the AIDS crisis is the attitude of resignation that many Africans seem to have taken toward it. In a continent where 12,000 children die every day of hunger and hunger-related causes, that attitude is depressingly easy to understand: what to us is some horrific medieval plague to them is simply one more deadly infectious disease.

In the West, newspaper editors came up with the term *compassion fatigue* to describe their audiences' reactions to Africa's problems (and, one suspects, to justify their meager coverage of them). A famine in Mozambique, civil war in the Sudan, a slaughter of innocents in Burundi, chaos in Zaire and elsewhere—in the post–Band Aid era, the litany of Africa's woes was buried in the back pages of the press. The impression to be gleaned from this coverage was that Africa's problems were monolithic, insurmountable, and utterly alien—and therefore, presumably, not of pressing concern to us.

"Why Africa Stays Poor, And Why It Doesn't Have To," by David Aronson first appeared in *The Humanist,* March/April 1993, pp. 9-14. Reprinted by permission.

Though examples of African success are admittedly few, they provide much-needed evidence that the continent's current woes are not insurmountable. A few African countries made it through the 1980s with their economies and political systems reasonably intact: the Gambia and Botswana are islands of democracy; the Cameroon's economy grew by 8 percent per year through much of the decade; and Mauritius' economy grew by over 5 percent per head. And though Africa's problems seem nearly universal, they are far from monolithic.

Zambia provides an instructive example of the continent's shifting fortunes. In 1964, when Zambia achieved independence under the leadership of Kenneth Kaunda, it was one of Africa's brightest prospects, with a highly developed infrastructure and abundant copper reserves that gave it some of the highest income levels on the continent. Kaunda himself was genial and charming, his political rhetoric an eloquent blend of socialism, pacifism, and pan-Africanism. Zambia became one of the world's largest per capita recipients of foreign aid. However, from 1965 to 1988, Zambia's average annual rate of growth was *minus* 2.1 percent, and today it is one of the poorest countries in the world—ranked eighteenth from the bottom in per capita GNP.

Belying his oft-proclaimed "humanistic philosophy," Kaunda officially declared Zambia a one-party state in 1972. Under the infamous Emergency Powers Act—a holdover from the colonial era—he regularly had dissenters jailed, beaten, or sent into exile. Kaunda also came to enjoy a life-style that few Zambians could have imagined, playing golf several times a week on a course adjoining his presidential estate and stocking his private zoo with peacocks and deer.

It was not, however, the gap between his professed idealism and his actual politics that brought Kaunda down—after all, any number of governments have survived telling more risible lies. Nor were Zambia's problems the result of outside pressures, though Zambia respected the African trade embargo against South Africa at considerable cost to its own economy. Rather, Kaunda's downfall was the result of economic policies that eventually boxed him in—policies that left him unable to fulfill the social contract he had established with his urban population.

Kaunda relied upon his nation's copper reserves to create a huge bureaucracy; at one point, nearly 40 percent of Zambia's wage earners received their checks from the government. At the same time, Kaunda poured money into the cities at the expense of the countryside. Schools, medical facilities, and subsidized housing were all available to the urban population; farmers had to make do with rigged, artificially low prices for

> *Mobutu pillaged his country with scarcely a murmur of dissent from his Western sponsors—particularly the United States.*

their produce.

The economic consequences were as devastating as they were predictable. The rural poor moved to urban areas, where they came to rely upon subsidized food as part of the social contract; declining food production forced the government to spend an ever-greater percentage of its earnings on imported grain. Zambia borrowed heavily in the 1970s and early 1980s on the strength of its copper reserves and used the money to support urban consumption, to import food, and to keep afloat its bloated civil service. But as copper reserves dwindled and international commodity prices dipped, the government could no longer afford its part of the social contract. In 1990, rioting erupted in several cities, forcing Kaunda to allow elections. The results were overwhelming: in Kaunda's home seat of Nchanga, the vote went 20,680 for his opponent to only 637 votes for Kaunda himself.

In Zaire, Zambia's giant neighbor to the north, the pro-democracy opponents of President Mobutu Sese Seko have had a far rougher time of it. Mobutu has ruled Zaire since 1965, when he seized power in a CIA-sponsored coup after the conflicts of Zaire's early years. It is frequently claimed that, whatever Mobutu's sins, he at least brought unity to a war-torn country; in fact, the government in place at the time had already made peace with all but one minor rebel group when Mobutu staged his coup. If Mobutu brought unity to Zaire, it is a unity born of Zairians' near-universal hatred of his regime.

Under Mobutu, Zaire's economy fell through the floor. In 1990, per capita annual income was $170; real wages had plummeted to one-tenth their 1960 levels. In Mobutu's 27 years in power, not a single hospital has been built. The road network that linked Zaire together in 1960 has crumbled; the country is larger than the United States east of the Mississippi, yet there are now fewer miles of paved roads in all of Zaire than there are in Toledo, Ohio. Only 3 percent of the central government's budget goes to health and education; 23 percent goes to the military, and 50 percent to lining the pockets of Mobutu and his ruling elite. Mobutu Sese Seko—his adopted name means "the cock that leaves no hen untouched"—has accumulated a dozen French and Belgian chateaux, a Spanish castle, and a 32-bedroom Swiss villa. His net worth is variously estimated at $3 billion to $7 billion. Meanwhile, one out of every two children born in Zaire dies before the age of five.

Mobutu pillaged his country with scarcely a murmur of dissent from his Western sponsors—particularly the United States. There were two reasons for this: first, Zaire is an important supplier of minerals, including industrial diamonds, copper, and cobalt (needed for the alloys used in military aircraft); and second, Mobutu allowed his country to be used as a base for covert American military and diplomatic expeditions into Angola and Chad. His stature as one of America's most important assets in the region was underlined in 1988 when Mobutu became Africa's first head of state to meet with President George Bush.

In 1990 and continuing through most of 1991, pro-democracy agitators put Mobutu on the defensive. Led by the popular long-time dissident Etienne Tshisekedi, they forced him to convene a national conference on the future of the country's leadership. Mobutu did all he could to disrupt the proceedings of the conference; in the ensuing riots, scores of people in various cities were killed by the military. Mobutu himself fled to a luxury yacht cruising the Congo River but retained control of the elite army troops and his security forces. Meanwhile, Zaire's economy and its political institutions have collapsed.

During the Cold War, diplomats and politicians in the West frequently justified their support for Mobutu by arguing that the alternative was a return to the chaos of Zaire's early years. This argument is founded, as I've indicated, on a willful misreading of history; it ignores the fact that the situation in Zaire was well past its crisis point when Mobutu took power, and it also implies that Western support for Mobutu was motivated, at least in part, by a humanitarian concern for Zairians. "We care," the diplomats seemed to be saying, "that Zairians not be forced to suffer again the internecine warfare that accompanied the birth of their nation." But since the confrontation between the national conference and Mobutu reached a stalemate, the West has done almost nothing to help resolve the situation; in the absence of Cold War exigencies, we seem to have lost interest in the country. What remains, as political scientist Rene Lemarchand has written, "is a country teetering on the brink of chaos—a devastated economy, an empty treasury, a peasant sector reduced to subsistence agriculture, an urban population at the edge of starvation, and hundreds of thousands of unemployed youth for whom life has nothing to offer." For years, we heard from our politicians and diplomats that it was "Mobutu or chaos." Now it is both.

Finally, there is Uganda, once known as the "pearl" of East Africa. No longer. The rolling green hills and fertile soil that so pleased the colonialists are still there, but since the 1971 coup that brought Idi Amin to power the country has lurched from one disaster to the next, bloodier one. Consider: there are half a million internally displaced people and over a quarter-million foreign refugees living in Uganda; it is probably the one country most hard-hit by AIDS in the world; its economy is in a shambles, with tea and other agricultural exports down to one-fifth or one-tenth of 1970 levels; and it has suffered from more or less continuous political unrest and warfare. Even by African standards, Uganda is a basket case.

The reasons for this startling decline are complicated. Before 1963, Uganda boasted a flourishing economy, a well-trained populace, extraordinarily fertile land, and a highly developed infrastructure. Makerere University was one of the best in Africa, and Kampala was among Africa's most cosmopolitan cities. For several years after its independence, Uganda seemed poised to fulfill the immense potential that was perceived for it. The economy flourished under the moderate leadership of Prime Minister Milton Obote, and with the kabaka, or king, of Buganda assuming the largely ceremonial role of president, the political situation seemed remarkably stable.

In fact, however, Uganda was built on the most fractured of foundations. Buganda was only one (albeit the principal one) of several traditional kingdoms the British had lumped together into one protectorate; and with independence came not only heightened ethnic rivalries but also religious, regional, and ideological tensions. To be sure, the challenges to nation-building represented by these divisions were to be found throughout the continent: the jigsaw puzzle of national boundaries the colonial powers left behind bore little relationship to pre-existing ethnic or linguistic identities. But in Uganda, these divisions were never overcome. No central authority ever managed to impose a national identity on the disparate groups competing for power, and the result was that no one group or party was ever seen as legitimate. The tensions within the polity could be submerged, but they could never be resolved.

Uganda's political history probably reached its nadir during the eight-year reign of Idi Amin, the notoriously brutal army sergeant turned international buffoon. But Uganda's problems neither began nor ended with Amin. Obote, Uganda's first prime minister, had spent most of his time in office trying to contain the forces that were tearing the country apart. In 1966, for example, he ousted the kabaka under pressure from radical youth groups in the capital; to keep the army happy, he tripled its share of the nation's budget. That did not prevent factions from developing within it, and eventually one of them—led by Amin—secured the patronage of Great Britain and Israel, who were dissatisfied with Obote's growing radicalism.

Amin's 1971 coup brought an end to Uganda's economic growth. During his bloody rule, coffee, tea, and cotton production fell to one-third their earlier levels, and industrial performance dropped 85 percent. Amin was finally overthrown in 1979, after an attempted invasion of Tanzania turned into a rout of the Ugandan army by the better-trained, better-led Tanzanian troops. But a succession of leaders proved unable to stop the anarchy that had characterized Amin's final years, and the economy continued to sputter. Under Uganda's current leader, Yoweri Museveni, there has been a modest economic recovery, and aid from the major donor nations has also begun to trickle in. But though a semblance of order prevails, it seems only a matter of time before the next political tremors erupt.

> *The national boundaries left behind by the colonial powers bore little relation to actual ethnic or linguistic identities.*

Are there any observations to be made, any lessons to be distilled, from the divergent experiences of these three countries? The first is that it pays not to live near one of Africa's hot spots. Zambia respected the boycott against South Africa at enormous cost to its own economy. Other countries, particularly Mozambique and

Angola, endured South African-sponsored terrorism and civil war that effectively blocked any possibility of their own development. And though guerrilla movements in these countries have now acquired a momentum of their own, support from the apartheid regime was critical in launching them. Much the same could be said of the Libyans in Chad and the Ethiopians in Djibouti.

The second observation to be made is that it was the misfortune of certain African nations to come of age during the intellectual heyday of a certain kind of nondemocratic socialism. It is true that, with few exceptions, Soviet and Chinese ties to Africa were always considerably weaker than the cold warriors would have had us believe. Thus, Herman Cohen, the outgoing American Assistant Secretary of State for African Affairs, has now blamed *European left-wingers* for implementing in Africa "the biggest socialist fantasies that they weren't able to implement in their own countries." But Cohen's is a simplistic rendering of history—a piece of Reaganite revisionism. African countries like Ghana, Tanzania, and Zambia "went" socialist not from any affection for the leaden Stalinists (Soviet advisers were despised wherever they went), nor from European left-wingers' denunciations of their own bourgeoisie (we should be so lucky, thought the Africans), but for complex reasons of their own. Socialism represented the opposite of capitalism, the system of Africa's colonizers. It offered a blueprint for social and political mobilization that seemed more coherent and plausible to some African leaders, and, with its hazy vision of a utopian future, it fed into a hallowed conception of Africa's past—a vision of a prelapsarian (read: precolonial) Eden whose traditions contained all the promises of Western political thought but none of its painful contradictions.

Nothing, alas, could have been more impractical for these fledgling nations. To work at all, socialism requires an educated bureaucracy and a vigorous, disciplined central government—precisely the conditions that did not exist in post-colonial Africa. After 30 years, the verdict is unequivocal: capitalism's record is mixed, but socialism—at least of the sort practiced by Kaunda in Zambia, Nyerere in Tanzania, or Nkrumah in Ghana—has been an utter failure.

The third observation to be made is that it was disastrous for Africa to come of age during the Cold War. Both the United States and the Soviet Union helped to install and maintain tyrants whose records are a blot upon the human race. America's support for Mobutu in Zaire and the Soviets' for Mengistu in Ethiopia may (or may not) have been justified by the superpowers' national-security needs, but the price that

> *It was disastrous for Africa to come of age during the Cold War. Both the U.S.A. and the U.S.S.R. helped to install tyrants whose records are a blot upon the human race.*

ordinary Zairians and Ethiopians paid for that security lies beneath the gravestones of countless children. It is important to be blunt about this: we have a moral duty to acknowledge how much of our own putative security has been purchased with the blood of others.

A fourth observation concerns the extent of corruption in Africa. From the governor who pockets half the state's budget for public works to the nurse at the local clinic who demands a bribe before changing a bandage, corruption is so deeply entrenched and so endemic that it requires an explanation. Peter Wanyande, a Kenyan political scientist at Kenyatta University in Nairobi, suggests that corruption flourishes in Africa in part because there are few institutions aside from the state through which the gifted and the capable can rise. But in contrast to the tribe or the extended family, the state remains an abstraction, commanding neither loyalty nor affection. The dispensation of favors and the profitable manipulation of alliances thus become the *modus vivendi* for Africans on the make. In Nigeria, in Zaire, and increasingly in Kenya, corruption is not so much the grease that keeps the wheel turning as it is the wheel itself.

This is not to justify the flagrant abuses of a Mobutu but, rather, to put them into context. The powerful in Africa steal because they can; the poor steal to survive. Corruption in Africa, then, was born of the structural weaknesses of African political and civil society, and it is now a cause as well as a consequence of African poverty.

A fifth observation is that, like the former Yugoslavia, many African countries are still wrestling with the demon of tribalism (a polluted term, no doubt, but more accurate than its euphemisms). Tribal conflicts simmer in Uganda, Rwanda, Burundi, Kenya, and Zaire, and they have boiled over in Chad and the Sudan. Tribalism hurts countries not only when blood is spilt; the energy expended on keeping the peace between rival ethnic groups represents a considerable distraction from the business of economic, political, and cultural development.

The culminating lesson is that, in Africa as elsewhere, the prerogatives of history are impossible to deny. Lumped arbitrarily into geographic units at the Berlin Conference of 1885 and launched toward independence 80 years later in entities that corresponded in no way to African realities or experience, the continent has limped behind the rest of the world—not because Africans are stupid or lazy or incapable of collective action, but because they are caught in structures that negate their strengths and frustrate their efforts. Africans remain poor because the states in which they live are artificial constructs, unmoored to the societies or people they govern, largely free of any constraint from civil institutions, and subject all too often to fracture along the substantive divisions of class, religion, and ethnicity.

Are there any grounds for hope, any reasons to believe, that Africa will do better in its next 30 years than it has in the last three decades? Certainly, there has been a veritable sea-change in African attitudes

215

toward development. Nigerian statesman Olusagen Obasanjo is among the most eloquent:

> I believe that for us in Africa, our salvation lies in our own hands and nowhere else. Only we can be the architects of our future; as we have been the architects of our misfortune by and large for the past quarter of a century.

Then, too, some countries—most notably Ghana—have embarked on ambitious economic restructuring programs to eliminate waste, corruption, and mismanagement and to encourage private industry. The World Bank has been the driving force behind much of this movement, and there is, naturally, considerable debate about how fair these programs are and whether or not they will work. To a certain extent, the programs represent a recolonization of Africa by Western technocrats. This is not necessarily bad, some argue: at the helms of central banks, it is better to have the gnomes of Zurich than the kleptocrats of Kinshasa. But these programs also represent the third or fourth generation of development initiatives sponsored by the World Bank, and they share with the Bank's earlier initiatives a refusal to acknowledge the political environments in which they operate. In fact, economic policies prescribed with little regard for such realities may exacerbate—as they have in the past—both the political and the economic problems confronting Africans. This is an age-old criticism of the Bank, but one that, despite an abundance of evidence, it seems constitutionally incapable of attending. The Bank's charter, argue its defenders, is properly economic and deliberately apolitical. Zairians—who have seen the results of such "aid" in the Mercedes-Benzes driven by the political elite through that country's corroded streets—might be excused for wondering whether the intention has not been improperly political and deliberately uneconomic. Their president could, after all, pay off the country's entire national debt.

The end of the Cold War ought to be a tremendous spur to African political and economic development; the West no longer needs to support dictators simply because they happen to be *our* dictators. Smith Hempstone, the American ambassador to Kenya and a conservative of long standing, has not hesitated to agitate on behalf of Kenya's fledgling democrats, even though he has irked Kenyan President Daniel arap Moi considerably in doing so. Significantly, in the past three years, a half-dozen African countries have followed Eastern Europe into democracy, and another dozen or so are struggling to achieve it. It is no exaggeration to say that the democracy movement represents the most important political shift in Africa since independence. It is clear, moreover, that the democracy movement has far deeper roots than many observers could have imagined. "Throughout the continent, independent political and social forces have emerged to challenge moribund, authoritarian-patrimonial regimes of many varieties," says Peter Lewis. "The result has been a succession of movements pressing for fundamental political change." At every level of government, indigenous institutions are evolving as a response to popular demands. The shape and future direction of these institutions are still difficult to determine; all that can now be said is that they belong to Africa in a way that the pre-fab parliaments and constitutions left behind by the colonialists never did. Kwame Nkrumah's famous injunction—"Seek ye first the political kingdom"—is at last being realized.

The news out of Africa has been grim for so long that one hesitates to conclude on a dispiriting note. However, the battle between democratic and despotic governments in Africa is by no means won: in Togo, for example, the former dictator is back on his throne, and in Kenya, President Daniel arap Moi seems to have survived the recent election. Moreover, the United States continues to send the bulk of its economic assistance to tyrannical regimes. Randall Robinson, executive director of Trans-Africa, points out that, in the last fiscal year, the United States gave $130 million to Kenya, Zaire, and Malawi—"states where there is massive corruption, broad repression, and little if any appreciation of democratic values." On the other hand, countries like Benin, Botswana, and Namibia, which are now fully democratic, received just $30 million. (Namibia, with the most liberal and democratic constitution on the continent, received $500,000.)

What is to be done? Let me conclude by listing several things the West can and should be doing to help Africa. I use the word *should* advisedly—not to insist on Western culpability but, rather, on Western responsibility, which should derive from a decent regard for humankind.

First, Western nations—particularly the United States—can send more aid more selectively. The U.S. budget for all of sub-Saharan Africa is about $800 million, roughly one-third of what we give to better-off (but more "strategic") countries such as Egypt or Pakistan. We ought, furthermore, to be giving our aid to those countries which are making genuine political and economic reforms; whatever else aid does, it bestows official approval on a Third World country's government. There's no longer any reason (if there ever was) for the United States to support tyrants.

The largest incentive the West can offer Africa's reformers is the prospect of debt relief. Compared to Latin America's debt burden, Africa's has been ignored. The numbers involved are much smaller and don't pose a threat to the world banking system; yet, according to the *Economist*, Africa's burden is

> *It is no exaggeration to say that the democracy movement represents the most important political shift in Africa since independence.*

50 percent greater as a percentage of gross domestic product than Latin America's, and the continent remains "profoundly, unsustainably indebted." Swapping debt for political and economic reform is one of the best and least costly ways for the West to help out.

Second, there should be more contact between Africa and the West: more student exchanges, more artistic conferences, more sister-cities and business conventions, and more professional colloquia. Ties between newspapers, universities, and community or religious groups should be strengthened. The West can and must find a way to encourage the institutions of African civil society; their continuing development represents the best hope for the continent's future.

Third, the early-warning system that is supposed to signal the potential for famine in Africa needs to be strengthened. Cobbled together by the United Nations after the Sahelian famine of the mid-1970s, it has not been an effective agency; it predicted major famine-related deaths in 1991, and when these failed to materialize, it lost much of its credibility. It needs a better intelligence network on the ground and more effective links to the wider community of Africanist scholars, diplomats, and journalists.

Fourth, U.N. troops ought to be empowered as a peace-making force, not just as peace-keepers. As in the case of Somalia and Yugoslavia, difficult decisions about the conflicting claims of various groups and issues of international law and human rights will need to be addressed. At what point does a regime so violate the accepted standards of international law that it forfeits its claims to sovereignty? How feasible or desirable is direct intervention in another country's affairs? What price is the West willing to pay to save the innocent victims of a murderous regime? My bias here is plain.

Finally, the press could do a far better job than it has covering African issues. It is one of the sadder ironies of our time that famines in Africa only make headlines when thousands have already died. Unlike earthquakes or typhoons, famines can be spotted months and even years in advance—when the rains fail, crops wither, and prices for grain shoot up. All these things take time and can be monitored. Given their pervasive influence, the media have a moral responsibility to alert the world to the potential of famine and to agitate aggressively, if need be, to ensure that the matter is attended to before people die.

Beyond the issue of famine, however, is the more mundane issue of poverty. In its quiet way, poverty—of the day-to-day sort that millions of Africans are mired in—is more destructive than the occasional spectacular famine. Every day, some 12,000 children in Africa die because their parents are too poor to buy them the food or medicine they require. Though the press duly takes note of Africa's rebellions, civil wars, coups d'etats, and corruption, the biggest story in Africa—the quiet struggle of ordinary people to survive against grim odds—has hardly been told at all. The media should find a way to bring this story home. In its global ramifications, the gap between the rich and the poor countries of the world is certainly one of the most important stories of our time—and yet, it is being largely ignored.

SOUTH AFRICAN APARTHEID: A SOCIO-SPATIAL PROBLEM

DeWitt Davis, Jr.

DeWitt Davis, Jr. (Ph.D., Lund University, Sweden, 1972) is professor of geography and teaches in the Department of Geography at the University of the District of Columbia, Washington, D.C. USA. His major research areas are population geography and urban geography.

Some form of apartheid has existed in South Africa for three quarters of a century, that is, since the founding of the Union in 1910. The various forms are all unquestionably types of segregation. Apartheid, which in 1949 was one of the main policies in the Malan regime, was created "to perpetuate the separateness of the population groups;" "separate development and multinational development." The Botha regime called it "good neighbourliness". The present government is negotiating reform with non-White political leaders. The purpose of this article is to demonstrate that apartheid has a socio-spatial framework and that geography has a role to play in understanding apartheid.

AFRICAN POPULATION OF INDEPENDENT HOMELANDS: 1991	
Transkei - 1976*	3,440,900 (11.8)**
Bophuthatswana - 1977	2,403,000 (8.3)
Venda - 1979	556,100 (1.9)
Ciskei - 1981	838,600 (2.9)
Sub-total	7,238,600

*Year of independence of the so-called homelands
**Percent of the total African population
Sources: Race Relations Survey: 1991/92. South African Institute of Race Relations, 1992

The South African government has distributed its most important human resource, its African population, over certain restricted space and has controlled most of the precious natural resources in the country. The Africans are most important in terms of absolute numbers and cheap labor for the government and private sector. This socio-spatial apartheid has prevented a more equal sharing of land and has impeded economic, physical, political, and social development for the entire country.

The intent since the Population Registration Act of 1950 has been to separate the non-White groups – the Africans, Asians, and Coloureds – and to exacerbate polarization further by dividing the African population into ten ethnic groups. The Act was repealed by the Tricameral Parliament in June 1991 and only the Conservative Party is in opposition to the repeal. Certainly there are cultural differences among these groups, but the differences are stressed and emphasized by government, making them more important than they need to be. The most important matter in abolishing apartheid is a socio-spatial problem, the unpleasant reality of how to resolve the group areas and homelands (Africans prefer to call them bantustans).

Color and location

The Group Areas Act of 1966 forced

This socio-spatial apartheid has prevented a more equal sharing of land

non-Whites, especially Africans, to live on the fringes of the city boundaries as well as the periphery of the entire urban region, and was repealed in June 1991. However, land has not been returned to non-Whites, nor have they been offered reparations for the urban residential land they were forced to vacate.

The creation of homelands (the government calls them independent and non-

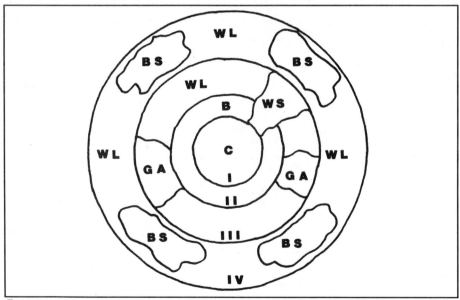

Zones

I	C-City
II	B-Buffer Zone
III	GA-Group Areas
	WS-White Suburbs
IV	BS-Bantustans
	WL-Land for the Whites

Schematic Socio-Spatial Model of South Africa

From *Focus*, Fall 1992, pp. 12-17. © 1991 by the American Geographical Society. Reprinted by permission.

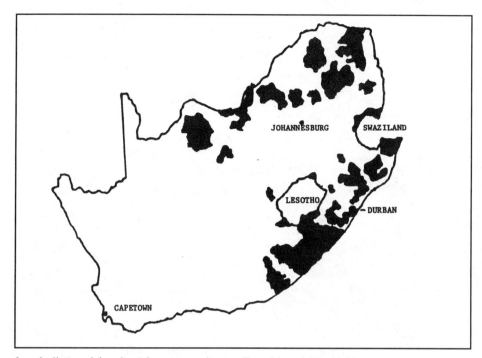

Land allocated for the African population, Republic of South Africa.

firmly in place. The "Schematic Socio-Spatial Model" depicts the zones of distribution of the four racial groups (African, White, Coloured, and Asian) in relation to group areas, homelands, and cities. In Zone I the cities were designated for the White population. Zone II was a buffer zone separating the cities from Zone III. The buffer was made up of open space, industry, or small neighborhoods of Coloureds and/or Indians. Zone III was composed of the group areas, townships and non-White residential areas for groups of color. The White suburbs extend out from a city or are separated from it without a buffer zone. The homelands are located far from cities in remote areas and are governed by Africans, within Zone IV. The remaining area is officially for White citizens of South Africa.

Approximately 58 percent of the African population and 0.2 percent of the other three racial groups own property and reside on 13 percent of the land allocated for homelands. (The remaining 42

independent homelands), another mechanism for apartheid, forced African people to live in different regions of the country according to their so-called ethnic affiliation. The Tomlinson Commission (1955), with the intention of creating an African commonwealth in South Africa, constructed this monstrous geographic structure. Today, there are various proposals on how South Africa should be restructured, put forth by the leaders of the different homelands and political factions.

Color and movement

An extremely important component to segregation or apartheid is to restrict population movement of particular groups to certain spaces. Although apartheid is slowly dismantling, the socio-spatial structure is

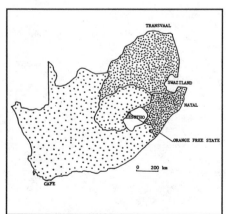

1 dot = 20000 persons
Gross population density of Africans in the respective provinces, including the bantustans' population.

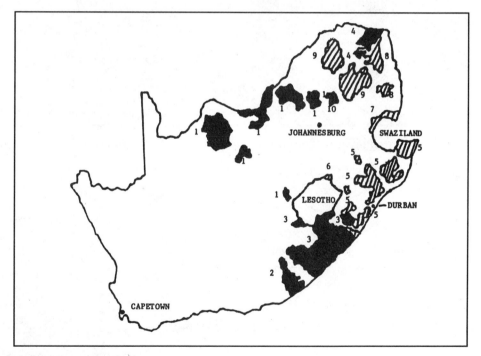

Indendent Republics
1. Bophuthatswana
2. Ciskei
3. Transkei
4. Venda

Self-Governing States
5. Kwazulu
6. Qwaqwa
7.Swazi
8. Gazankulu
9 Lebowa
10 Ndebele
The ten independent republics and self-governing states, Republic of South Africa.

D. Davis, Jr.

Rural residential area outside Umtata, the capital city of Transkei, a so-called independent state.

percent of the African population, 13 percent of the White population, and 11 percent of the Asian and Colored groups are located on 87 percent of South Africa's white designated areas.)

The government decided to segment the

D. Davis, Jr.

WHITE ETHNIC GROUPS 1980

GROUPS	POPULATION
Afrikaans	2,465,328
Afrikaans and English	14,137
Dutch	12,095
English	1,610,265
French	7,316
German	41,057
Greek	17,597
Italian	14,358
Jewish	117,963
Portuguese	56,601
Total	4,356,717

Sources: Department of Statistics and South Africa Official Yearbook 1986.

Residential area and hospital in Soweto, a township outside of Johannesburg.

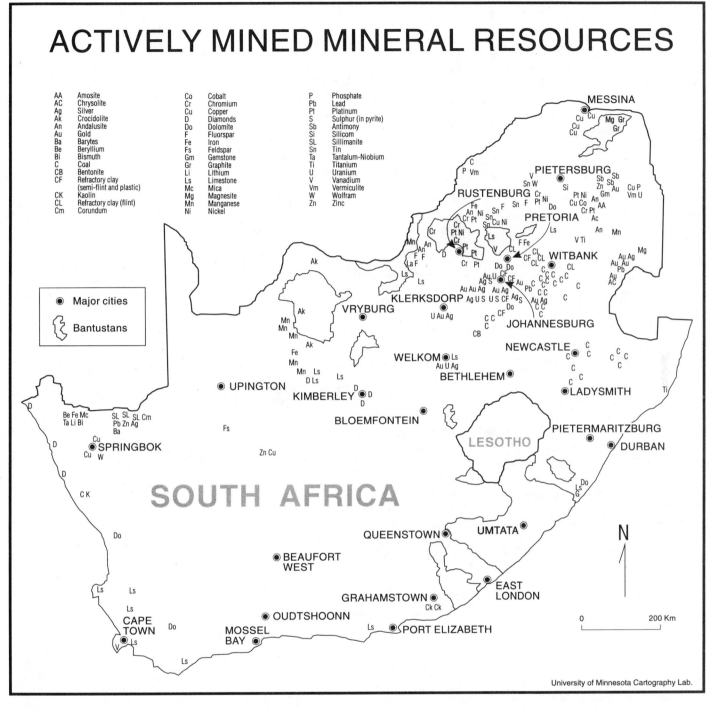

ACTIVELY MINED MINERAL RESOURCES

AA	Amosite	Co	Cobalt	P	Phosphate
AC	Chrysolite	Cr	Chromium	Pb	Lead
Ag	Silver	Cu	Copper	Pt	Platinum
Ak	Crocidolite	D	Diamonds	S	Sulphur (in pyrite)
An	Andalusite	Do	Dolomite	Sb	Antimony
Au	Gold	F	Fluorspar	Si	Silicom
Ba	Barytes	Fe	Iron	SL	Sillimanite
Be	Beryllium	Fs	Feldspar	Sn	Tin
Bi	Bismuth	Gm	Gemstone	Ta	Tantalum-Niobium
C	Coal	Gr	Graphite	Ti	Titanium
CB	Bentonite	Li	Lithium	U	Uranium
CF	Refractory clay (semi-flint and plastic)	Ls	Limestone	V	Vanadium
		Mc	Mica	Vm	Vermiculite
CK	Kaolin	Mg	Magnesite	W	Wolfram
CL	Refractory clay (flint)	Mn	Manganese	Zn	Zinc
Cm	Corundum	Ni	Nickel		

University of Minnesota Cartography Lab.

13 percent allocated for homelands, and divided it among the ten ethnic groups to create 10 homelands, with a total African population of about 16,796,900; as shown in the two tables of African population in independent and non-independent homelands. The map "Land Allocated for the African Population" shows the segmented 13 percent, and the map of "Ten Independent Republics and Self-Governing States" depicts that proportion of the African land subdivided into the different homelands. Only four of these are so-called independent homelands:

Bophuthatswana, Transkei, Venda, and Ciskei. The homelands are not totally African because Asians, Coloureds and Whites also live in them.

As a matter of social fact, when people migrate into foreign territory they eventually mix

The table of African population in white designated areas provides the absolute numbers. The map of "Gross Population Density of Africans in the Respective Provinces" indicates that there is extremely high gross density in Kwazulu, Qwaqwa, Lebowa, Ciskei, and Gazanzulu, but somewhat lower in the other five homelands. The 1991 census clearly shows that there are also approximately ten White cultural groups. They intermingle and intermarry; they do not compete among themselves as they would if the government had convinced them to maintain their language, cultural

5. POPULATION, RESOURCES, AND SOCIOECONOMIC DEVELOPMENT

ELECTRICITY GRID

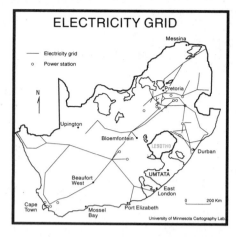

Electricity grid
o Power station

N

Messina
Pretoria
Upington
Bloemfontein
LESOTHO
Durban
UMTATA
Beaufort West
East London
Cape Town
Mossel Bay
Port Elizabeth
0   200 Km
University of Minnesota Cartography Lab.

heritage, and provided them with specific geographic spaces within the boundaries of South Africa. Two tables list the different cultural groups among the African and White populations in the country.

Color, planning, and development

It is clear from the maps that the land segmentation and population distribution were not well conceived and planned by the government, because spatial discontinuity and lack of compactness in a political unit or country can lead to many development problems.

For example, it is essential for the labor pool to be located conveniently close to major mineral resources, transportation, and communication activities. In this case the apartheid regional and urban planners deserve very low marks. The industrial border areas called "growth points" are located near homelands but actually in the white designated areas. If the industries were located within the homelands, there would be greater economic benefit to the labor pool from which the industries obtain a large proportion of employees.

Further, many of the border area industries are located at the mineral resources, which for the most part are outside the homelands (a table is provided of South Africa's major mineral resources; and see

MAJOR HIGHWAYS

N

Messina
Pietersburg
Pretoria
Witbank
Vryburg
Johannesburg
Klerksdorp
Newcastle
Kimberley
Bethlehem
Ladysmith
Upington
Bloemfontein
Pietermaritzburg
LESOTHO
Durban
Beaufort West
Queenstown
UMTATA
Oudtshoorn
East London
Cape Town
Mossel Bay
Port Elizabeth
0   200 Km
University of Minnesota Cartography Lab.

map of "Actively Mined Mineral Resources"). The South African government deliberately created these so-called homelands to be near, but not containing, important mineral resources: thus the homelands could provide mining labor, but could have no territorial power over what they were mining.

A close look at the country's infrastructure shows that the homelands were never to be actively connected with other regions of South Africa. In the highway network, national roads and tarred roads are routed outside the homelands, as shown on the map of "Major Highways". This routing impedes industrial and economical development within the major urban areas and towns of the homelands. Even though the tarred roads have a greater network and are connected to the national roads, they still bypass the homelands.

The railway system shows the same pattern of isolating the African population located within the designated 13 percent

of South Africa's bounded space as shown on the map of "Major Railways". On the map of "Major Airways" the large and middle sized White cities are directly connected with airline circuits, but there are few African cities within the homelands tied into this civil airway network.

Electricity is the basic resource for better communication. The "Electricity Grid" map displays, again, that even with communication, the homelands and townships, at the periphery of the white cities, are obviously neglected by the South African planners. Economic and industrial growth cannot be expected with such drawbacks. These economic activities must be more evenly distributed if a country is to develop without social and political chaos and eventual revolution.

AFRICAN ETHNIC GROUPS IN 1980

GROUPS	POPULATION
Lemba-Venda	182,034
North Ndebele	265,977
Northern Sotho	2,372,522
Shangana-Tsonga	1,024,160
South Ndebele	394,856
Southern Sotho	1,780,511
Swazi	848,749
Tswana	1,356,067
Xhosa	2,927,377
Zulu	5,769,718
Total	16,921,971

Excluding Bophuthatswana, Ciskei, Transkei, and Venda
Source: South Africa Official Yearbook 1986.

Four ways to maintain segregation

After analyzing the maps on population, structure and location of bantustans, min-

AFRICAN POPULATION IN WHITE DESIGNATED AREAS BY PROVINCES: 1991

Cape Province	2,325,100 (8.0)*
Natal Province	1,079,300 (3.7)
Transvaal Province	7,169,100 (24.7)
Orange Free State Province	1,692,100 (5.8)
Sub-total	12,265,600
Total	29,062,500**

*Percent in parenthesis
**The total includes the Africans in the white designated areas and Africans in the ten Bantustans.
Source: Race Relations Survey: 1991/92. South African Institute of Race Relations, 1992

eral resources, transportation connectivity, and communication connectivity, it becomes very apparent that the South African government with its grandiose plans does not and never has had intentions for the majority of its population to realize equality at the township and group areas level, nor at the homeland level.

There are four government ploys to establish and maintain inequality and segregation. Firstly, *population:* the government legally divided groups on the basis of skin color and in a hierarchical order of color. Further, the government subdivided the African group with a strongly convincing model that they are ten separate peoples and should maintain their individual cultures and languages and live separately. As a matter of social fact, when people migrate into foreign territory they eventually mix, provided there are no legal restrictions or extreme racial and cultural taboos. As these different groups migrate to urban areas, they tend to intermingle. If the government really had a strong social conscience about maintaining cultural heritage and language, it should have convinced the ten white groups in the same manner and with the same intensity.

Secondly, *land allocation:* the South African government, certainly cognizant of land segmentation and fragmentation and the importance of contiguity, connectivity, compactness, and bounded space, created a physically disruptive homeland system.

group areas, townships, and homelands to control and maintain the important natural resources that produce great wealth. Planners certainly were aware of the importance of compactness; but valuable mineral resources are not found everywhere equally. As a result, the government allocated homelands in such a spatial arrangement in order to maintain control of major mineral resources within white designated areas.

Abandoning the apartheid system will also mean abandoning the spatial and legal functioning of group areas and homelands. As a result, Zones II, II, and IV will be

MAJOR AIRWAYS

AFRICAN POPULATION OF NON-INDEPENDENT HOMELANDS: 1991	
KwaZulu	5,211,000 (17.9)*
KaNgwane	496,100 (1.7)
Ndebele	516, 100 (1.8)
Gazankulu	607,000 (2.1)
Lebowa	2,276,300 (7.8)
Qwaqwa	451,800 (1.6)
Sub-total	9,558,300
Total	29,062,500**

*Percent in parenthesis of so-called non-independent homelands.
**The total includes the Africans in the ten Bantustans and Africans in the so-called white designated areas.
Source: Race Relations Survey: 1991/92. South African Institute of Race Relations, 1992.

This is economically, politically, and socially costly.

Thirdly, *the economic impact:* the government has organized this discontinuity of

MINERAL RESOURCES OF SOUTH AFRICA	
	WORLD RANK
Gold	1
Manganese	1
Vanadium	2
Chrome	2
Platinum Group Metals	3
Ferrochromium	1
Alumino-silicates	1
Vermiculite	2
Manganese ore	2
Zirconium minerals	2
Titanium minerals	2
Uranium	n.a.
Coal	5
Iron ore	2
Lead	3
Copper	3
Phosphate rock	2
Ferrosilicon	2
Zinc	2

Source: Minerals Bureau, June, 1985, South Africa.

totally eliminated. This is perhaps an impossible task, to return and/or rearrange land and boundaries, by negotiation and peacefully. Its completion will allow the total population of South Africa, Black (non-White) and White, to have access to and benefit from the gross national product generated by the country's valuable natural resources. Every person of voting age within the boundary of South Africa will be allowed to vote. This will lead to a more democratic socio-spatial distribution in the country.

After the apartheid government has been totally dismantled, industrial development and population redistribution should take place. Development should be encouraged around several large industrial growth poles in the sparsely populated areas of the four provinces, especially in the northern and northwestern parts of the Cape Province. People of all ethnic groups and from the two more densely populated provinces, the Transvaal and Natal, should be encouraged to relocate to these new centers.

Geography is important to understanding the inequalities of apartheid, and helps

us realize how long it may take for South Africans to no longer think in terms of the mental map of apartheid.

Further Readings

Best, A. and deBlij, H. 1977. *African Survey.* New York: John Wiley.

Davis, D. and Sanders, R. 1986. "Segmentation and Fragmentation of Land in South Africa." Paper presented at AAG Annual Meeting, Minneapolis, MN.

DeCrespigny, A. 1980. "South Africa: the Case for Multiple Partition," *Journal of Racial Affairs* 31:50-57.

Leach, G. 1987. *South Africa: No Easy Path to Peace.* Suffolk, Great Britain: Methuen.

Stephenson, G.V. 1986. "Pakistan: Discontiguity and the Majority Problem," *Geographical Review* 58 (2):195-213.

South African Institute of Race Relations, 1992. "Race Relations Survey: 1991/92." Johannesburg, South Africa.

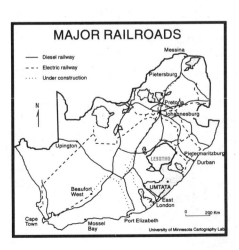

MAJOR RAILROADS

A Water and Sanitation Strategy for the Developing World

Daniel A. Okun

DANIEL A. OKUN is Kenan Professor of Environmental Engineering, emeritus, at the University of North Carolina at Chapel Hill. This article was drawn from a report prepared for the United Nations Development Programme, under the direction of Frank Hartvelt, deputy director of the agency's Division for Global and Interregional Programmes and with the help of Donald T. Lauria of the University of North Carolina at Chapel Hill. The text of the article was delivered by the author on 13 May 1991 at the National Academy of Sciences in Washington, D.C., as the second Abel Wolman Distinguished Lecture sponsored by the Water Science and Technology Board of the National Research Council.

This question, asked by some friends about to embark on a trip abroad, has been put to me more frequently than any other in my almost 50 years in the water resources field. If they are going to Asia, Africa, or Latin America, I have almost always been obliged to answer "no," and the answer has remained unchanged over the years. The reason is clear; there is an inadequate supply of safe drinking water in many of the cities on these continents. This lack of potable water is attributable to the cities' limited water resources and the poor facilities used to treat and distribute water and is compounded by the absence of proper sewerage. Despite ef-

Some residents of Surabaya, Indonesia's second largest city, must buy water from vendors at a price that is much greater than that of piped water.

From *Environment*, Vol. 33, No. 8, October 1991, pp. 16-20, 38-43. Reprinted with permission of the Helen Dwight Reid Educational Foundation. Published by Heldref Publications, 1319 Eighteenth St., NW, Washington, DC 20036-1802.

forts over the last half-century by industrialized countries and international agencies, such as the United Nations Development Programme (UNDP), the World Bank, and the World Health Organization, the situation in these cities is actually worsening.[1]

When travelers fly to these cities, they may see an impressive skyline with modern hotels, offices, and apartment buildings. What they do not see is the absence of physical infrastructure, such as sewers and water lines, to serve these buildings. In a neighborhood in Alexandria, Egypt, for example, sewers were not built when residential buildings were being constructed. As a result, sewage collects in pools a few feet away from the doors to these buildings. (A description of conditions in Alexandria may be found in the box on this page.) In cities where the public water supply is inadequate and developers sink wells, which lower the water table, land subsidence often ensues, and, in coastal areas, saltwater intrusion ultimately fouls the wells. The absence of sewerage is even more serious. The wastewater is discharged with little if any treatment into drainage channels and urban streams so that the surficial groundwater and the soil become badly contaminated. In Bangkok, for example, household wastewater is discharged into the klongs, or canals. Pipes containing water destined for homes run through the klongs, and, when the water pressure in the pipes is low, wastewater in the klongs seeps into the pipes. Such unsanitary situations are aggravated along the perimeters of urban areas, where increasing numbers of migrants to the cities assemble in densely populated communities that are often without any basic services whatsoever.

This lack of basic water and sanitation services leads to a number of problems in the less developed countries. For example, although water-borne diseases, particularly those that cause diarrhea among infants and children, are completely manageable in the industrialized world, they exact a heavy toll in the less developed countries.[2] Also, women and children in these countries must spend hours each day fetching water for the home—time that might

be put to more socially and economically productive uses.

In 1977, the United Nations held a water conference in Mar del Plata, Argentina, to address these problems. The conference's resultant action plan gave rise to the declaration of the 1980s as the International Drinking Water Supply and Sanitation Decade.[3] However, the plan's mandate extends far beyond a concern for just water supply and sanitation; it also recommends the sound management of water resources for all purposes, including agricultural irrigation, industry, hydroelectric power, fisheries, and inland navigation. There was an explicit call for integrated water management.

Some progress was made in providing water and sewer services during the 1980s, as an additional 1.3 billion people, mostly those in the rural areas, received water services (see Tables 1 and 2).[4] However, the growth of the urban population caused a 15-percent increase in the number of city dwellers not receiving water services and almost a 30-percent increase in the number of those residing in neighborhoods without sewer systems.[5] The greatest impact of the International Drinking Water Supply and Sanitation Decade resulted from the higher priority given to water supply and sanitation at national and international levels. Most external support agencies are now committed to assisting water supply and sanitation programs and projects.

The traditional approach to supplying water, whether for household or agricultural use, is to estimate the demand for water and to assume that the resource is available for the taking. Whoever can afford it is generally free

SANITATION PROBLEMS IN ALEXANDRIA, EGYPT

Alexandria, Egypt, known as the "Paris of the Middle East," still has the appearance of a lovely city—but only when viewed from afar. The turquoise waters of the Mediterranean Sea that once laved the sands along the Corniche are now brown and smelly and often transmit enteric diseases to those who dare to swim in them. The spray that a car sends up when driven through residential neighborhoods near the coast is not rainfall or water from leaking mains; it is wastewater from the multifamily dwellings that are provided with water but not with a sewer line. Residents walk on boards to get to their homes and the neighborhood shops, and children play in the sewage-filled pools outside their homes. Even the cosseted visitor to Alexandria cannot escape the refuse, the flies, and the foul puddles that are ubiquitous in the city.

The impacts on public health can be seen through the city's infant mortality rate, which is in the 100-per-1,000 live births range (as compared to the infant mortality rate of 10 per 1,000 live births in more developed countries). Egyptians suffer the highest rates of water-borne diseases in the Middle East, and Alexandrians have the highest rates in all of Egypt. When cholera struck Egypt in the summer of 1970, Alexandria was hardest hit, with an attack rate of 100 per 100,000 persons, compared to attack rates of 25

per 100,000 persons in Cairo and 16 per 100,000 persons in the country as a whole. When cholera hit again in 1974, all districts in Alexandria registered attack rates of more than 200 per 100,000 persons, and some districts had attack rates in the 1,000-per-100,000 range.[1]

The reason for such high rates is poor sanitation generally and the absence of an adequate sewer system specifically. A glance at the skyline of Alexandria, and at those of many metropolitan areas in developing countries, reveals the active construction of buildings, including highrise residential and commercial buildings, without the necessary sanitation infrastructure. Developers are permitted to build and install their own facilities for supplying water but are not obligated to invest in the public sewer system. Although low-cost solutions to sanitation problems are available for rural areas, the costs of building urban sewer systems are high. Thus even Egypt, which has been a favored nation for loans and grants by external support agencies for the last two decades, has only now, with massive infusions of financial aid, begun to improve many of its life-threatening sanitation problems.

1. These data are from a memorandum prepared by the author on 11 August 1977 for the study of the requirements for wastewater collection and disposal in Alexandria.

5. POPULATION, RESOURCES, AND SOCIOECONOMIC DEVELOPMENT

to extract this resource from groundwater or from streams virtually at will. In the more developed countries, local water shortages have been alleviated by gigantic projects that transfer water from one area to another. This approach encounters political obstacles because watersheds do not respect governmental boundaries. Moreover, the prospect of communities being flooded by major reservoirs, the loss of flora and fauna because of excessive diversions from streams, and the land subsidence caused by excessive groundwater abstractions inhibit, as they should, the easy acquisition of water from "new" sources and the overexploitation of existing sources.

Although water is a renewable resource, freshwater is being depleted from surface and groundwater sources faster than it is being replaced by rainfall. Except for desalinized seawater, which is beyond the financial capabilities of all but a few rich countries, there are no new sources of freshwater. All countries will have to live with what they have or with what they can negoti-

ate from others. Countries with arid and semi-arid regions are already facing serious water shortages. Even cities in rain-rich countries are now experiencing water shortages in certain areas.

Until now, few measures have been instituted to husband existing water resources or to devise mechanisms for the effective allocation of these resources. Even in the industrialized countries where these problems have already been widely recognized, institutions and regulations do not offer much guidance. The challenge is exacerbated by the inherent conflict over water resources not only between agricultural and urban demands but also between different regions and countries. Accordingly, the development of policies, legislation, and regulations for the management of this resource and the creation of institutions capable of establishing mechanisms for allocating water must be a priority.

Urban Water Management

During the second half of the 20th

century, the world population will grow 150 percent, but the urban population will grow 300 percent, so that, by the end of the century, almost half of the total population will be living in cities.[6] By 2000, the number of people living in cities will be about 50, 40, and 85 percent of the populations of Asia, Africa, and Latin America, respectively. Moreover, the number of cities with more than 1 million inhabitants will have increased to more than five times the number in 1950. In addition, 18 out of a total of 22 "giant cities" (urban areas with more than 10 million inhabitants) will be in less developed countries by 2000, compared to only 1 out of a total of 4 such cities in 1960.

The explosive growth of cities in the less developed countries is generally not accompanied by the provision of the necessary water and sewerage. Because urbanization increases and concentrates the demand for water, cities throughout the world suffer from water shortages. In the less developed countries, such shortages also result from inadequate facilities to distribute

TABLE 1
EXTENT OF WATER SERVICES AND SEWER SYSTEMS IN DEVELOPING COUNTRIES

Service	Provision	Number of people in 1980 (in millions)	Percentage of population	Number of people in 1990 (in millions)	Percentage of population	Change in number from 1980 to 1990 (in millions)	Percentage change from 1980 to 1990
Water supply	Served	1,411	44	2,758	69	+1,347	+95
	Unserved	1,825	56	1,232	31	−593	−32
Sanitation	Served	1,502	46	2,250	56	+748	+50
	Unserved	1,734	54	1,740	44	+6	0
Total population		**3,236**		**3,990**		**+754**	**+23**

SOURCE: Adapted from data collected by the World Health Organization.

TABLE 2
EXTENT OF WATER SERVICES AND SEWER SYSTEMS IN URBAN AREAS OF DEVELOPING COUNTRIES

Service	Provision	Number of people in 1980 (in millions)	Percentage of urban population	Number of people in 1990 (in millions)	Percentage of urban population	Change in number from 1980 to 1990 (in millions)	Percentage change from 1980 to 1990
Water supply	Served	720	77	1,088	82	+368	+51
	Unserved	213	23	244	18	+31	+15
Sanitation	Served	641	69	955	72	+314	+49
	Unserved	292	31	377	28	+85	+29
Total urban population		**933**		**1,332**		**+399**	**+43**

SOURCE: Adapted from data collected by the World Health Organization.

water. Urban water shortages are accompanied by agricultural demands for water. In the less developed countries, institutions for managing water resources effectively are only now beginning to be developed. However, even in the more developed countries, such institutions are not always adequate.[7]

With a 40-percent increase in the urban population over the last decade, almost all cities in the less developed countries are experiencing water supply and sewerage problems. The water supply problems arise because of inadequate quantity and are exacerbated by leakage, poor accounting, and contamination at the source. Available and economical sources have already been exploited. Population growth, increases in per-capita domestic consumption with higher standards of living, and growing industrialization combine to increase demand. Moreover, the use of water for agriculture near cities must be recognized as temporary because the cities will eventually expand. Studies in China, as well as experience in many other countries including the United States, have shown that the most economical source of additional water is the reclamation of wastewater to be reused for agricultural, industrial, and nonpotable urban purposes.[8]

A major problem with water reclamation in developing countries, however, is that much of the population is not yet served by sewerage, and only a small portion of municipal wastewater receives treatment. Water reuse requires that wastewaters be collected and adequately treated. The reuse of wastewater is attractive because many uses for water are not consumptive, and the wastewaters generated can be reclaimed again. Urban irrigation and evaporative cooling are consumptive uses of water, but they offer an opportunity for wastewater disposal that does not pollute rivers, lakes, and other areas into which wastewaters are discharged.

In cities where high-rise residential and commercial buildings have already been provided with potable water, the high cost of retrofitting water lines for reclaimed water is daunting. Hence,

the use of reclaimed water is most easily initiated for large users in or near urban areas and for newly developing urban areas where lines can be laid out during construction. In this regard, cities without fully developed sewer systems have an advantage; sewage lines and treatment plants can be built to accommodate markets for reclaimed water.

The potential for reusing water highlights the need for research on the planning and design of urban sewer systems to reduce their cost. In the more developed countries, sewer systems are designed that require high building costs. In this case, heavy capital investments are justified because they reduce maintenance problems. But, because of the backlog in building sewer systems in less developed countries, capital costs should be minimized. Studies of this approach have already begun.[9]

The reuse of municipal wastewater has been extensive in the United States for, among other things, landscape irrigation in urban areas and cooling towers in power plants and for many commercial uses in the arid areas of the Southwest and in Florida. The use of reclaimed water for flushing toilets has recently been introduced in commercial buildings in California and has been common for many years in residential buildings in Singapore and Japan. Water reclamation for urban, industrial, and agricultural uses is an option that will become increasingly attractive wherever the costs are comparable with those for obtaining alternative sources of water.[10]

Capacity Building

The less developed countries face

WATER MANAGEMENT IN NORTHERN CHINA

Beijing and Tianjin, the principal port city in northern China, are two of the country's largest and most important cities.[1] The Beijing-Tianjin region, an extensively industrialized area with 18 million people, sits at the bottom of the Hei River basin where little flow remains in the river after water is drawn for the household, industrial, and agricultural needs of the approximately 100 million people living in the upper reaches of the basin. About 40 percent of this region's population lives in rural areas, and agricultural irrigation requires about 65 percent of the available water. The situation here is typical of urban centers throughout the world: Falling groundwater tables and increasing land subsidence, saltwater intrusion, and heavy pollution are rendering much of the region's water unusable.

Heroic efforts have succeeded in bringing water into the region from outside the basin. Now, plans are being made for a $1-billion project to bring water in from the silt-laden Yellow River about 100 miles away. If constructed, this project would meet the current needs of the region. When asked about the future, however, officials rest their hopes on bringing water from the Yangtze, more than 500 miles away.

Studies of the water resources in the region have concluded that the reduction of agricultural water use through more efficient irrigation practices and the

reclamation of wastewater for nonpotable urban and industrial needs should have the highest priority in the region's water management plans. These approaches could meet the region's water needs at a far lower cost than those of the proposed massive capital projects. However, although these approaches are technically and financially feasible, they suffer from the same problems that face major urban areas in all developing countries: the inability to allocate scarce water resources effectively because of the many authorities responsible for water management, the absence of a rational policy for pricing water, and the difficulty of initiating urban water reclamation and reuse projects because of institutional inflexibility at the municipal level. This situation, so common throughout the world, reveals that the solutions to water management problems do not founder from a lack of technology or even from a lack of funds but from the lack of a capacity to effect change.

1. Over the past four years, the author has studied the water resources of the Beijing-Tianjin region for the World Bank and the China State Science and Technology Commission. See R. A. Carpenter, *Final Report: UNDP-SSTC International Workshop on Control of Environmental Pollution in China* (Beijing: UN Development Programme, February 1990); and East–West Environment and Policy Institute, *Summary Report: Water Resources and Management for the Beijing-Tianjin Region* (Honolulu, Hawaii: East–West Environment and Policy Institute, 1988).

two closely related problems in water management. First, the performance of water supply and sanitation systems has been inadequate, in part, because of exploding urban growth, the use of inappropriate technology, and the lack of institutions for sound water management. Second, insufficient attention has been paid to water resources management as a whole and, in particular, to the competing demands for water by the agricultural sector and by urban populations.

UN agencies have responded by agreeing to develop a comprehensive strategy for the 1990s.[11] The consensus is that failures in meeting objectives for water supply and sanitation services and in achieving effective water resources management in developing countries are not caused by a lack of technology or the unavailability of funds. Even in cases where funds are available and the required technology involves only well-established practices, projects have not been sustained. (For more on this topic, see the box on previous page.)

The capacity of a country to receive developmental assistance from external support agencies, such as the World Bank, the U.S. Agency for International Development, and the African Development Bank, among others, must be built up so that programs and projects can later be sustained with indigenous resources. The United Nations Development Programme was charged with developing a strategy for "capacity building."[12] The building of a country's capacity to receive foreign assistance requires the establishment of a favorable policy environment and adequate institutional development. The latter includes the development of sound management systems, incentive structures, and the personnel skills that are needed for sustainable development of water-related programs and projects. In this context, the objective of capacity building is the integrated management of the different subsectors of the water resources sector, such as water supply, sanitation, and agricultural irrigation.

Elements of capacity building have been pursued over the years, but with

In Alexandria, Egypt, wastewater from a sewer plugged with solid wastes overflows into Montaza Canal, the city's source of drinking water.

indifferent success and often outright failure. The less developed countries are littered with water supply and sanitation projects that failed to meet their objectives or were abandoned. Policy, institutional, and personnel development have traditionally enjoyed little if any priority when a project included the construction of facilities or the provision of equipment. Part of the problem results from widespread perception that the external support agencies themselves are not interested in capacity building. They are perceived as being driven by the need to meet lending or granting quotas and seem far more concerned with the completion of work on a project than with its sustainability. Although water supply and sanitation projects frequently fail because of institutional inadequacies, external support agencies are loathe to make a loan or grant contingent upon the assurance that a national or local capacity exists to use the funds because any such condition might cause a delay.[13]

To external support agencies and national leaders, investments in capacity building do not appear as rewarding as do investments in capital projects, which move larger blocks of money.

Moreover, capacity building is not as visible as capital projects are. Capacity building cannot be used for dedication ceremonies and plaques to memorialize national leaders.

Capacity building requires a two-point approach. First, external support agencies must improve their ability to assist capacity building in the less developed countries, and they must be perceived by those countries as giving capacity building high priority. Second, a variety of approaches to build capacity must be introduced into the less developed countries at both the national and the utility, or service, levels for water supply and sanitation and for agricultural irrigation.

External Support Agencies

In industrialized countries, applicants for loans or grants, whether from commercial banks, from the central government, or from private foundations, must show that they have the capacity to use the funds effectively. The potential lender or grantor then examines the applicant's affairs to ascertain whether the capacity does, in fact, exist.[14] Problems arise when the lender or grantor has objectives other than just assistance to the applicant. Among external support agencies, quotas for lending or granting may have been established to justify the existence and continued funding of the agency itself or to enhance the status of the agency. The driving force for granting or lending may be political, diplomatic, or commercial. Thus, certain countries may be targeted for grants and loans irrespective of their expressed needs. Competition within agencies to achieve high levels of lending or granting, which may enhance the role and remuneration of the responsible staff members, sometimes results in "pushing" the approval of loans and grants without too much concern for whether the country can use the funds effectively. Although loans or grants may be evaluated for their effectiveness in meeting their avowed objectives, it may not happen until many years later when the staff responsible for them will have long departed the scene.

The integration of water management is very important to capacity building. Capacity building requires that efforts be made to overcome barriers among different aspects of water management so that different policies for water supply and sanitation and irrigation, for example, will not be advocated by an external support agency through separate governmental agencies. Such a process has already begun in the World Bank, and its integrated management may well be a focus for capacity-building activity. The ultimate objective should be to encourage countries to undertake the integration of water management themselves.[15]

Domestic Efforts

The initiative to build capacity in the water sector of a less developed country must come from within the country. Investments in water management are not likely to be sustained if the countries themselves are not fully committed to capacity building at all appropriate levels. One measure of such a commitment is a country's readiness to inaugurate and participate in an assessment of its own water management practices, which would identify the pol-

icy environment, the institutional resources, the demand for and the availability and quality of water, and the availability of human, material, educational, and financial resources.

Implementing activities that would contribute to the capacity building of a country may be approached at two levels. At the national level, issues for capacity building include the development of a favorable policy environment; the establishment of legal and regulatory frameworks; the integrated management of water resources, river basins, and water demand; the use of information systems and organizational models; and the promotion of the roles played by professional associations. National governmental agencies are often involved in local projects, and their relationships with local agencies affect the country's capacity to use funds effectively. Furthermore, international and bilateral assistance is generally provided through federal agencies or, in large countries, provincial or regional agencies. At the local level, most projects are implemented through municipal utilities or private companies that provide water and wastewater disposal services and sewerage. The issues to be examined at the local level include institutional resources and alternative management models; local water availability, demand, conservation, and quality; financial viability, including funds for operation and maintenance and capital cost recovery; human resources development; and consumer organizations. Integrated water management is entirely appropriate between water supply and sanitation agencies, and the inclusion of local irrigation districts is advantageous where water reclamation for residential, urban, industrial, and agricultural nonpotable reuse is feasible.

The Policy Environment

A tradition of providing free water for agricultural irrigation and heavily subsidized water for household use, often embodied in policy and law in the less developed countries, has placed heavy constraints on sound water management and on the funding of water projects. However, even where such

traditions hold sway and national governments are in economic and political disarray, local initiatives have sometimes been able to create institutions that plan, finance, construct, and manage high-quality water enterprises. Institutional development does not absolutely require the establishment of sound policies and laws. Indeed, the creation of strong local institutions can often be a stimulus to their establishment.

Certain policy issues in urban water management must be addressed. For example, although water resources are renewable, they are nonetheless being depleted. Floods and droughts cannot be controlled, but their impacts on the water supply should be ameliorated. Also, water that falls as rain and then flows into rivers, lakes, and underground aquifers may be free, but making it available on-site for agricultural, residential, industrial, or other uses is as costly as controlling water pollution. Commitments to meet these costs are essential. Finally, to encourage conservation, utilities must charge users at a price that reflects the amount of water used.[16] Accordingly, the metering of water production and use is essential. At the very least, charges should be adequate to cover the costs of managing, operating, and maintaining water service facilities. Because funds from external support agencies are not likely to meet all the capital costs for developing water resources and distributing water for agricultural, industrial, and household uses in the less developed countries, charges must include capital cost recovery.

Institutional Development

The essential ingredient in capacity building is institutional development. According to a 10-year evaluation of the U.S. Agency for International Development's Water and Sanitation for Health project, "local institution-building is the key to transferring sustainable skills."[17] Sustainability of a project is cited as the most important measure of its success.

Institutions in water management include national ministries and regulatory agencies and local organizations, such as municipal departments of public works and utilities. Many types of institutions have been successful; indeed, there is no universally suitable model that can be prescribed. Institutions are the products of a country's history, society, and economy. The choice of which institutions are developed is a local prerogative. A major problem arises when too many institutions with overlapping authorities affect one sector of a country. New institutions should not necessarily be created; those that already exist should be assessed, and people should be open to and aware of other models that have been successful and may be appropriate. There have been many different models of institutions, each with its own advocates.[18] They include

- government administrative, regulatory, and operating agencies at the national and local levels;
- national and local quasigovernmental agencies or authorities (known as "quangos" in the United Kingdom) that are financially self-sustaining and enjoy their own personnel practices;
- local public utilities for water supply (many advantages accrue when separate municipal sanitation agencies that manage sewerage and wastewater treatment plants are combined in a local or regional public utility);
- private companies that own and operate water utilities, which are usually regulated by national governmental agencies (privatization of sewerage and irrigation agencies is less common);
- publicly owned agencies that contract with private companies for the operation and management of water and sanitation facilities; and
- river basin organizations, which have considerable potential for implementing effective water resources management (they were highly successful in England and Wales for 15 years).[19]

Of course, because institutions are essentially made up of people, major changes are often difficult to implement. All that external support agencies can do, where institutional structures are perceived as constraints to effective management, is to offer a variety of options for change and to support those promising changes that government officials may elect to pursue.

Laws

A country's institutions are inextricably tied to its laws and its political system. A common complaint in industrialized countries is that legislation and regulations have overwhelmed water management. Consumers of water are in conflict, and resolution increasingly rests with the courts. In less developed countries, laws and regulations are often inadequate, and, where they do exist, they are seldom enforced.

Institutions have a responsibility for the promulgation of appropriate laws and regulations. They must have the professional expertise, both administrative and technical, to advise on appropriate and implementable legislation and regulations.

A 1990 Food and Agriculture Organization/World Health Organization working group on the legal aspects of water supply and wastewater management "emphasizes (*inter alia*) the need to insure that governments have the legal power to allocate and reallocate water reuse rights," and it suggests a "water rights administration."[20]

Laws are particularly important today in the less developed countries where the demand for water for urban and industrial growth and for agricultural irrigation exceeds available resources. Water for household use has traditionally enjoyed priority. However, because agriculture uses about 80 to 85 percent of water resources, agricultural interests hold dominion over water resources in most countries, including the United States.

Personnel

Institutions depend on people. The boxes and lines of an institution's organizational chart are less important than the people who occupy the boxes. An ideal institutional structure with poor personnel has less potential than a poor one with high-quality people.

Education and training are essential elements for developing the competence of personnel. The nine members of the International Training Network, part of the UN Development Programme's and the World Bank's Water and Sanitation Program, can assist countries in their training efforts. Bilateral aid for financing training is extensively available from the Nordic countries, the Netherlands, the United Kingdom, Japan, and Germany, but there is shamefully little available today from the United States. The U.S. Agency for International Development has no identifiable program directed at such assistance in water management. It is a very short-sighted posture.

Training and education at professional levels within the country are also necessary. Much water-related education, research, and other work can be commissioned to local universities and other local institutions. This arrangement helps educational institutions obtain and retain staff in water management. The university, its students, and the utility all benefit. Such "twinning" arrangements may be encouraged between industrialized and less developed countries, thus providing for an exchange of up-to-date technical material and information on the specialized needs of less developed countries.

Professional Associations

Nongovernmental professional associations have long played an important role in enhancing the capacity of industrialized countries in the water sector.[21] They provide a mechanism for reporting and updating technical knowledge about water management and make it available through publications, conferences, and short courses. They work with national decision makers in defining water policy and setting standards. They provide a link among public agencies, consultants, manufacturers, industry, and the public. Finally, they promote national and international exchanges and cooperation in training, research, technology, and the development of water management. Although they are well established in the more developed countries, professional associations are either nonexistent or weak in most less developed countries. Initiatives by professional associations from the industrialized countries can contribute to the capacity building of less developed countries.[22] Members of these associations can make special arrangements with colleagues in the less developed countries to distribute their publications, which are generally not easily available because of hard-currency restrictions. Also, external support agencies can sponsor professionals to attend conferences that they could not otherwise afford to attend.

Private Companies

Engineering firms from industrialized countries are major actors in the developing world. They are generally employed by clients in the developing countries, but an external support agency is usually involved in the selection of consultants when it is the source of funds. Consultants are selected for their technical competence; little attention is paid to the role they should play in capacity building. The "joint-venture" of a foreign firm and a local one on a project is often a marriage of convenience. The local firm helps to secure the contract and afterward is involved only in "housekeeping" rather than fulfilling professional obligations. A foreign consulting firm whose selection is based on its commitment to the development of the less developed country's personnel, as well as on its technical expertise, will be eager to use its professional staff to assist in training and supporting local personnel. They already exercise such functions in their own countries as an inherent professional obligation.

Because multinational manufacturing corporations have production facilities throughout the developing world that require services for water supply and wastewater and solid waste management and disposal, they have a stake in the quality of the water supply and sanitation institutions of those countries. These corporations also have close working relationships with the utilities that serve them in industrialized countries. They can thus promote cooperation between utility personnel in industrialized countries and those in the less developed countries.

Consumer Organizations

Decisions about local water management policies are too often made among staff members of an external support agency and high-level government officials, neither of whom are knowledgeable about the local population. There have been cases where water supply systems have been built with foreign assistance but were later abandoned because customers chose to have their homes connected to these systems. If customers choose, for whatever reason, to continue to purchase water from vendors at exorbitant prices, the water supply project eventually will be abandoned. (For a case study of what happened to such a project in Nigeria, see the box on this page.)

Projects must include funds to support one or more consumer organizations. These organizations could assess the needs of the customers, their ability and willingness to pay, and their preferred level of service. Such information can help officials avoid costly missteps that are common today. Also,

the role of women in consumer organizations has been shown to be instrumental in their success and in the sustainability of water supply and sanitation projects.

A Water Strategy

Despite major commitments by many countries and the extensive assistance of external support agencies, water services, including water supply and sanitation for urban areas and water for irrigation in agricultural areas, are not keeping up with demand. Furthermore, many investments in water services have not been sustained. The major constraint has been the less developed countries' lack of capacity to develop and utilize the resources available.

The challenges faced by the less developed countries are enormous and are made more difficult by the increasing numbers of people migrating to urban areas. As these populations grow, the demand for safe drinking water and for reliable sewerage will grow as well.

THE WATER-VENDING SYSTEM OF ONITSHA, NIGERIA

The poor in developing countries have long been perceived as not being able to pay for household water service. The fact is, however, that, because the poor in the urban areas of these countries are generally not provided with a public water service, they are obliged to buy water, often of questionable quality, from private, unregulated vendors at a cost per liter that is some 30 times greater than that paid by the well-to-do who live in homes that have connections to the municipal water system. The monthly cost of buying small amounts of water from vendors may exceed by several times the monthly cost of buying water from municipal sources.[1]

A study of the water-vending system in Onitsha, Nigeria, found that these private vendors were responsible for more than 95 percent of the sales in water to the city's residents.[2] The poor were annually paying water vendors twice the operational and maintenance costs and 70 percent of the annual capital costs of the new municipal water system.

The Onitsha situation illustrates the importance of community involvement in the planning and management of water systems. The new municipal water system was planned without any participation from the city's residents. As a result, these residents were unsure of the reliability and the quality of the service that they would receive and thus were reluctant to connect to the municipal system. Instead, they continued to pay high prices for a much inferior service. Clearly, funding for the development of consumer organizations and for their participation in the planning and execution of water and sanitation projects is almost as essential as the money to pay the engineers' fees.

---

1.   B. Zaroff and D. A. Okun, "Water Vending in Developing Countries," *Aqua* 5 (1984):289–95.
2.   D. Whittington, D. T. Lauria, and X. Mu, "A Study of Water Vending and Willingness to Pay for Water in Onitsha, Nigeria," *Water Development* 19 (1991):179–98.

5. POPULATION, RESOURCES, AND SOCIOECONOMIC DEVELOPMENT

To provide water and to build and maintain sewers for their growing populations, the less developed countries must adopt strategies that will allow the efficient use of available assistance and lead to a sustainable water supply and sanitation system. Such water strategies should focus on capacity building through improvements in the policy environment and institutional and human resources development. Specific actions should include the enhancement of information, education, and training programs for personnel and the establishment of cooperative programs with professional associations, multinational corporations, consulting engineering firms, and consumer organizations. Particular attention must be paid to the integration of water management for water supply, sanitation, and irrigation at both the national and local levels. Moreover, water reclamation should be developed and made available in industrialized and less developed countries alike.

NOTES

1. During the 1980s (the International Drinking Water Supply and Sanitation Decade), the number of people who received water services increased by about 370 million people in Asian, African, and Latin American cities. The number of those who did not, however, increased by about 30 million. Sanitation services fell even further behind. An additional 85 million urban dwellers were living in neighborhoods not served by a sewer system.

2. The infant mortality rate in the less developed countries is 10 times greater than that in the industrialized world. The vulnerability of these populations to endemic enteric diseases is revealed by the explosive outbreak of cholera in Peru earlier this year.

3. United Nations, *Report of the United Nations Water Conference* (New York: United Nations, 1977).

4. Adapted from data collected by the World Health Organization.

5. Adapted from data collected by the World Health Organization.

6. UN Department of International Economic and Social Affairs, *World Population Prospects* (New York: United Nations, 1989).

7. These problems are as common in the United States as they are in the less developed countries. See W. Viessman, Jr., "A Framework for Reshaping Water Management," *Environment*, May 1990, 10. For example, sustaining barge traffic on the Mississippi River during the 1988 drought required decisions for regulating the river's flow. But these decisions were stymied because the responsibility was fragmented among 18 federal agencies and several states. Also, in the United States, there are no coherent federal water policies. This absence, which now most heavily affects the country's burgeoning cities, is pointed out in L. B. Leopold, "Ethos, Equity, and the Water Resource," *Environment*, April 1990, 16. He states that water rates are heavily subsidized for irrigation in the western United States, where 33 to 45 percent of the irrigated acreage is devoted to surplus crops that must also be subsidized.

8. See R. A. Carpenter, *Final Report: UNDP–SSTC International Workshop on Control of Environmental Pollution in China* (Beijing: UN Development Programme, February 1990); and East–West Environment and Policy Institute, *Summary Report: Water Resources and Management for the Beijing-Tianjin Region* (Honolulu, Hawaii: East–West Environment and Policy Institute, 1988).

9. The UN Development Programme's and the World Bank's Water and Sanitation Project in Washington, D.C., is engaged in such studies.

10. D. A. Okun, "Realizing the Benefits of Water Reuse in Developing Countries," *Water Environment and Technology* 2 (1990):78-82.

11. The Administrative Committee on Coordination Intersecretariat Group for Water Resources represents all UN agencies concerned with water management, including the UN Development Programme, the World Bank, the World Health Organization, the Food and Agriculture Organization, the UN Educational, Scientific, and Cultural Organization, the UN International Children's Emergency Fund, the World Meteorological Organization, and the UN Department of Technical Cooperation and Development, and HABITAT.

12. See D. A. Okun and D. T. Lauria, *Capacity Building for Water Resources* (New York: UN Development Programme, August 1991). Under the UN Development Programme, the author, D. T. Lauria, and Frank Hartvelt, deputy director of the agency's Division for Global and Interregional Programmes, prepared the draft document "Capacity Building for Water Resources Management: An International Initiative for Sustainable Development in the 1990s," which served as the background for the agency's international symposium A Strategy for Water Resources Capacity Building, held in Delft, the Netherlands, in June 1991. The final document contains a "Delft Declaration" and conclusions and recommendations that will be presented at the International Conference on Water and the Environment, to be held in Dublin in January 1992. Recommendations from this conference will be forwarded to the June 1992 UN Conference on Environment and Development in Rio de Janeiro.

13. In the 1990 report "Major Issues in Bank-Financed Water Supply and Sanitation Projects," based on 125 World Bank water supply and sanitation projects, J. B. Buky of the bank's Operations Evaluations Department asserted that capacity building must occur prior to, rather than during, project implementation. He noted that success was achieved "where the Bank had the fortitude and the patience to prepare the ground . . . by insisting on pre-project action for build-up of institutional competence." External support agencies seldom have the patience to delay capital or technical assistance loans and grants pending the development of a country's reasonable capacity to use the funds effectively.

14. A major departure from this prudent policy may be the case of the savings and loan fiasco of the 1980s in the United States.

15. The U.S. government may establish a central water agency, such as the old National Water Council, to avoid some of the problems caused by conflicting pressures on limited resources. See Viessman, note 7 above.

16. See United Nations, *Water Resources: Progress in the Implementation of the Mar del Plata Action Plan*, doc. no. E/C 7/1991/8 (New York: United Nations, 1991).

17. The Water and Sanitation for Health (WASH) project was instituted in 1981 as the U.S. Agency for International Development's contribution to the International Drinking Water Supply and Sanitation Decade. WASH provides technical assistance to developing countries on behalf of the agency. The project, which is continuing into the 1990s, is being operated under contract by Camp, Dresser, and McKee International of Cambridge, Mass.

18. Okun and Lauria, note 13 above.

19. The creation of 10 public water authorities in England and Wales was successful, as described in D. A. Okun, *Regionalization of Water Management: A Revolution in England and Wales* (London: Applied Science Publishers, 1977). The authorities oversaw regions whose boundaries were based on hydrologic boundaries, and they were responsible for virtually all water-related activities in their areas. They were so efficient, in fact, that Prime Minister Margaret Thatcher, in pursuit of privatization and for the benefit of the Exchequer, elected to sell off selected assets of the water authorities—namely, those responsible for water supply and for wastewater collection and disposal. The water companies that were formed are now under considerable attack for using their financial resources to buy properties around the world unrelated to water, while neglecting their pollution control obligations. "Rivers Get Dirtier as Water Firms Divert Their Profits," *The Sunday Times*, 16 June 1991.

20. Food and Agriculture Organization and World Health Organization Working Group on Legal Aspects of Water Supply and Wastewater Management, *Legal Issues in Water Resources Allocation, Wastewater Use and Water Supply Management* WHO/CWS/90.19 (Geneva: World Health Organization, 1990).

21. C. Rietveld, "Building National Capacities for Sustainable Water Supply and Sanitation Coverage" (Paper presented at the Foundation for the Transfer of Knowledge, International Water Supply Association Conference, Copenhagen, May 1991).

22. In the United States, professional associations active in the water supply and sanitation field include the American Water Works Association, the Water Pollution Control Federation, the American Public Health Association, and the American Society of Civil Engineers. All have some interest in the developing world. The American Water Works Association has just inaugurated a Water for People program. Some of the international associations are the London-based International Water Supply Association, with its Foundation for the Transfer of Knowledge; the International Association for Water Pollution Research and Control, also based in London; the International Water Resources Association in Urbana, Illinois; and regional associations such as the Inter-American Association of Sanitary Engineers (AIDIS) in Mexico City.

50 Trends Shaping the World

The nuclear threat that kept antagonists at bay for the last four decades has largely been removed. While much turmoil has resulted, trends toward new alliances and cooperation to solve global problems bode well for sustaining peace.

Marvin Cetron and Owen Davies

About the Authors

Marvin Cetron is president of Forecasting International, Ltd., 1001 North Highland Street, Arlington, Virginia 22210. He is co-author (with Owen Davies) of *American Renaissance: Our Life at the Turn of the 21st Century* (St. Martin's Press, 1989) and co-author (with Margaret Gayle) of *Educational Renaissance: Our Schools at the Turn of the Twenty-First Century* (St. Martin's Press, 1991).

Owen Davies is co-author of *American Renaissance* and former senior editor of *Omni* magazine. His address is P.O. Box 355, Hancock, New Hampshire 03449.

This article is adapted from their book, *Crystal Globe: The Haves and Have-Nots of the New World Order* (St. Martin's Press, 1991).

The world will be a more peaceful and prosperous place in the 1990s than it has been in the decades since World War II, because the premise by which it operates has changed. In the coming years, it will no longer be influenced by the needs of ideological and military competition, but instead by the need to promote international trade and the well-being of the trading nations. Major military conflicts will be all but unthinkable, because they are contrary to the mutual interests of nations that are interdependent in the global economy. Wars will not suddenly disappear, but they will be primarily small and regional in nature. These conflicts will stem from local antagonisms and the ambitions of Third World rulers, and peace will be restored by the joint effort of the entire world community.

This fundamental change will be the guiding theme of the 1990s.

Politically, this will be an interesting era. Nations will increasingly band together, however briefly, with traditional enemies to further their short-term interests.

No single nation will have the power to dominate in this new global order. World leaders will be military powers as well as leaders from the three powerful regional economic blocs now coming to dominate international commerce: the European Community, the Pacific Rim, and the North American alliance. Each group will be heavily influenced by its largest members but will act primarily by consensus in all matters of common interest.

Vast regions of the world will be left out of this interlocking arrangement, save on the occasions when they can serve the interests of the major powers. The Middle East will retain much of its wealth and influence, thanks to the continued importance of oil. Africa, the Indian subcontinent, and Southeast Asia will remain much as they are now, doomed to poverty largely by their own leaders and used by the industrialized nations as little more than stockpiles of raw material.

Yet, even these nations should benefit from the new global structure. In a more peaceful and prosperous world, the developed nations will have a better opportunity to help their less-fortunate neighbors deal with economic and social problems, to whatever extent local politics allow it. Progress will come slowly in the Third World, but it will move more quickly under the new commercial priorities than it did under ideological and military domination.

In the pages that follow, we will outline many of the trends that are emerging from today's ferment to form tomorrow's new world order.

Population

1. In the industrialized countries, the "birth dearth" has cut growth almost to nothing, while in the developing world, the population bomb is still exploding.

• The rich get richer, the poor have children: Throughout the industrialized world, workers can look forward to national retirement programs or social security. In the developing lands, those too old for labor rely on their children to support them — so they have as many as they can.

• Thanks to better health care, children have a greater chance to survive into adulthood and produce children of their own. This will tend to accelerate population growth, but contraceptive use is increasing, with an opposite effect on growth.

• In the developed world, the vast Baby Boom generation is approaching middle age, threatening to overwhelm both medical and social-

From *The Futurist*, September/October 1991, pp. 11-21. Adapted from *Crystal Global: The Haves and Have-Nots of the New World Order.* © 1991 by Marvin J. Cetron and Owen Davies (St. Martin's Press) and reprinted with permission by the World Future Society.

5. POPULATION, RESOURCES, AND SOCIOECONOMIC DEVELOPMENT

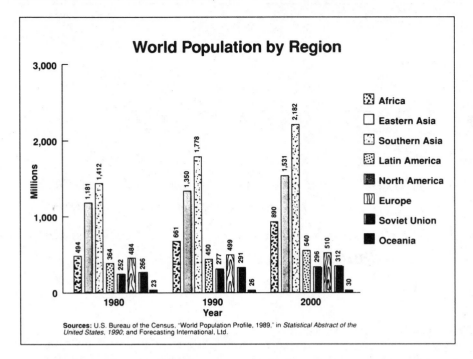

World Population by Region

Sources: U.S. Bureau of the Census, "World Population Profile, 1989," in *Statistical Abstract of the United States, 1990*; and Forecasting International, Ltd.

security programs. These costs will consume an increasing portion of national budgets until about 2020.

2. The AIDS epidemic will slaughter millions of people worldwide, especially in Africa.

• According to the World Health Organization, the AIDS-causing human immunodeficiency virus will have infected up to 40 million people by 2000.

• By 1990, some 5 million people in sub-Saharan Africa already carried the disease — twice as many as just three years earlier. In some cities, as much as 40% of the population may be infected.

3. A host of new medical technologies will make life longer and more comfortable in the industrialized world. It will be many years before these advances spread to the developing countries.

4. As the West grows ever more concerned with physical culture and personal health, developing countries are adopting the unhealthy practices that wealthier nations are trying to cast off: smoking, high-fat diets, and sedentary lifestyles. To those emerging from poverty, these deadly luxuries are symbols of success.

• In the United States, smokers are kicking the habit. Only 35% of American men smoke, down from 52% twenty years ago; 29% of women smoke, down from a peak of 34%.

• However, the developing world continues to smoke more each year. Even Europe shows little sign of solving this problem.

5. Better nutrition and the "wellness" movement will raise life expectancies.

• In developed countries, children born in the 1980s will live to an average age of 70 for males, 77 for females. In developing countries, the average life expectancies will remain stalled at 59 years for males and 61 for females.

Food

6. Farmers will continue to harvest more food than the world really needs, but inefficient delivery systems will prevent it from reaching the hungry.

• According to the World Bank, some 800 million people are chronically malnourished by U.N. standards. As the world population grows, that number will rise.

7. The size and number of farms are changing.

• In the United States, the family farm is quickly disappearing. Yet, giant agribusinesses reap vast profits, while small, part-time "hobby" farms also survive. This trend will begin to affect other developed nations during the 1990s and will even-

tually spread to the rest of the world.

• Former Iron Curtain countries will find it difficult to turn their huge, inefficient collective farms back to private owners; progress in this effort will be uneven.

• Land reform in the Philippines and Latin America will move at a glacial pace, showing progress only when revolution threatens. Most of the vast holdings now owned by the rich and worked by the poor will survive well into the twenty-first century.

8. Science is increasing the world's supply of food.

• According to the U.S. Office of Technology Assessment, biotechnology and other yield-increasing developments will account for five-sixths of the growth in world harvests by 2000; the rest will come from newly cultivated croplands.

• Biotechnology is bringing new protein to developing countries. Bovine growth hormone can produce 20% more milk per pound of cattle feed, while genetic engineering is creating fish that grow faster in aquafarms.

9. Food supplies will become healthier and more wholesome.

• Most nations will adopt higher and more-uniform standards of hygiene and quality, the better to market their food products internationally. Consumers the world over will benefit.

10. Water will be plentiful in most regions. Total use of water worldwide by 2000 will be less than half of the stable renewable supply. Yet, some parched, populous areas will run short.

• The amount of water needed in western Asia will double between 1980 and 2000. The Middle East and the American West are in for dry times by the turn of the century. Two decades later, as many as 25 African nations may face serious water shortages.

• We already know how to cut water use and waste-water flows by up to 90%. In the next decade, the industrialized countries will finally adopt many of these water-saving techniques. Developing countries reuse little of their waste water, because they lack the sewage systems required to collect it. By 2000, building this needed infrastructure will become a high priority in many parched lands.

• Cheaper, more-effective desalination methods are on the horizon. In the next 20 years, they will make it easier to live in many desert areas.

Energy

11. Despite all the calls to develop alternative sources of energy, oil will provide more of the world's power in 2000 than it did in 1990.

• OPEC will supply most of the oil used in the 1990s. Demand for OPEC oil grew from 15 million barrels a day in 1986 to over 20 million just three years later. By 2000, it will easily top 25 million barrels daily.

12. Oil prices are not likely to rise; instead, by 2000 they will plummet to between $7 and $9 a barrel. A number of factors will undermine oil prices within the next 10 years:

• Oil is inherently cheap. It costs only $1.38 per barrel to lift Saudi oil out of the ground. Even Prudhoe Bay and North Sea oil cost only $5 per barrel.

• The 20 most-industrialized countries all have three-month supplies of oil in tankers and storage tanks. Most have another three months' worth in "strategic reserves." If OPEC raises its prices too high, their

customers can afford to stop buying until the costs come down. This was not the case during the 1970s oil shocks.

• OPEC just is not very good at throttling back production to keep prices up when their market is glutted. They will not get any better at doing so in the 1990s.

13. Growing competition from other energy sources will also help to hold down the price of oil:

• Natural gas burns cleanly, and there is enough of it available to supply the world's entire energy need for the next 200 years.

• Solar, geothermal, wind-generated, and wave-generated energy sources will contribute where geographically and economically feasible, but their total contribution will be small.

• Nuclear plants will supply 12% of the energy in Eastern Europe and the Soviet Union by the end of the century.

Environment

14. Air pollution and other atmospheric issues will dominate eco-policy discussions for years to come.

• Soot and other particulates will be more carefully scrutinized in the near future. Recent evidence shows

that they are far more dangerous than sulfur dioxide and other gaseous pollutants formerly believed to present major health risks. In the United States alone, medical researchers estimate that as many as 60,000 people may die each year as a direct result of breathing particulates. Most are elderly and already suffering from respiratory illness.

• By 1985, the concentration of carbon dioxide in the atmosphere had increased 25 times since preindustrial days. By 2050, the concentration is likely to increase 40% over today's levels if energy use continues to grow at its current pace. Burning fossil fuel will spew about 7 billion tons of carbon into the air each year by 2000, 10–14 billion in 2030, and 13–23 billion in 2050.

• Blame global warming for at least some of the spread of Africa's deserts. Before the process runs its course, two-fifths of Africa's remaining fertile land could become arid wasteland. Up to one-third of Asia's non-desert land and one-fifth of Latin America's may follow. Global warming will not only hurt agriculture, but will also raise sea levels, with consequent impacts on habitation patterns and industries.

• Brazil and other nations will soon halt the irrevocable destruction of the earth's rain forests for very temporary economic gain. Those countries will need economic help to make the transition. The World Bank and the International Monetary Fund (IMF) will help underwrite alternatives to rain-forest destruction.

• Acid rain such as that afflicting the United States and Canada will appear whenever designers of new power plants and factories neglect emission-control equipment. Watch for it in most developing countries.

15. Disposal of mankind's trash is a growing problem, especially in developed nations. Within the next decade, most of the industrialized world will all but run out of convenient space in its landfills.

• The U.S. Environmental Protection Agency estimates that existing technologies could reduce the total amount of hazardous waste generated in the United States by 15%–30% by 2000.

• For now, recycling is a necessary nuisance. By 2000, recyclables will

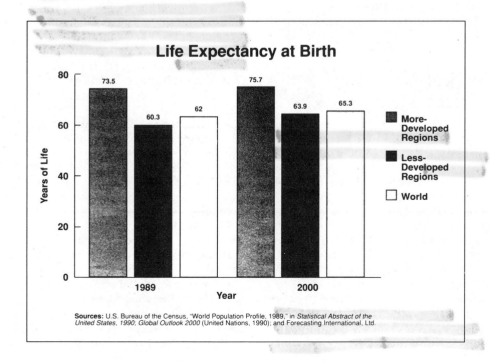

Life Expectancy at Birth

Years of Life

1989: 73.5 (More-Developed Regions), 60.3 (Less-Developed Regions), 62 (World)

2000: 75.7 (More-Developed Regions), 63.9 (Less-Developed Regions), 65.3 (World)

■ More-Developed Regions
■ Less-Developed Regions
□ World

Sources: U.S. Bureau of the Census, "World Population Profile, 1989," in *Statistical Abstract of the United States, 1990*; *Global Outlook 2000* (United Nations, 1990); and Forecasting International, Ltd.

become valuable resources, as research finds profitable new uses for materials currently being discarded. Recycling will save energy as well: Remanufacturing requires less energy than does the full iron-ore-to-Cadillac production process.

Science and Technology

16. High technological turnover rates are accelerating.

• All the technological knowledge we work with today will represent only 1% of the knowledge that will be available in 2050.

17. Technology has come to dominate the economy and society in the developed world. Its central role can only grow.

• For some economists, the numbers of cars, computers, telephones, facsimile machines, and copiers in a nation define how "developed" the country is.

• Personal robots will appear in homes in the developed world by 2000. Robots will perform mundane commercial and service jobs and environmentally dangerous jobs, such as repairing space-station components in orbit.

18. The technology gap between developed and developing countries will continue to widen.

• Developed countries have 10 times as many scientists and engineers per capita as the developing world. The gap between their spending on research and development grew threefold from 1970 to 1980.

• Technologically underdeveloped countries face antiquated or nonexistent production facilities, a dearth of useful knowledge, ineffective organization and management, and a lack of technical abilities and skills. Under these conditions, underdevelopment is often self-perpetuating, which weakens the country's ability to compete in international markets.

• The widening technology gap will aggravate the disparity in North–South trade, with the developed nations of the Northern Hemisphere supplying more and more high-tech goods. The less-developed countries of the South will be restricted to exporting natural resources and relatively unprofitable low-tech manufactured products.

19. Nations will exchange scientific information more freely, but will continue to hold back technological data.

• Basic research is done principally in universities, which have a tradition of communicating their findings.

• Fifty-three percent of Ph.D. candidates in U.S. science and engineering programs are from other countries. Anything they learn will return to their homelands when they do.

• The space-faring nations — soon to include Japan — will share their findings more freely.

• Technological discoveries, in contrast, often spring from corporate laboratories, whose sponsors have a keen interest in keeping them proprietary. More than half of the technology transferred between countries will move between giant corporations and their overseas branches or as part of joint ventures by multinationals and foreign partners.

20. Research and development (R&D) will play an ever-greater role in the world economy.

• R&D outlays in the United States have varied narrowly (between 2.1% and 2.8% of the GNP) since 1960 and have been rising generally since 1978.

• R&D spending is growing most rapidly in the electronics, aerospace, pharmaceuticals, and chemical industries.

Communications

21. Communications and information are the lifeblood of a world economy. Thus, the world's communications networks will grow ever more rapidly in the next decade.

• A constellation of satellites providing position fixing and two-way communication on Earth, 24 hours a day, will be established in the 1990s. A person equipped with a mini-transceiver will be able to send a message anywhere in the world.

22. The growing power and versatility of computers will continue to change the way individuals, companies, and nations do their business.

• Processing power and operating speeds for computers are still increasing. By 2000, the average personal computer will have at least 50 times the power of the first IBM PCs and 100 or more times the power of the original Apple II.

• Computers and communications are quickly finding their way into information synthesis and decision making. "Automatic typewriters" will soon be able to transcribe dictation through voice recognition. Computers will also translate documents into various languages. Today's best translation programs can already handle a 30,000-word vocabulary in nine languages.

• The revolution in computers and communications technologies offers hope that developing countries can catch up with the developed world. However, few have yet been able to profit from the new age of information. In 1985, developing countries owned only 5.7% of the total number of computers in the world; most of these computers are used mainly for accounting, payroll processing, and similar low-payoff operations.

Labor

23. The world's labor force will grow by only 1.5% per year during the 1990s — much slower than in recent decades, but fast enough to provide most countries with the workers they need. In contrast, the United States faces shortages of labor in general, and especially of low-wage-rate workers.

• Multinational companies may find their operations handicapped by loss of employees and potential workers to the worldwide epidemic of AIDS, especially in Africa, since many firms rely on indigenous workers.

24. The shrinking supply of young workers in many countries means that the overall labor force is aging rapidly.

• Persons aged 25 to 59 accounted for 65% of the world labor force in 1985; almost all growth of the labor force over the next decade will occur in this age group.

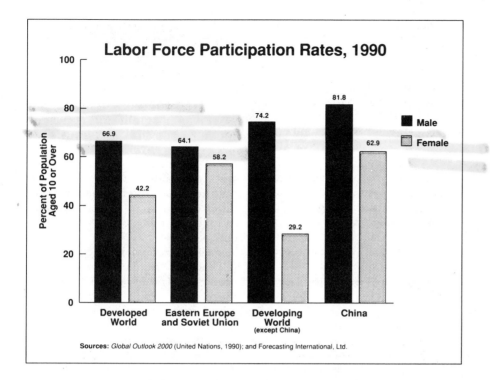

Labor Force Participation Rates, 1990

Sources: *Global Outlook 2000* (United Nations, 1990); and Forecasting International, Ltd.

moderate rate, with developed nations showing the fastest increases.

Industry

29. Multinational and international corporations will continue to grow, and many new ones will appear.

• Companies will expand their operations beyond national borders. For example, Marconi Space Systems (a British General Electric company) and Matra Espace (of France) got together to form Matra Marconi Space, "the first international space company."

• Many other companies will go international by locating new facilities in countries that provide a labor force and benefits such as preferential tax treatment, but that do not otherwise participate in the operation. Ireland pioneered this practice with U.S. companies in the insurance, electronics, and automobile industries. It found that when companies leave, for whatever reason, the country loses revenue and gains an unemployed labor force.

30. Demands will grow for industries to increase their social responsibility.

• A wide variety of environmental disasters and public-health issues (e.g., the *Exxon Valdez* oil spill and Union Carbide's accident at Bhopal, India) have drawn public attention to the effects of corporate negligence and to situations in which business can help solve public problems not necessarily of their own making.

• In the future, companies will increasingly be judged on how they treat the environment — and will be forced to clean up any damage resulting from their activities.

• Deregulation will be a thing of the past. There will be increased government intervention: Airlines will be compelled to provide greater safety and services; the financial-service industry will be regulated to reduce economic instability and costs; electric utilities will be held responsible for nuclear problems; and chemical manufacturers will have to cope with their own toxic wastes.

31. The 1990s will be the decade

25. Unions will continue to lose their hold on labor.

• Union membership is declining steadily in the United States. It reached 17.5% in 1986. According to the United Auto Workers, it will fall to 12% by 1995 and to less than 10% by 2000.

• Unionization in Latin America will be about the same as in the 1980s; unionization in the Pacific Rim will remain low; unionization in the developing world as a whole will remain extremely low.

• Increased use of robots, CAD/CAM, and flexible manufacturing complexes can cut a company's work force by up to one-third.

• Growing use of artificial intelligence, which improves productivity and quality, will make the companies adopting it more competitive, but will reduce the need for workers in the highly unionized manufacturing industries.

26. People will change residences, jobs, and even occupations more frequently, especially in industrialized countries.

• High-speed MAGLEV trains will allow daily commutes of up to 500 miles.

• The number of people who retrain for new careers, one measure of occupational mobility, has been increasing steadily.

• The new information-based organizational management methods — nonhierarchical, organic systems that can respond quickly to environmental changes — foster greater occupational flexibility and autonomy.

27. The wave of new entrepreneurs that appeared in the United States during the 1970s and 1980s is just the leading edge of a much broader trend.

• In 1986, the number of new businesses started in the United States hit a record 700,000. In 1950, there were fewer than 100,000 new business incorporations. A similar trend has appeared in Western Europe, where would-be entrepreneurs were until recently viewed with suspicion. And a new generation of entrepreneurs is growing throughout Eastern Europe and even in Japan.

• From 1970 to 1980, small businesses started by entrepreneurs accounted for most of the 20 million new jobs created in America. In 1987, small businesses accounted for 1 million new jobs, compared with 97,000 in larger companies.

28. More women will continue to enter the labor force.

• In both developed and developing regions, the percentage of working women has increased since 1950. Women represented 36.5% of the world's labor force in 1985. This growth is expected to continue at a

of microsegmentation, as more and more highly specialized businesses and entrepreneurs search for narrower niches.

Education and Training

32. Literacy will become a fundamental goal in developing societies, and the developed world will take steps to guard against backsliding toward illiteracy. Throughout the world, education (especially primary school for literacy) remains a major goal for development as well as a means for meeting goals for health, higher labor productivity, stronger economic growth, and social integration. Countries with a high proportion of illiterates will not be able to cope with modern technology or use advanced agricultural techniques.

• Most developed countries have literacy rates of more than 95%. The increasing levels of technological "savvy" demanded by modern life, however, often are more than people are prepared to meet, even in the most modern societies.

• The proportion of illiterates among the world's adult population has steadily decreased, although the absolute number has grown. In developing countries, the proportion of illiterates will drop from 39% in 1985 to 28% by 2000, while the number of illiterate adults will have climbed by 10 million.

• Worldwide, the proportion of children not enrolled in school will fall from 26% in 1985 to 18% by 2000. Primary-education enrollment has risen dramatically in most of the developing world except for Africa. In 31 sub-Saharan countries reporting their enrollment rates, the rates had fallen for boys in 13 countries and for girls in 15.

• Useful, job-oriented knowledge is becoming increasingly perishable. The half-life of an engineer's professional information today is five years.

33. Educational *perestroika* is changing American schools. In the long run, this will repair the nation's competitive position in the world economy.

• The information economy's need for skilled workers requires educational reform.

• Science and engineering schools will be actively recruiting more students.

• Foreign-exchange programs will grow markedly in an attempt to bolster the competence of American students in international affairs.

34. Higher education is changing as quickly as primary and secondary schools.

• The soaring cost of higher education may force program cuts. If so, developing countries face an ultimate loss of foreign exchange, as their industries fall further behind those of cheaper, more-efficient competitors.

• There are too few jobs for liberal arts college graduates in many developing countries. For instance, Egypt cannot keep its promise to give a job to every graduate; the civil service is grossly overstaffed already.

• The concept of "university" is changing. Increasingly, major corporations are collaborating with universities to establish degree-granting corporate schools and programs. Examples include the General Motors Institute, Pennsylvania State University's affiliation with a major electronics company, and Rutgers University's affiliation with a major pharmaceutical house.

• More private companies will market large electronic databases, eventually replacing university libraries.

World Economy

35. The world economy will grow at a rapid rate for the foreseeable future, but the gap between rich and poor countries will widen.

• World trade will grow at a brisk 4.5% annually in the next decade. As one result, international competition will continue to cost jobs and income in the developed market economies.

• The gross domestic products (GDPs) of the developed market economies will grow at 3.1% on average in the 1990s as investment demand increases and the economic integration in Europe introduces capital efficiency.

• The economies of Eastern Europe and the Soviet Union may recover with a GDP growth rate of 3.6%.

• The developing economies will fall further and further behind the industrialized nations, largely because their populations will continue to rise faster than their incomes. GDPs in the developing economies will grow by 4.3% a year (well below the 5.1% rate they enjoyed in the 1970s). In the 1970s, their per capita GDP was one-tenth that of the developed countries. By 1985, it had fallen to one-twelfth. By 2000, it will be one-thirteenth.

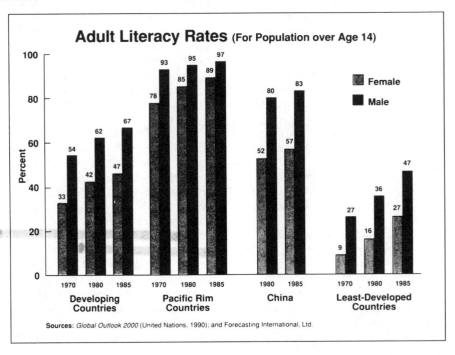

Adult Literacy Rates (For Population over Age 14)

Sources: *Global Outlook 2000* (United Nations, 1990); and Forecasting International, Ltd.

- By reducing military budgets, the fabled "new world order" will make more money available for business.

36. The world economy will become increasingly integrated.

- There is a "ripple effect" among closely linked national stock exchanges. The impact of a major event on one exchange perturbs all the others. Stock markets will become more fully connected and integrated.

- By 2000 or so, all national currencies will be convertible, following a model similar to the European Community's Exchange Rate Mechanism.

- It will become increasingly difficult to label a product by nation (e.g., "Japanese cars") since parts often come from several countries to be assembled in others and sold in yet others. Protective tariffs will become obsolete — for the good of the worldwide economy.

37. The world is quickly dividing itself into three major blocs: the European Community, the North American free-trade zone, and Japan's informal but very real Pacific development area. Other regions will ally themselves with these giants: Eastern Europe with the EC, Mexico with the United States and Canada. The nations of Latin America will slowly build ties with their neighbors to the North. The Australia–New Zealand bloc is still trying to make up its mind which of these units to join — the Pacific Rim, where its nearest markets are, or Europe and North America, where its emotional bonds are strongest.

- The economic structure of all these regions is changing rapidly. All but the least-developed nations are moving out of agriculture. Service sectors are growing rapidly in the mature economies, while manufacturing is being transferred to the world's developing economies.

- Within the new economic blocs, multinational corporations will *not* replace the nation-state, but they will become far more powerful, especially as governments relinquish aspects of social responsibility to employers.

38. The European Community will become a major player in the world economy.

- By 1992, the EC will represent a population of 325 million people with a $4-trillion GDP.

- By 1996, the European Free Trade Association countries will join with the EC to create a market of 400 million people with a $5-trillion GDP. Sweden, Norway, Finland, Austria, and Switzerland will join the founding 12.

- By 2000, most of the former East Bloc countries will be associate members of the EC.

39. The 25 most-industrialized countries will devote between 2% and 3% of their GDP to help their poorer neighbors.

- Much aid to poorer countries will be money that formerly would have gone to pay military budgets.

- The World Bank and IMF will help distribute funds.

- Loans and grants may require developing nations to set up population-control programs.

40. Western bankers will at last accept the obvious truth: Many Third World debtors have no hope of ever paying back overdue loans. Creditors will thus forgive one-third of these debts. This will save some of the developing nations from bankruptcy and probable dictatorship.

41. Developing nations once nationalized plants and industries when they became desperate to pay their debts. In the future, the World Bank and the IMF will refuse to lend to nations that take this easy way out. (Debtors, such as Peru, are eager to make amends to these organizations.) Instead, indebted nations will promote private industry in the hope of raising needed income.

42. Washington, D.C., will supplant New York as the world financial capital. The stock exchanges and other financial institutions, especially those involved with international transactions, will move south to be near Congress, the World Bank, and key regulatory bodies.

- Among the key economic players already in Washington: the Federal Reserve Board, the embassies and commercial/cultural attachés of nearly every country in the world, and the headquarters of many multinational and international corporations.

- In addition, several agencies cooperating with the United Nations, including the International Monetary Fund and the General Agreement on Tariffs and Trade, have their headquarters or routinely conduct much of their business in Washington.

Warfare

43. The world has been made "safer" for local or regional conflicts. During the Cold War, the superpowers could restrain their aggressive junior allies from attacking their neighbors. With the nuclear threat effectively gone, would-be antagonists feel less inhibited. Iraqi President Saddam Hussein was only the first of many small despots who will try to win by conquest what cannot be achieved by negotiation.

- The United States and the Soviet Union will sign a long procession of arms treaties in the next decade. The two countries will make a virtue of necessity, but both will act primarily to cut expensive military programs from their budgets.

- The Warsaw Pact has already disintegrated. NATO, seeking a new purpose, will eventually become an emergency strike force for the United Nations. The number of guns, tanks, and military planes in Europe will fall to little more than half their peak levels.

- Terrorist states will continue to harbor chemical and biological weapons until the international community finally takes a firm stand.

44. Brushfire wars will grow more frequent and bloody. Among the most likely are:

- Israel vs. the Arab countries. We foresee one last conflict in this region before the peace that now seems near actuality becomes a reality. Israel will win this one, too.

- India vs. Pakistan. The two have feuded with each other since the British left in 1947; religious differences, separatism in Kashmir, and small stocks of nuclear weapons make this a hot spot to watch carefully.

- Northern Ireland vs. itself. This perpetually troubled land will remain its own worst enemy. In trying to keep Ireland under control, the British face an increasingly unpleasant task.

45. Tactical alliances formed by common interests to meet immediate

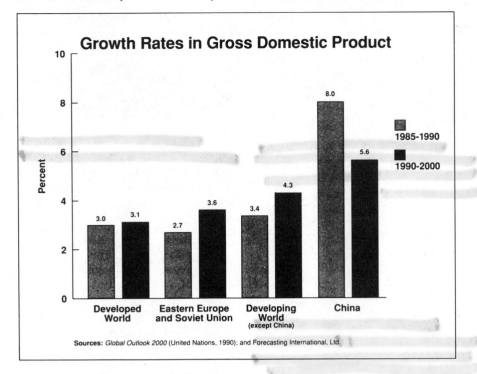

Growth Rates in Gross Domestic Product

Legend: 1985-1990, 1990-2000

Developed World: 3.0, 3.1
Eastern Europe and Soviet Union: 2.7, 3.6
Developing World (except China): 3.4, 4.3
China: 8.0, 5.6

Sources: *Global Outlook 2000* (United Nations, 1990); and Forecasting International, Ltd.

needs will replace long-term commitments among nations.

• In the Middle East, "the enemy of your enemy is your friend." Iran and Iraq will tolerate each other in their stronger hatred for the West. The United States and Syria will never be friends, but both dislike Iraq.

• Turkey and Greece will be hard-pressed to overlook their differences about Cyprus, but may do so in an effort to counter terrorism.

International Alignments

46. The Information Revolution has enabled many people formerly insulated from outside influences to compare their lives with those of people in other countries. This knowledge has often raised their expectations, and citizens in many undeveloped and repressed lands have begun to demand change. This trend can only spread as world telecommunications networks become ever more tightly linked.

• East Germans learned of reforms elsewhere in Eastern Europe via West German television; Romanians learned through Hungarian media.

• International broadcasting entities such as Voice of America, the British Broadcasting Corporation, and Cable News Network disseminate information around the world, sometimes influencing and inspiring global events even as they report on them.

47. Politically, the world's most important trend is for nations to form loose confederations, either by breaking up the most-centralized nations along ethnic and religious lines or by uniting independent countries in international alliances.

• Yugoslavia will soon split into a loose confederation based on the region's three dominant religions: Greek Orthodoxy, Roman Catholicism, and Islam. Czechoslovakia is already loosening the ties between its Czech and Slovak regions. And, following a brief, unsuccessful attempt at new repression by the right wing of the Communist Party, the Soviet Union will reorganize itself as a confederation of 15 largely independent states.

• Quebec will secede from Canada, probably in 1996. The four eastern Canadian provinces will be absorbed into the United States by 2004, and the other Canadian provinces will follow suit by 2010.

• Hong Kong and Macao will rejoin China, through previously made

agreements, by 1997. Taiwan will seek to join Mainland China shortly thereafter. The two Koreas will reunite before 2000.

48. The role of major international organizations will become extremely important in the new world order.

• The United Nations will finally be able to carry out its mission. The World Court will enjoy increased prestige. UNESCO's food, literacy, and children's health funds will be bolstered. The World Health Organization will make progress in disease eradication and in training programs. The Food and Agriculture Organization will receive more funding for starvation relief and programs to help teach farming methods.

• More countries will be willing to reform internally to meet requirements for International Monetary Fund loans and World Bank programs that provide development and education funds and grants.

• More medical aid from developed countries will be provided, frequently under the auspices or coordination of the United Nations or Red Cross/Red Crescent, to countries devastated by plagues, famine, or other natural disasters. Red Cross and Red Crescent will step up activities in such areas as natural-disaster relief and blood programs.

• Cooperation will develop among intelligence agencies from different countries (e.g., Interpol, the CIA, and the KGB) in order to monitor terrorism and control antiterrorism programs and to coordinate crime fighting worldwide.

49. International bodies will take over much of the peacekeeping role now being abandoned by the superpowers. The Conference on Security and Cooperation in Europe (CSCE) — a group of 35 nations (including the United States and the Soviet Union) — will pick up where NATO and the Warsaw Pact left off by creating a pan-European security structure.

• CSCE will transform the diplomatic process into an institution.

• The methods of operation for voting on CSCE matters will likely be revised (currently, each of the member nations holds veto power).

50. The field of public diplomacy will grow, spurred by advances in communication and by the increased importance and power of international organizations.

Credits/ Acknowledgments

Cover design by Charles Vitelli

1. Geography In a Changing World
Facing overview—Courtesy of NASA.

2. Land-Human Relationships
Facing overview—United Nations photo.

3. The Region
Facing overview—United Nations photo by Derek Lovejoy.
135—Map by Joe LeMonnier.

4. Spatial Interaction and Mapping
Facing overview—Department of Public Works, State of California. 185-189—Graphics by Ian Worpole.

5. Population, Resources, and Socioeconomic Development
Facing overview—United Nations photo by J. P. Laffont.

ANNUAL EDITIONS ARTICLE REVIEW FORM

■ NAME: _____ DATE: _____

■ TITLE AND NUMBER OF ARTICLE: _____

■ BRIEFLY STATE THE MAIN IDEA OF THIS ARTICLE: _____

_____

_____

_____

_____

■ LIST THREE IMPORTANT FACTS THAT THE AUTHOR USES TO SUPPORT THE MAIN IDEA:

_____

_____

_____

_____

_____

_____

■ WHAT INFORMATION OR IDEAS DISCUSSED IN THIS ARTICLE ARE ALSO DISCUSSED IN YOUR
TEXTBOOK OR OTHER READING YOU HAVE DONE? LIST THE TEXTBOOK CHAPTERS AND PAGE
NUMBERS:

_____

_____

_____

_____

_____

_____

■ LIST ANY EXAMPLES OF BIAS OR FAULTY REASONING THAT YOU FOUND IN THE ARTICLE:

_____

_____

_____

_____

■ LIST ANY NEW TERMS/CONCEPTS THAT WERE DISCUSSED IN THE ARTICLE AND WRITE A
SHORT DEFINITION:

_____

_____

_____

_____

*Your instructor may require you to use this Annual Editions Article Review Form in any number of ways:
for articles that are assigned, for extra credit, as a tool to assist in developing assigned papers, or simply
for your own reference. Even if it is not required, we encourage you to photocopy and use this page;
you'll find that reflecting on the articles will greatly enhance the information from your text.

ANNUAL EDITIONS: GEOGRAPHY 94/95

Article Rating Form

Here is an opportunity for you to have direct input into the next revision of this volume. We would like you to rate each of the 40 articles listed below, using the following scale:

1. **Excellent: should definitely be retained**
2. **Above average: should probably be retained**
3. **Below average: should probably be deleted**
4. **Poor: should definitely be deleted**

Your ratings will play a vital part in the next revision. So please mail this prepaid form to us just as soon as you complete it.
Thanks for your help!

Rating	Article	Rating	Article
1	1. The Four Traditions of Geography	1	22. Lash of the Dragon
1	2. The American Geographies: Losing Our Sense of Place	3	23. Low Water in the American High Plains
	3. Spatial Environment and Social Adaptation in Japan—A Traveler's Perspective	2	24. The Key to Understanding the Former Soviet Union
2	4. The Balkans	2	25. Transportation and Urban Growth: The Shaping of the American Metropolis
2	5. Will Russia Disintegrate Into Bantustans?	3	26. Hispanic Migration and Population Redistribution in the United States
2	6. Beaches on the Brink		
2	7. What's Wrong With the Weather?	2	27. Rhine-Main-Danube Canal: Connecting the Rivers at Europe's Heart
1	8. The Green Revolution	3	28. The Chunnel: The Missing Link
2	9. Geography and Gender	2	29. An Alternative Route to Mapping History
2	10. Troubled Waters	1	30. Cultural Commitments and World Maps
1	11. A Current Catastrophe: El Niño	1	31. A Sense of Where You Are
2	12. The Deforestation Debate	1	32. The New Cartographers
1	13. Climate Change: The Threat to Human Health	1	33. The Aging of the Human Species
4	14. Exploring the Links Between Desertification and Climate Change	2	34. The Global Threat of Unchecked Population Growth
2	15. Water Tight	3	35. The Landscape of Hunger
2	16. The Rise of the Region State	3	36. We're All Minorities Now
1	17. Africa's Geomosaic Under Stress	1	37. Why Africa Stays Poor, And Why It Doesn't Have To
4	18. Middle East Geopolitical Transformation: The Disappearance of a Shatterbelt	2	38. South African Apartheid: A Socio-Spatial Problem
1	19. Welcome to the Borderlands		
	20. Reclaiming Cities for People	2	39. A Water and Sanitation Strategy for the Developing World
3	21. The Aral Sea Basin: A Critical Environmental Zone	2	40. 50 Trends Shaping the World

(Continued on next page)

...OUT YOU

...me_ Jennifer Willis _____ Date_ 4-6-95 _

...e you a teacher? ☐ Or student? ☑

...ur School Name_ Olivet Nazarene University _____

...epartment _____

Address_ ONU Box 8193   P.O. Box 592 _____

City_ Kankakee _____ State_ Il _ Zip_ 60901 _

School Telephone #_(815) 937-6522 _____

YOUR COMMENTS ARE IMPORTANT TO US!

Please fill in the following information:

For which course did you use this book?_ Physical + Cultural Geography ___

Did you use a text with this Annual Edition? ☑ yes ☐ no

The title of the text?_ Customized Social Science 201, Physical + Cultural Geography _

What are your general reactions to the Annual Editions concept?

I really like it.

Have you read any particular articles recently that you think should be included in the next edition?

The Aging of the Human Species

Are there any articles you feel should be replaced in the next edition? Why?

Welcome to the Border lands - Boring
Exploring the Links between Desertification + Climate change. - Boring

Are there other areas that you feel would utilize an Annual Edition? Information

May we contact you for editorial input? No

May we quote you from above? Yes

ANNUAL EDITIONS: GEOGRAPHY 94/95

BUSINESS REPLY MAIL

First Class   Permit No. 84   Guilford, CT

Postage will be paid by addressee

The Dushkin Publishing Group, Inc.
Sluice Dock
DPG **Guilford, Connecticut 06437**

No Postage
Necessary
if Mailed
in the
United States